# Alzheimer's Disease

## Biology, Biophysics and Computational Models

# Alzheimer's Disease

## Biology, Biophysics and Computational Models

**Don Kulasiri**
*Lincoln University, New Zealand*

**Jingyi Liang**
*UiT The Arctic University of Norway, Norway*

World Scientific

NEW JERSEY • LONDON • SINGAPORE • BEIJING • SHANGHAI • HONG KONG • TAIPEI • CHENNAI • TOKYO

*Published by*

World Scientific Publishing Europe Ltd.

57 Shelton Street, Covent Garden, London WC2H 9HE

*Head office:* 5 Toh Tuck Link, Singapore 596224

*USA office:* 27 Warren Street, Suite 401-402, Hackensack, NJ 07601

**Library of Congress Cataloging-in-Publication Data**
Names: Kulasiri, Don, author. | Liang, Jingyi, author.
Title: Alzheimer's disease : biology, biophysics and computational models / Don Kulasiri, Lincoln University, New Zealand, Jingyi Liang, UiT The Arctic University of Norway, Norway.
Description: New Jersey : World Scientific, [2022] | Includes bibliographical references and index.
Identifiers: LCCN 2021021071 | ISBN 9781800610118 (hardcover) | ISBN 9781800610125 (ebook) | ISBN 9781800610132 (ebook other)
Subjects: LCSH: Alzheimer's disease. | Alzheimer's disease--Research.
Classification: LCC RC523 .K85 2022 | DDC 616.8/311--dc23
LC record available at https://lccn.loc.gov/2021021071

**British Library Cataloguing-in-Publication Data**
A catalogue record for this book is available from the British Library.



For any available supplementary material, please visit
https://www.worldscientific.com/worldscibooks/10.1142/Q0297#t=suppl

Desk Editors: Vishnu Mohan/Michael Beale/Shi Ying Koe

Typeset by Stallion Press
Email: enquiries@stallionpress.com

*Don Kulasiri dedicates this book to his late mother Jayawathie Sinhababu, a teacher par excellence.*

*Jingyi Liang dedicates this book to her mother, Jihong Zhu* (朱继宏)*, and her father, Hongde Liang* (梁红德).

# Preface

This book has been written during the COVID-19 global pandemic when countries and communities around the world are faced with very difficult choices between economic activities and the well-being of their populations. Elderly around the globe are highly susceptible to COVID-19, and during the first outbreak in early 2020, we saw the spikes in the deaths among elderly populations compared to other age groups. Many governments have taken action to take care of the elderly by sacrificing the economic activities through a series of lockdown measures. Different cultures have diverse ways of looking after their grandparent generation: as extended families, elderly care communities, separate facilities and more support through governmental programs. Just like the COVID-19 pandemic, dementia affects millions of elderly people throughout the world, and Alzheimer's disease (AD) in particular is devastating at the individual, familial and social levels.

AD still does not have a cure and 47 million people suffer from it worldwide at present. Immeasurable social, emotional and financial consequences of AD increase as decades go by, and AD poses one of the biggest scientific challenges of our time. As life expectancy improves because of medical research and vastly enhanced healthcare across the world in comparison to the previous centuries, we are faced with the challenge of understanding the causes of aging-associated neurodegenerative diseases. We hope that this book would help the reader to understand AD from mechanistic and biochemical perspectives at intra- and inter-cellular levels.

In this book, we explain some of our recent research on biophysics and computational models associated with AD. We take a heavily context-driven modeling approach using mathematics and computation; hence, a great amount of biological details are necessary to understand the modeling we have done. The first two chapters are about AD and its associated biology, and then in Chapters 3 and 4 we discuss a region where action potential, the fundamental signaling motif in brain, is generated. This small region is called the axon initiation segment (AIS), dysfunction of which has a strong relationship to AD. In Chapter 5, we discuss the biomolecular dynamics within AIS providing a mathematical model of related protein interactions. We discuss presynaptic mechanisms and biophysics in Chapter 6, and Chapters 7–9 are based on our modeling related to biochemical pathways of AD. Finally, in Chapter 10, we attempt to quantify synaptic plasticity to understand AD better.

We acknowledge the first author's former research students for their contributions: Piyush Bhardwaj's contributions in Chapters 3 and 4 are gratefully acknowledged; and Dr. Yao He, Sun Yat Sen University, China, contributed immensely to Chapters 5 and 10. Thank you Piyush and Yao for your help in writing these chapters. We also acknowledge Professor Sandhya Samarasinghe at Lincoln University, New Zealand for her support in many ways. Last but not least, the first author acknowledges the encouragement and facilitation he received from Professor Philip Maini during his visits to Wolfson Centre for Mathematical Biology, Mathematical Institute, Oxford University, UK, over the years.

December 2020

Don Kulasiri
*Lincoln University, New Zealand*

Jingyi Liang
*UiT The Arctic University of Norway, Norway*

# Contents

# Chapter 1

# Introduction to Alzheimer's Disease

## 1.1. Introduction

Alzheimer's disease (AD) is the leading cause of dementia and, unfortunately, is still incurable. About 47 million people worldwide are affected by AD or some other dementia [1] and the social, emotional and financial implications of AD are immeasurable. As more and more people live longer due to improved healthcare around the world, the aging-associated neurodegenerative diseases are posing the greatest challenges to science and it is becoming imperative for us to understand the causes of these diseases, AD in particular, as it is the focus of this book, at molecular and cellular levels. Here we aim to summarize the main facts as we know them now so that we can develop a mechanistic understanding of dementia and AD in the later chapters; it is important to see the broader picture of the problem before we embark on our journey through this book: the fascinating biology and dynamics within and outside the neuron, the basic building block of brain.

Memory formation, retention and retrieval are central functions of the human brain, without which we would not have created the diversity in our cultures, human relationships, arts and sciences, in short, human civilizations. When we pause and think about our memories, we see that they form our identities, the personal sense of self, and pleasant, unpleasant and neutral pasts. What if one happens to lose these memories? The effects are devastating not only to the individual concerned but for the family and also for the society at large. We have to develop care giving practices and institutions for the patients at enormous costs to society not to mention the damage these diseases such as AD inflict on the wider fabric of society. So it becomes ever more important for us to understand the neurodegeneration in an attempt to cure these diseases even though we do not yet have effective remedies to AD. But the good news is that the sciences involved in the search of understanding AD have advanced tremendously during the last several decades.

While humans may have suffered from AD for the entirety of human history, it was Dr. Alois Alzheimer (1864–1915) who reported the case of his female patient Auguste Deter who was suffering from presenile dementia early last century. In 1895, Dr. Alzheimer became the director of the asylum in Frankfurt mainly to conduct clinical studies on manic depression and schizophrenia. During his stay in Frankfurt, Alzheimer met 51 years old Auguste, who was suffering from disorientation, impaired memory and disabilities in cognitive functions such as reading and with symptoms ranging from hallucinations to the loss of higher mental functions. After Deter's death in 1906, her brain and all the medical records he had made were in his possession, and he systematically recorded neuropathology of the case. Deter was later recognized as the first documented case of AD. Alzheimer published his work on the histopathology of the patients for whom he was the psychiatrist and he showed that the histological diagnosis can be used to answer clinical questions. Alzheimer also noticed that the degenerative process that underlies AD was independent from the inflammatory reactions. Alzheimer's writings even after 100 years remain valid even though the degenerative processes were poorly understood until the 1980s. Alzheimer presented the clinical and neuropathologiacal features of Auguste Deter as "a peculiar disease of the cerebral cortex" because pathological and anatomical investigation of the brain showed a thinner cerebral cortex which was quite normal among the elderly. Further, the neurofibrillary tangles and amyloid plaques were observed. The neurofibrillary tangles represent the changes in the cytoskeleton of nerve cells leading to cell death, and amyloid plaques consist of deposited extracellular neurotoxic substance. For the first time, Alzheimer related the dementia to the lesions in Auguste's brain. This presentation is the beginning of AD research: today the neurofibrillary tangles and amyloid plaques are important topics in neuroscience research. Alzheimer's second case, a demented 56 year old man, Johann F., who died of AD in 1910, showed similar lesions of his brain as were in the first case. For decades following Alzheimer's research, the hypothesis that the lesions associated with neurons were the cause of dementia was accepted with neuropathological and psychological assessments with limited understanding of neurogenerative aspects of AD.

As mentioned before, AD is characterized by progressive and irreversible loss of memory and cognitive function and the symptoms of AD worsen over time with variable rates of AD progression. Currently, an AD patient can live four to eight years after diagnosis but in some cases this could be extended up to 20 years. The changes in the brain of an AD patient start occurring years before any signs of AD appear (named preclinical AD). After the onset of AD, the progression of the disease can be categorized into three stages: mild, moderate and severe AD, but these stages can overlap with each other. In mild AD (early stage), a person may be functionally independent but he or she may have memory lapses such as forgetting words regularly used or the location of everyday objects. He or she may still drive, work and be part of social activities. Close family members begin to notice these moments more regularly, and it is advisable to seek medical help at this stage. Moderate AD is the longest stage which lasts for many years during which appropriate care regimes need to put in place for patients. Moderate stage is characterized by more pronounced symptoms exposing the vulnerabilities associated with the inability to attend to normal routine activities. In severe AD, individuals lose the ability to respond to their environment, and their bodily movements are compromised. Communication could be painfully difficult with others, and significant personality changes occur during this stage. Extensive help would be needed at this stage.

The progress of understanding neurogenerative diseases has been slow until a few decades ago mainly due to the complexity associated with the functions of brain. The brain consists of 100 billion neurons which are interconnected within the specialized segments of brain, and now only we begin to understand the system dynamics of all these interconnections which are responsible for memory formation, retention and retrieval. Further, neurons as cells are highly specialized over millions of years of evolution and they do not divide into daughter cells like other cells in the body. They are postmitotic and largely cannot be recovered once lost even though neurogenesis occurs in a human adult brain to a marginal level [2]. Therefore, most research into finding a therapeutic target for AD focused on the processes associated with the neurons and their surrounding environments.

## 1.2. Brain and Neurons

In this section, we give a very brief overview of brain and the changes that occur when AD sets in. More details can be found in excellent text books that are available (see [3] for example) and we intend to cover only the essentials of the brain as an aid to understand the subsequent chapters. There are very many neuroscience texts that expound all the intricate facts thus far scientifically discovered of the human brain, and this book is not intended to be such an exhaustive treatment; the field is expanding rapidly, so our references would hopefully provide more in-depth coverage of a wide range of topics associated with brain science.

The complexity of the human brain is unparalleled. We often marvel at the quintessentially human pursuits on a wider spectrum: creative writings of exceptional novels, timeless poetry and artworks, dazzling mathematical theorems, latest technologies, and space exploration to name a few, but often we take for granted the organ that is responsible for these wonders. Now only we begin to understand how the multifarious and often multitasking brain interacts with its dynamically varying environments. The complexity of the brain can be attributed to the ways it has evolved over millenia to comprise of a multifaceted self-organization of cellular units called neurons in a structured manner. The sheer number of neurons is mind-boggling, around 100 billion, and they are surrounded by glial cells.

The brain has many levels of organization starting from single neurons to separate regions with specific functions associated with them, and intriguing communicating mechanisms within individual cells (intra-cellular), in-between cells (inter-cellular), across the regions of the brain connecting central nervous systems (CNS), and brain to muscles to name some of the few involved. Biomolecules play very important roles in the brain through biochemical pathways [4], and the roles of biomolecules in neurons are an active area of research in neuroscience and computational biology. The next level of organization are neurons which interact with their neighbors through the connections that are called "synapses" and they play an important role in memory formation. The networks in the brain facilitate the information transfer among cells, and the adaptive nature of synapses and networks significantly contribute to flexibility that is seen in response to environmental changes. This flexibility, "plasticity", is a main subject of research among scientists. At a macro-level, the brain has several regions with unique physiological characteristics (Fig. 1.1): forebrain, midbrain and hindbrain, and with four lobes: frontal, temporal, parietal and occipital. Reasoning, motor function, emotions and cognitive functions are associated with the frontal lobe; the parietal lobe, located behind the frontal lobe, handles information processing [5]; the temporal lobe at the side of the head supports hearing, memory and emotions; the occipital lobe, found at the back of the brain, contains the primary visual cortex [3].

It is fair to say that neurons are one of the most, if not the most, highly evolved cells to perform specialized functions including simultaneous chemical and electrical signaling. Neurons have somato-dendritic and axonal compartments [6–8]. Like many other cells, a neuron has soma containing the nucleus, mitochondria, polyribosomes, Golgi apparatus and endoplasmic reticulum (ER). Neurons come in different shapes, unipolar, multipolar and bipolar (Fig. 1.2), and multipolar neurons are a very common type: they have

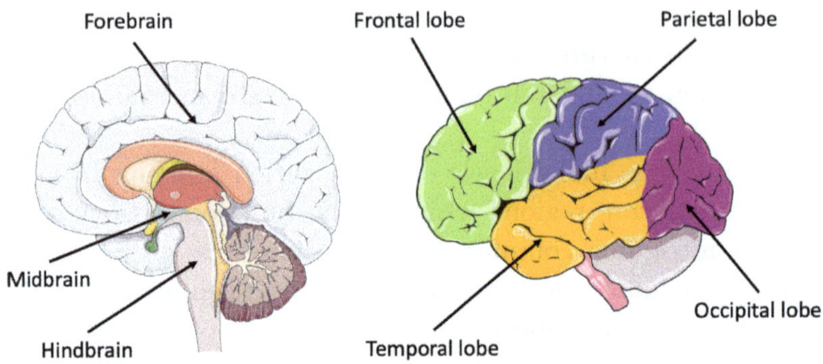


Fig. 1.1. The brain and its parts can be divided into three main categories with four lobes. This figure is produced using Servier Medical Art (https://smart.servier.com).

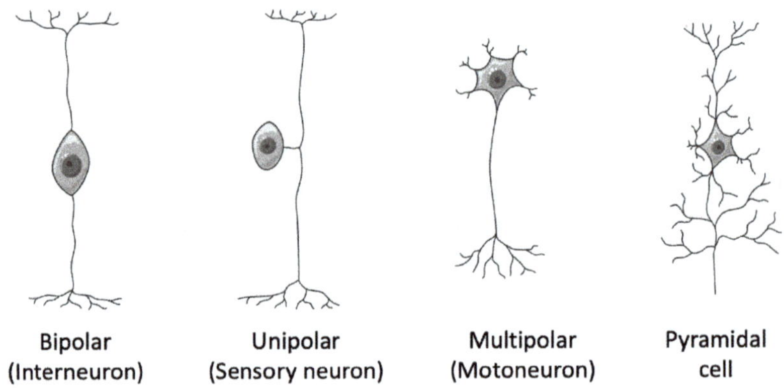


Fig. 1.2. Basic neuron types. This figure is produced using Servier Medical Art (https://smart.servier.com).

several dendrites stemming from their cell body followed by a long axon covered with a myelin sheath terminating in an axon terminal.

What happens in the neuron is a unique phenomenon in terms of biological signal transduction: the dendrites act as receptors of input signals and they are transmitted to the soma which is the integrator of the information, and the integrated signals are transmitted through the axon. The outputs travel along the axon ending up in the axon terminals and these outputs get transferred to next neuron in the form of neurotransmitters such as dopamine, acetylcholine, serotonin, glutamate and aspartate. The myelin sheath protects the signal transmission along the axon, and the speed of signal slows down greatly if the sheath is absent. The weakening of the myelin sheath causes several neurological disorders [3].

There are a multitude of biological processes associated with the flow of information from the dendrites to the axons, and neuronal polarity, which refers to the one directional travel of signals within a neuron due to the morphology, internal molecular structures, and functions of the axons and dendrites [8]. The vehicles for the signal transmission between neurons are action potentials (AP), which are caused by the change in membrane potential due to the efflux and influx of cations. The axon initial segment (AIS) is where AP begins, and the AIS divides the somato-dendritic and axon compartments [9]. The AIS is a very important region in the neuron and we discuss complex structure and its associated protein later in Chapter 3 because the experimental evidence suggests strong link between some AIS proteins and AD. It is quite logical to think that if APs are not properly formed in AIS, then faulty signals may not form memories in the synapses [10, 11].

There are different types of connections between neurons, and the diverse array of connections (synapses) based on where they get connected are shown in Fig. 1.3; some transfer information through electrical signals and others through chemical signals, so the synapses are categorized based on how they transfer information: electrical and chemical synapses. We pay more attention to chemical synapses which use neurotransmitters, as most neurogenerative diseases are related to their malfunction.

The chemical synapses between neurons play a vital role in synaptic transmission, which is a neurotransmitter-driven process (Fig. 1.4). The process involves a few sequential steps, including neurotransmitter synthesis; neurotransmitters release into the synaptic cleft; binding between neurotransmitters and postsynaptic ligand-gated receptors, as well as induction of postsynaptic response; and clearance of neurotransmitters out of synaptic cleft. Among the different types of neurotransmitters, amino acids and amines are involved in fast-synaptic transmission. Both amino acids and amines

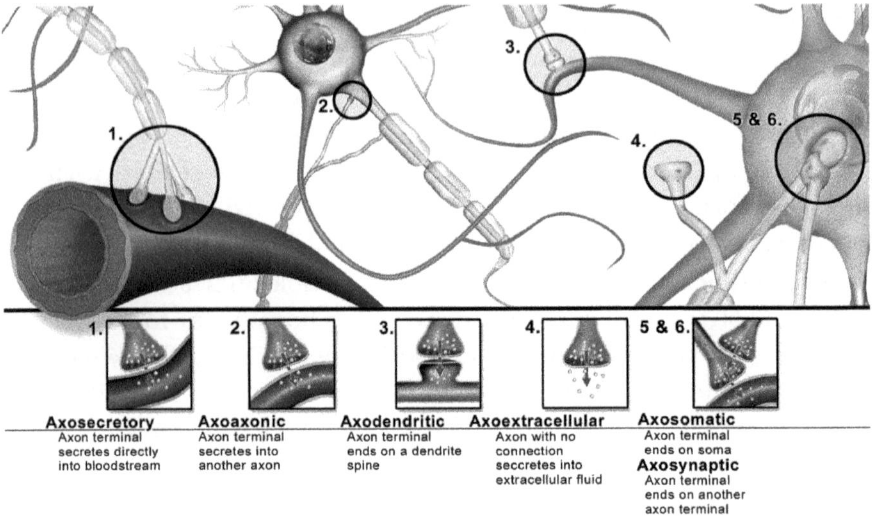


Fig. 1.3. Types of synapses [12].

are released from synaptic vesicles in presynaptic cells, and both of them are synthesized in the axon terminal by their synthesizing enzymes and are taken up shortly after synthesis into the synaptic vesicle [13]. However, amino acids, as the elementary blocks of proteins, are abundant in all cells; while amines are only synthesized by the neuron which releases them.

Action potential triggers the neurotransmitter release through depolarizing the presynaptic membrane. Action potential can be generated in many ways: by sensory nerves in response to environmental stimuli or by inter-neurons in response to synaptic transmission from their presynaptic neurons. Action potential depolarizes the presynaptic membrane at the active zone, where many voltage-gated calcium ($Ca^{2+}$) channels exist [13]. The depolarization opens these $Ca^{2+}$ channels and triggers $Ca^{2+}$ influx to increase the intracellular $Ca^{2+}$ level. The increased intracellular $Ca^{2+}$ level further triggers the neurotransmitter release [14].

The neurotransmitter release occurs rapidly through exocytosis, which fuses the synaptic vesicle membrane with the presynaptic membrane and releases its contents, neurotransmitters, into the

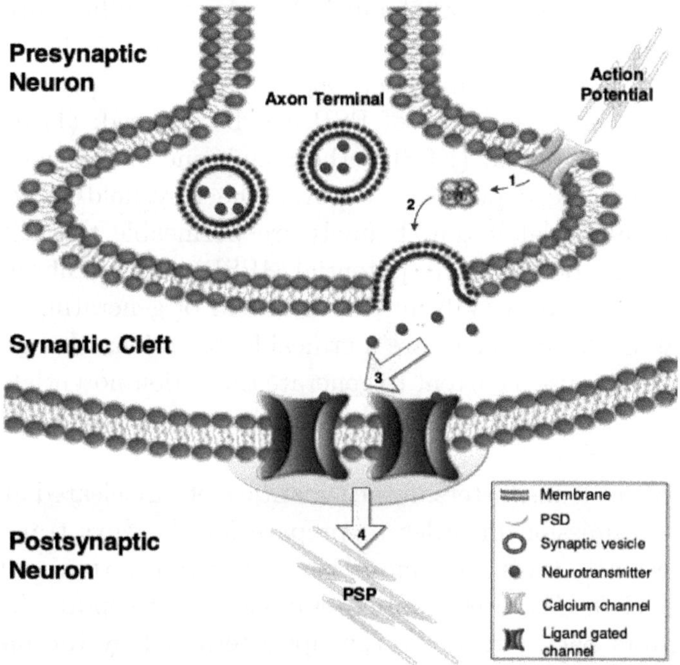


Fig. 1.4. Synaptic transmission. (1) Action potential depolarizes the presynaptic membrane to open the calcium channel, which allows an influx of $Ca^{2+}$ ions. (2) $Ca^{2+}$ ions trigger the release of neurotransmitters into the synaptic cleft by exocytosis at the active zone. (3) The released neurotransmitters diffuse across the synaptic cleft and bind to ligand-gated receptors at PSD. The binding opens the embedded ion channels of the receptors. (4) The opening of the ligand-gated ion channels allows an influx of ions to induce a postsynaptic potential (PSP). Also acknowledgment to [4]: Kulasiri, D. and He, Y. (2017) *Computational Systems Biology of Synaptic Plasticity: Modelling Of Biochemical Pathways Related To Memory Formation And Impairement,* Vol. 10 (World Scientific).

synaptic cleft. The release is triggered by a vesicle protein called synaptotagmin 1, which senses $Ca^{2+}$ and triggers the exocytosis [13, 15]. After the release, the empty synaptic vesicle re-enters the axonal terminal.

The released neurotransmitters bind to ligand-gated receptors in PSD and induce post synaptic potential (PSP). The ligand-gated receptor is composed of 4 or 5 subunits with an embedded ion channel. With a neurotransmitter bound, the subunits are twisted to expose the ion channel [16]. An ion influx through the exposed ion channel induces PSP as a result of the great ion gradient across the membrane. Depending on ion permeability of the ion channel, two types of PSP can be induced: (1) excitatory postsynaptic potential (EPSP): the membrane of the postsynaptic neuron would be depolarized towards the threshold of generating an action potential if the channels are permeable to cations; and (2) inhibitory postsynaptic potential (IPSP): the membrane would be hyperpolarized away from the threshold of generating an action potential if the channels are permeable to anions. Large enough EPSP would cause a current to generate the action potential in spike-initiation zone, which is usually located in the axon hillock, of the postsynaptic cell.

The neurotransmitters in the synaptic cleft are cleared at the end of the transmission in order to prepare for the next transmission. There are two ways to remove the neurotransmitters out of the synaptic cleft: (1) presynaptic reuptake to transport the neurotransmitters back into the presynaptic terminal by the membrane transporter proteins in the presynaptic membrane. Once back into the presynaptic terminal, the neurotransmitters are either degraded or reloaded into the synaptic vesicle; and (2) enzymatic destruction in the synaptic cleft to inactivate or destroy the neurotransmitters. Once the neurotransmitters are cleared, one cycle of synaptic transmission is completed. This system of signal transmission in synapses is the subject of active research in neuroscience, and it plays a significant function in memory formation and impairment, hence we discuss this system at length in Chapter 5.

## 1.3. AD and Morphology of the Brain

As we discussed earlier, AD affects all the organizational levels of the brain starting from single neurons. The onset of AD within a single neuron cannot be detected using molecular markers yet, but we can gain insights into the biochemical processes by developing the descriptions of biochemical pathways in a neuron, and then by constructing computational and mathematical models of

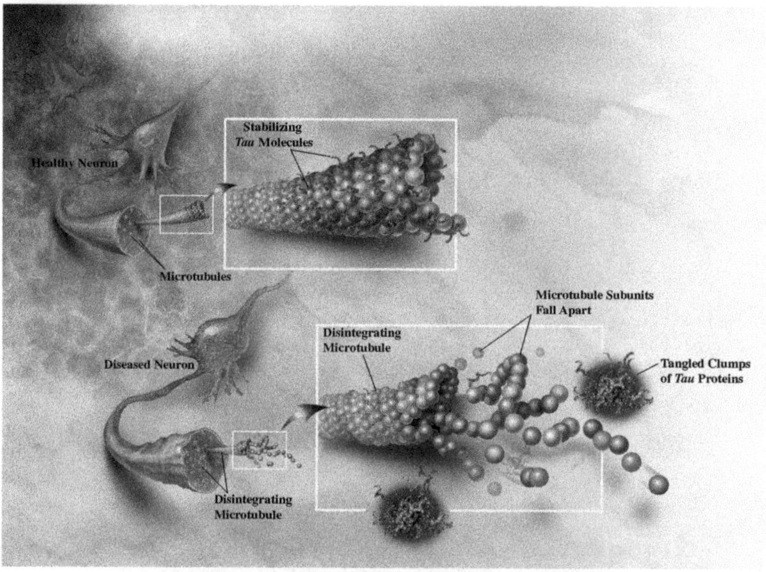


Fig. 1.5. Alterations in tau protein lead to the disintegration of microtubules in AD brain (public domain in US).

the pathways. (We discuss these pathways and models in the later chapters.) During the later stages of AD, we see that neurofibrillary tangles get formed in the inter-neuronal spaces (Fig. 1.5) and amyloid plaques are present within neurons. Compared to healthy neurons, inter- and intra-neuronal spaces are physically clogged by these toxic substances blocking the food and nutrient supply to the cells and tissues, and eventually leading to neuronal demise. This overly simplified picture is easy to understand, but genetic and molecular level details of the processes involved in leading to shrinkage of neurons are not completely understood, but we know a lot more now than the days of Dr. Alzheimer. The loss of neurons slowly but surely shrivels the regions of brain impairing memory and the functions of the body. Figure 1.6 shows the morphological changes of the brain as AD progresses through the different stages. Starting from the hippocampus, AD spreads from the medial temporal lobe to the whole brain with differing severity of dementia symptoms, eventually killing the patient.

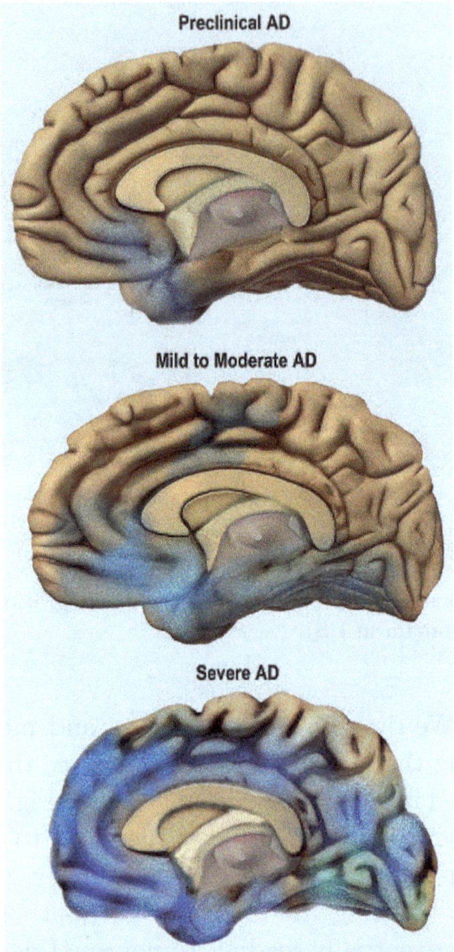


Fig. 1.6. Stages of AD brain (public domain in US). As the progression of AD, the disease spreads from the hippocampus and entorhinal cortex to the whole brain with different severity of dementia symptoms. The morphology of the AD brain is also changing during disease progression. The AD brain shows cortical shrinkage, ventricle enlargement, and hippocampus shrinkage as the disease progresses from mild to severe.

Because the progress of AD can be seen by observing the morphology of brain, medical scientists extensively use brain images for diagnosis of the disease. However, we are far from having an effective drug to intercede the progression of AD and to grow the affected neurons back to a healthy state. Nevertheless, there are reasons to be optimistic: neuroscience has advanced so much during the last three decades, especially in cellular and molecular levels, that it is quite conceivable that we may have successful therapeutic interventions in near future.

## 1.4. Conclusion

In this chapter, we briefly introduce AD, the physiology of the brain and how AD affects its morphology. We selectively discuss some of the background and this book is not a comprehensive one in neurobiology. Our intention is to discuss AD from mechanistic and functional points of view as much as possible paving the way to understand the biophysical aspects of AD in the later chapters. There is extensive literature about the brain and some of them are listed in the references.

## References

[1] Prince M., Comas-Herrera A., Knapp M., Guerchet M. and Karagiannidou M. (2016). World Alzheimer report 2016: Improving healthcare for people living with dementia: Coverage, quality and costs now and in the future, https://www.alzint.org/u/WorldAlzheimerReport2016.pdf.

[2] Bergmann O., Spalding K.L. and Frisén J. (2015). Adult neurogenesis in humans. *Cold Spring Harbor Perspect. Biol.*, *7*(7), p. a018994.

[3] Bear M., Connors B. and Paradiso M.A. (2020). *Neuroscience: Exploring the Brain* (Jones & Bartlett Learning, LLC).

[4] Kulasiri D. and He Y. (2017). *Computational Systems Biology of Synaptic Plasticity: Modelling of Biochemical Pathways Related to Memory Formation and Impairement* Vol. 10. (World Scientific).

[5] Ackerman P.L. (1992). Predicting individual differences in complex skill acquisition: Dynamics of ability determinants. *J. Appl. Psychol.*, *77*(5), p. 598.

[6] Kole M.H. *et al.* (2008). Action potential generation requires a high sodium channel density in the axon initial segment. *Nature Neurosci.*, *11*(2), pp. 178–186.

[7] Kole M.H. and Stuart G.J. (2008). Is action potential threshold lowest in the axon? *Nature Neurosci.*, *11*(11), pp. 1253–1255.

[8] Leterrier C. *et al.* (2015). Nanoscale architecture of the axon initial segment reveals an organized and robust scaffold. *Cell Rep.*, *13*(12), pp. 2781–2793.

[9] Kole M.H. and Stuart G.J. (2012). Signal processing in the axon initial segment. *Neuron.*, *73*(2), pp. 235–247.

[10] Jones S.L. and Svitkina T.M. (2016). Axon initial segment cytoskeleton: Architecture, development, and role in neuron Polarity. *Neural Plasticity*, 2016, 6808293.
[11] Huang L., Hu W. and Shao Z. (2005). Study on techniques of decoding output data packages from AIS. *J. Jimei Univ. Natural Sci.*, *10*(1), pp. 37–41.
[12] Blaus B. (2014). Medical gallery of Blausen medical 2014. *Wiki J. Med.*, *1*(2), pp. 1–79.
[13] Südhof T.C. (2004). The synaptic vesicle cycle. *Annu. Rev. Neurosci.*, *27*, pp. 509–547.
[14] de Jong A.P. and Verhage M. (2009). Presynaptic signal transduction pathways that modulate synaptic transmission. *Curr. Opinion Neurobiol.*, *19*(3), pp. 245–253.
[15] Stevens C.F. (2004). Presynaptic function. *Curr. Opinion Neurobiol.*, *1*(3), pp. 341–345.
[16] Wollmuth L.P. and Sobolevsky A.I. (2004). Structure and gating of the glutamate receptor ion channel. *Trends Neurosci.*, *27*(6), pp. 321–328.

# Chapter 2

# Neuron and Its Environment as Complex Dynamic Systems

## 2.1. Introduction

The functional aspects of the brain have many levels of interactions, and therefore, one way to understand its functioning is to adopt the systems approach as in physics, especially in thermodynamics. During the last two decades, "systems biology" has emerged to reflect the complexity of life, and there are many approaches to understanding living cells and organs as systems having components and interactions which are dynamic in various scales of time. It is important to briefly overview the ideas that propelled the field of systems biology as a convergence of biology, mathematics, computing, and physics so that we can develop ways to understand the brain and AD. The conceptual foundations of systems biology emanate from various disciplines and are still active areas of research. In this chapter, we adopt approaches and concepts which are quite common in physics to discuss the biology of the brain. We begin by understanding systems in general and their inputs and outputs. There are many excellent papers and books on systems biology and the reader may want to read them to broaden the understanding, and we include a few references at the end of this chapter, which is not an exhaustive list, so we have missed many more excellent references.

Systems thinking started in thermodynamics in the 18th century to understand the relationships between work and heat during the industrial revolution. Most sciences have been reductionist in the sense that their experiments are focused on understanding individual components and their behavior rather than the interactions within a system among the components. This is especially true in biology where specificity is the central focus of an organism rather than understanding the general nature of similar organisms. In contrast, physics is about developing general theories using mathematics to understand the observed phenomena: Newtonian mechanics has been partly successful in ushering science and technology to what we have in modernity. However, the desire to have systems-level understanding has been a spasmodic theme in biology, but due to complexity and the evolutionary nature of organisms, this has not been an easy task. Thanks to revolutionary advances in measurement technologies during the last few decades, especially in molecular biology, there has been a resurgence in multi-faceted attempts at a systems-level understanding of biology, which is becoming more and more a quantitative science. We can amass large quantities of data rapidly in genomics for example, and bioinformatics software packages and programs are used to understand the phenomena measured. These "high-throughput" robotic technologies pave ways for us to develop a theoretical understanding of biological systems as in physics. But there are serious challenges in developing system approaches like those in physical and engineering systems to living systems, and we intend to discuss some of these challenges showing a specific example of a system in neurons in the next chapter.

Encouraged by the success in molecular biology for insights into the functions of intra-cellular entities such as genes, proteins and mRNAs and their interactions, systems biology as a scientific field has been around for the last two decades. Primary conceptual drivers for the emergence of systems biology are (1) network theory, (2) modularity as in electronics, (3) ideas around emergent properties, robustness and fragility, (4) use of mathematics and physics, and lately (5) data sciences. However, most of systems biology literature has a strong specificity flavor, i.e., has been strongly influenced by experimental biology as it should be, which is its strength and the limitation. Systems biology as it stands now is not a well-defined field because of its recent inception, and one of the key characteristics of systems biology is its emphasis on understanding the dynamic nature of interactions among the components in a system. However, living systems not only change within a multitude of timescales but also, they are capable of self-replication and annihilation which are not seen in engineered systems. Developing theories which encapsulate these aspects and emergent properties of systems is currently the biggest challenge in systems biology.

In thermodynamics, there are three kinds of systems depending on the nature of boundaries: isolated, closed and open systems. In isolated systems, there are no energy and matter moving across the boundaries; in closed systems, there is only energy moving in and out of the systems; and, in open systems, both energy and matter move in and out of the systems. Boundaries are very important aspects of biology: cells simply cannot exist without highly evolved membranes, and biological membranes have very complex functionalities, and living, breathing, and evolving membranes are one very extraordinary feature that distinguishes biological systems from other systems. Therefore, while biological systems are open systems in the thermodynamic sense, the "openness" here is not as simple as in engineered systems but is highly sophisticated and evolutionarily well-adapted for changing environments. We see that the boundaries associated with neurons are highly specialized membranes such as myelin sheath in neurons. In biological systems, the boundaries, i.e., membranes, are active participants of systems dynamics.

## 2.2. Components of a Neuron

As we have already seen in the previous chapter, neurons are amazingly important cells in our body. They have very special functions such as informing each other (signaling), keeping short and long term memories, and being responsible and supportive of the other numerous cognitive aspects of the brain. Neurons have also evolved structural features to enable these functions which are an area of active research and we find that an ever-growing body of experimental evidence is available to us. There are also organelles which are quite common to all the cells and in this chapter, we aim to introduce the functionalities and structures of neurons to the readers from physical sciences and mathematical backgrounds so that they would understand the later chapters. There are pertinent references at the end of the chapter for further reading.

Neurons are supported and surrounded by glial cells, which fill the spaces between neurons, consisting of astrocytes, oligodendrocytes, ependymal cells, and microglial cells (Fig. 2.1). Oligodendrocytes form the myelin sheath around axon which acts as an insulator of a wire conducting electrical current, and in the peripheral nervous system, the Schwann cells form the myelin sheath [1]. The myelin sheath is functionally important because its deterioration leads to demyelinating diseases such as multiple sclerosis and affects the muscle movements, sensations and cognition depending on the axons that are involved. There are numerous other functions that have

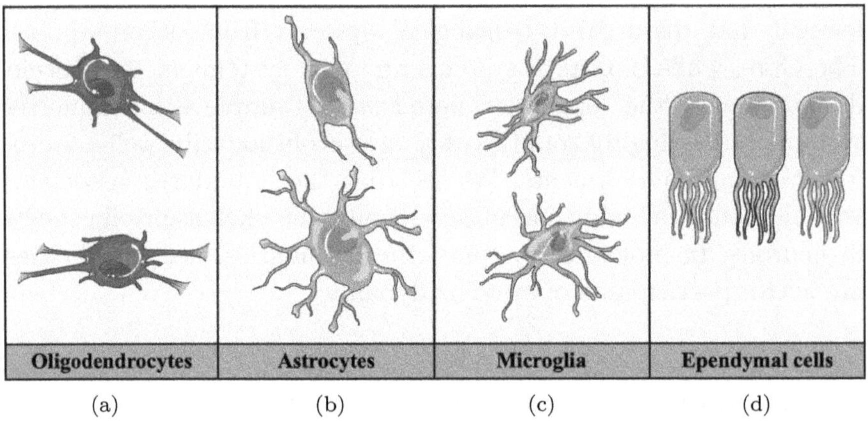


Fig. 2.1. Four types of neuroglia in the central nervous system (CNS). (a) Oligodendrocytes: glial cells that produce the myelin sheath to provide support and insulation to neuronal axons in CNS. (b) Astrocytes: star-shaped glial cells which are the most abundant cells in the human brain that maintain the blood-brain barrier and preserve the chemical environment by recycling ions and neurotransmitters. (c) Microglia: glial cells that act as the first and main form of active immune defense by removing cell debris, wastes and pathogens via phagocytosis. (d) Ependymal cells: simple cuboidal cells that line ventricles (in the brain) and the central canal (in the spinal cord) and are involved in the production of cerebrospinal fluid. This figure is produced using Servier Medical Art (https://smart.servier.com).

been discovered for which glial cells are responsible [2]: glial cells participate in the uptake of the neurotransmitters in the postsynaptic cells for intercellular signaling; they act as a buffer for ions in the extracellular environment; they remove the dead material from dying neurons; they metabolize the neurotransmitters; glial cells, like the myelin sheath, insulate and ghettoize different groups of neurons from each other; they are scaffolds for axonal growth and neuronal migration; and they nurture neurons providing necessary proteins and metabolites. We are beginning to understand the important roles glial cells play in the brain and there is some evidence that they may have a role in memory formation as well. Hence, glial cells are active participants in the brain instead of being passive supporting actors, according to recent research.

Even though neurons appear different from the cells in plants and animals, there are many components in neurons which are the same as those in other cells. The neuronal cell body or soma contains all the organelles which are common to other cells: the nucleus containing the genome which is identical to the genomes in other cells in the organism with some exceptions; giant neurons found in some invertebrates may have a large number of copies of the genome which may be due to division of genome without cell division. In these situations, a soma may have as many as 50,000 copies of the genome in the nucleus, and the functional implications of having this many copies of the genome are not yet understood. Apart from these kinds of situations, the nucleus is a system with its own complex dynamics, with the genome being responsible for gene expression specific to the neuron, i.e., specialized proteins are produced for the functions required in the neuron, just like in any other cell. The ways the genes are expressed are different in many different types of neurons and glia found in the brain.

As mentioned in Chapter 1, the presence of neurofibrillary tangles and amyloid-$\beta$ (A$\beta$) senile plaques is the neuropathological hallmark of AD [3]. It is also known that inheritance of specific genes is responsible for predisposition to onset of AD and some genes influence the disease progression. It is now recognized that certain genetic and biochemical pathways are instrumental in the onset and progression of AD and the mutational studies show the critical roles these pathways play in many aspects of AD. Many neurogenerative diseases including AD could have either familial, which is quite rare, or non-familial (common) forms, but there is enough evidence now to indicate that these diseases have genetic causes and we need to know more about them. Early-onset familial AD is an example of mutations due to APP, PSEN1, and PSEN2 genes, but these mutations are very rare, in less than 5% of AD patients. Late-onset AD, on the other hand, has only a single gene (APoE) that is responsible for increasing the risk of AD; however, the complexity is not only in the genetic networks, but also in the protein–protein interaction networks. We discuss these networks in later chapters, and despite the tremendous progress in understanding AD at the molecular level, we still cannot confidently say which factors contribute to the onset of AD to prevent its progress in the beginning. Therefore, our expectation is that if we understand the networks involved in enough detail, our understanding would lead to prevention, maybe through pharmaceutical intervention.

Most of the mechanisms that are responsible for the intracellular functions such as gene regulation and protein transcription are conserved across many cell types including neurons. However, one exception that stands out in neurons is its form consisting of different shapes of soma and the existence of dendrites and axons as parts of neurons, and structural stability of the morphology. The proteins that are associated with the structure of neurons play very crucial roles in not only maintaining the form, growth, and shrinkage of neurons but also in signaling right across different clusters of neurons in the brain. The onset of AD starts the process of contracting the affected neurons and therefore the structural proteins may have relationships with the causes of AD at molecular level.

The cytoskeleton is a collection of structural proteins which form networks of filaments within a neuron, and the main constituents of these networks, the microfilaments, the neurofilaments, and the microtubules. The microfilaments are highly ordered protein structures made of the proteins actin and myosin, and their role in skeletal muscle in producing muscle contraction is well understood. The neurofilaments are long filaments with a diameter of approximately 10 nm, and their disorganization is associated with AD. The disordered tangles of neurofilaments visible in an AD brain and senility may have a correlation, but the nature of it is not clear. The microtubules are present in axons and dendrites playing a significant role in neuronal development. We discuss the axon initial segment (AIS) in detail in Chapter 3 in which we discuss the role of structural proteins in relation to action potential (AP) initiation.

The dynamics of actin filaments contribute to the structural plasticity during the early phase of memory formation, and actin filaments act as a scaffold to recruit various postsynaptic proteins. Scaffolding proteins such as actin maintain the mechanical stability of the spine [4–7]. The two forms of actin, the monomeric globular form (G-actin) and the filament form (F-actin), are in dynamic equilibrium. This brief snapshot shows the complexity of structural proteins (microfilaments, neurofilament and microtubules) which support the morphology of neurons by dynamically interacting with many other proteins to maintain the crucial functions such action potential generation.

Neurons come in different shapes and sizes, and the lengths of axons are especially highly variable [8]; some axons are 1 m long, and if the proteins are synthesized in soma, how would they get transported to the terminals, i.e., presynaptic cells? Protein synthesis is known to occur in dendrites but not in axons, and the transport of necessary proteins and metabolites to distal regions of axons cannot be explained by diffusion as it would take days to move the vesicles of proteins and neurotransmitters to distal tips. For example, the time taken for passive diffusion of a protein to the tip of a 1 cm-axon starting from soma is about 10 days, which is contrary to the experimental observations. Neurons have a roster of mechanisms evolved over millions of years to address this transport problem which are highly sophisticated, and only now do we know some of the details of these mechanisms called axonal transport mechanisms (ATMs) [9]. ATMs consist of structural proteins and molecular motors, and they are a fascinating subject which is beyond the scope of this book.

## 2.3. AP and Synaptic Transmission

APs propagate the signals from one neuron to the other [10]. APs originate in the region between the axon hillock and the axon (Fig. 2.2), AIS [11], and travel along the axon to the presynaptic terminals. Then the signals get transferred to the postsynaptic cells of another neuron. APs are very important signaling vehicles in the brain, so we introduce them in this section, and we discuss the AIS in some detail in the next chapter to show the complexity of a system associated with the neuron. Some of the structural proteins play a major role in the AIS, dysfunction of which would be a factor in many neurogenerative diseases.

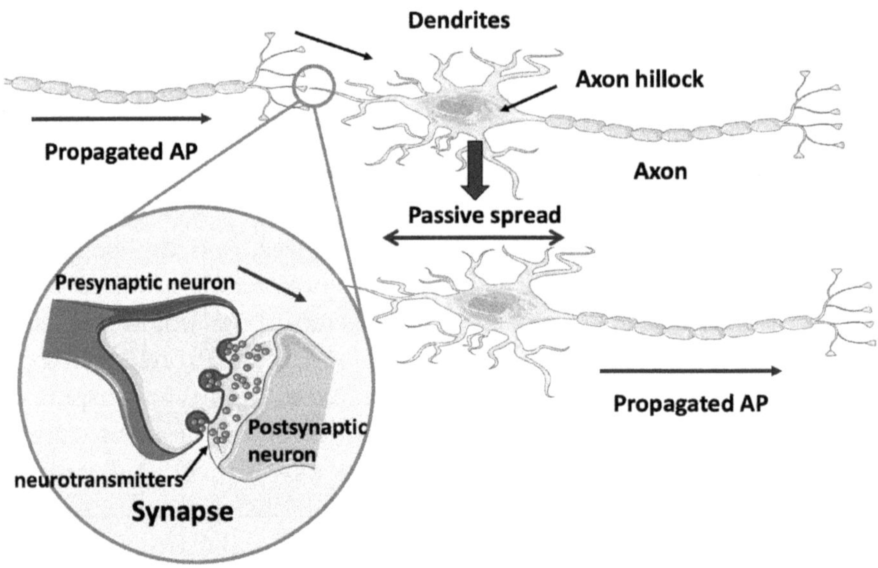


Fig. 2.2. Neuronal signal transmission requires propagation of action potential (AP) within a neuron and neurotransmitter passing between neurons. AP is generated at the axon hillock of a neuron after receiving the incoming signal from the previous neuron and then propagated along the axon. When an AP reaches the end of the axon, the presynaptic terminal, it induces the release of neurotransmitters, which then travel across the synaptic cleft to the postsynaptic neuron, passing the signal to the next neuron. This figure is produced using Servier Medical Art (https://smart.servier.com).

### 2.3.1. *Generation of APs*

The interior of axon membranes is negatively charged (approximately −65 mV at rest) [12]. At rest, there is a higher concentration of $Na^+$ outside of the cell membrane than inside, and vice versa for $K^+$. If a threshold of electrical charge reaches the axon hillock, voltage-sensitive $Na^+$ channels are made to open. Because there is a higher concentration of $Na^+$ as well as an electron gradient outside of the cell membrane, the opening of the V-gated channel causes a rapid influx of $Na^+$ into the cell. This causes the inside of the membrane to become more positively charged. The interior continues to get positively charged until it reaches approximately +40 mV. This voltage is the threshold at which $K^+$ channels open. Because the concentration of $K^+$ outside of the cell is lower as well as there is a positive electron gradient, $K^+$ moves out of the membrane from the $K^+$ channels. The $K^+$ channels overshoot their mark, however, and hyper depolarize the inner membrane such that its voltage is less than −65 mV. The axon then recovers back to its resting potential in what is known as the "refractory period" [13].

If a powerful electrical stimulation is imposed on the cell body, multiple action potentials can be sent down the axon [14]. For example, if there is a strong depolarization current sent to the neuron, it can be made to fire several action potentials. A weak current may produce a single action potential per second (denoted as 1 Hz) while a powerful current may produce 100 Hz or even up to 1000 Hz. This is known as the "firing frequency" of a neuron. These limits are caused by the effects of the refractory period. Finally, an ATPase $Na^+/K^+$ pump pumps $Na^+$ back out of the membrane and $K^+$ back in until the concentration is brought back to −65 mV, and the concentration of $Na^+$ and $K^+$ inside are as they were at the resting state.

There are two phases of the refractory period, the absolute refractory period and the relative refractory period (Fig. 2.3) [15]. The absolute refractory period is defined as a period when no more action potentials can be sent. This is because $Na^+$ channels get inactivated when the membrane is strongly depolarized. The $Na^+$ channels cannot be made to open again until the membrane

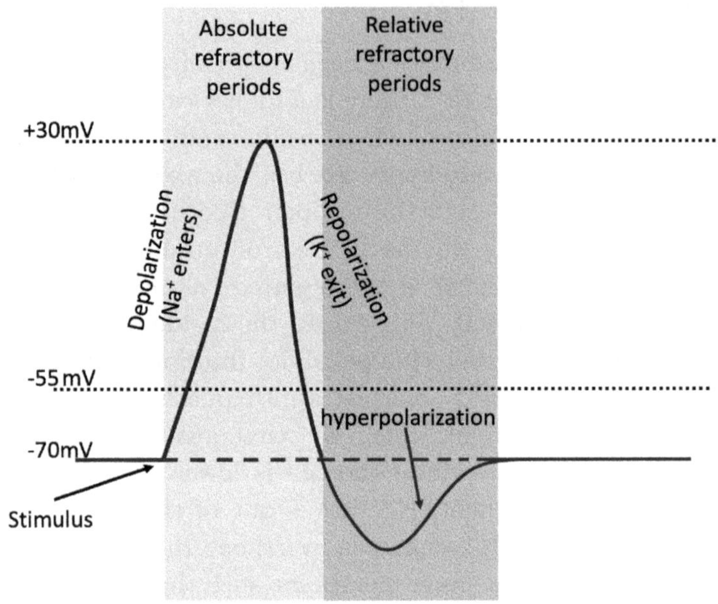


Fig. 2.3. Absolute and relative refractory periods of an action potential (AP). Absolute refractory period is the time from the opening of the $Na^+$ activation gates till the closing of inactivation gates. During this period, the neuron cannot respond to another stimulus. Relative refractory period follows the absolute refractory period. $Na^+$ gates are closed whereas $K^+$ gates are open, and repolarization is occurring. Only a strong stimulus can generate an AP.

becomes sufficiently negative. The relative refractory period refers to hyperpolarization of the membrane caused by $K^+$ effluxes. During this period, it is possible to produce another action potential, but extra depolarization is required to reach the threshold to induce an action potential.

### 2.3.2. *Propagation of APs along the axon*

Another feature of action potentials are that once they are sent down axonal bodies, the message is propagated along the entire length of the axon (Fig. 2.2) [16]. This is because when a section of an axon is depolarized, the charge spreads down to the next patch and causes it to depolarize, and thus open the V-gated channels. This pattern continues until action potential has completed its run down the axon.

Importantly, this is normally a one-way process since the previous patch of the membrane is in refractory. However, in certain neurons such as cerebellar Purkinje cells, APs can possibly travel backwards to the dendrite as a dendritic spike [17].

### 2.3.3. *AP induced synaptic transmission*

In the nervous system, a chemical synapse is a junction that comprises a presynaptic terminal, a postsynaptic component and a synaptic cleft between them (Fig. 2.2). When the AP reaches the end of the axon, the presynaptic terminal, the electrical signal is converted into a chemical one. Once the AP reaches the presynaptic terminal, it will promote the release of neurotransmitter vesicles into the synaptic cleft [18]. Glutamate is the major excitatory neurotransmitter in the central nervous system and each vesicle contains thousands of glutamate molecules [19]. The released glutamate molecules diffuse across the synaptic cleft to the postsynaptic cell or spill over into the extrasynaptic space. At the distal dendritic spine, a region called the postsynaptic density (PSD) attached to the postsynaptic site of the spine head is enriched with large numbers of proteins, including neurotransmitter receptors, cytoskeleton proteins, ion channels and other signaling proteins [20]. These proteins function collaboratively to translate the chemical signal from the presynaptic terminal into a postsynaptic response that we have already discussed in Chapter 1.

### 2.3.4. *Electrical stimulation protocols (stimulation frequency)*

Neuron systems are intrinsically noisy, natural-like stimulation signals are used as inputs to reproduce responses *in vitro*, *in vivo*, and *in silico*. A proper stimulation pattern protocol can lead to improved predictability and efficiency [21, 22]. Neuronal signal transmission is suggested to be frequency dependent. A variety of electrical stimulation protocols that produce certain stimulation patterns are applied to study the propagation of APs and synaptic transmission. In this section, we focus on discussing the common options of stimulus patterns that are widely used in neuroscience research.

#### 2.3.4.1. *Presynaptic stimulation*

When the AP reaches the presynaptic terminal, it triggers the release of glutamate vesicles to the synaptic cleft. The neurotransmitter receptors in the membrane of the postsynaptic spine are activated by the neurotransmitter, resulting in the postsynaptic potential. In Fig. 2.4, four typical types of presynaptic stimulation patterns used in research are shown: (1) single stimulation (or 1 s of stimulation at 1 Hz; 1 pulse); (2) 1 s of low-frequency stimulation (LFS) at 10 Hz (10 pulses); (3) 1 s of high-frequency stimulation (HFS) at 100 Hz (100 pulses); and (4) theta-burst stimulation (TBS). In a train of TBS, pulses are grouped into several bursts, and the time duration (200 ms) between two bursts allows desensitized regulators (such as NMDA receptors) to partially recover. TBS is considered to be a more physiologically relevant stimulus, which is close to the frequency of the endogenous hippocampal rhythm that triggers Long-Term Potentiation [LTP] [23, 24]. One train of TBS consists of 10 stimulus bursts at 5 Hz (200 ms separation between bursts) and each burst consists of four pulses at 100 Hz. Four trains of TBS are delivered at 0.1 Hz (10 s separation between trains), which is used to induce LTP experimentally [24].

#### 2.3.4.2. *Pairing stimulation*

When an AP travels along the axon, it has been found that AP can propagate back to the dendrites of the pyramidal neurons. This phenomenon is called backpropagation of action potential (bAP). bAP can create strong depolarization on the dendritic membrane, for a brief duration. It activates voltage-dependent channels to induce ion flux across the membrane [25]. We simulate the bAP by injecting a potential into the spine head and dendrite shaft to generate the amplitudes of bAPs of 66.4 and 66.7 mV, respectively [26].

Experimental evidence shows that neither postsynaptic membrane depolarization triggered by presynaptic stimulation nor bAP

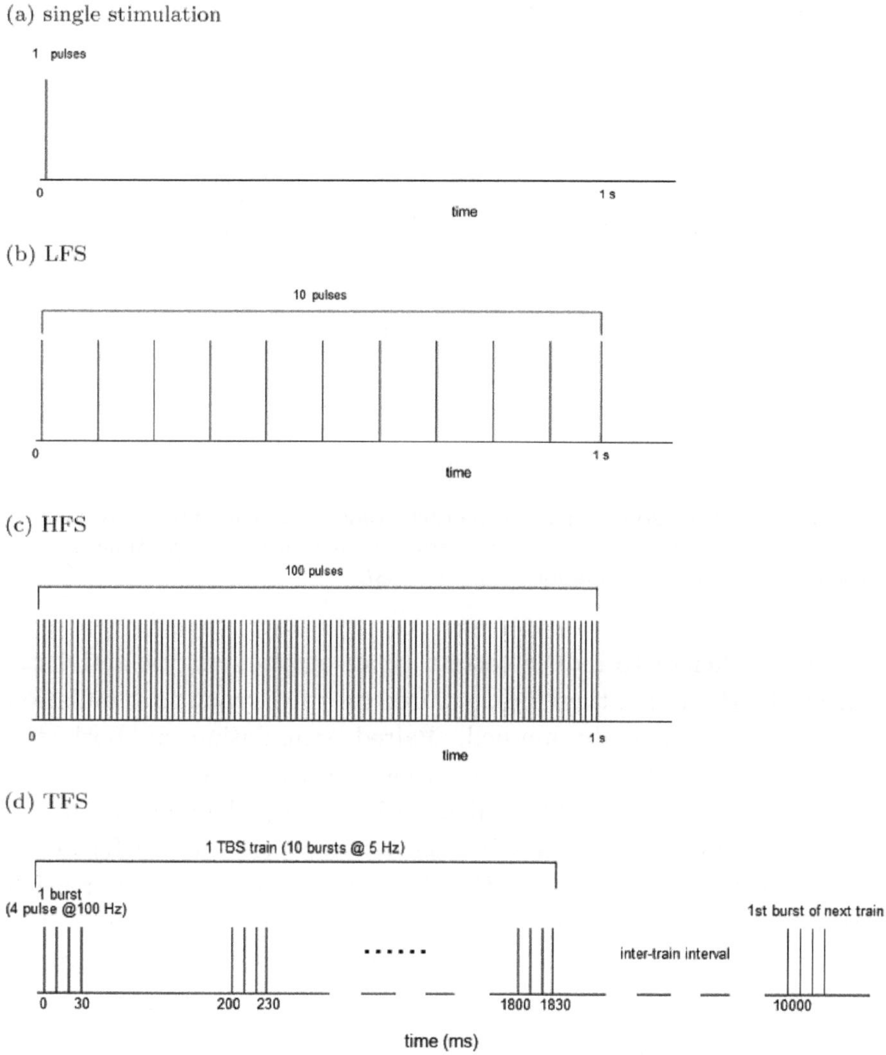


Fig. 2.4. Stimulation patterns. (a) A single pulse stimulation (or 1 s of stimulation at 1 Hz; 1 pulse); (b) 1 s of low-frequency stimulation (LFS) at 10 Hz (10 pulses); (c) 1 s of high-frequency stimulation (HFS) at 100 Hz (100 pulses); and (d) theta-burst stimulation (TBS). One TBS train consists of 10 stimulus bursts at 5 Hz and each burst consists of four pulses at 100 Hz. The TBS trains are delivered at 0.1 Hz (10 s separation between trains) [24].

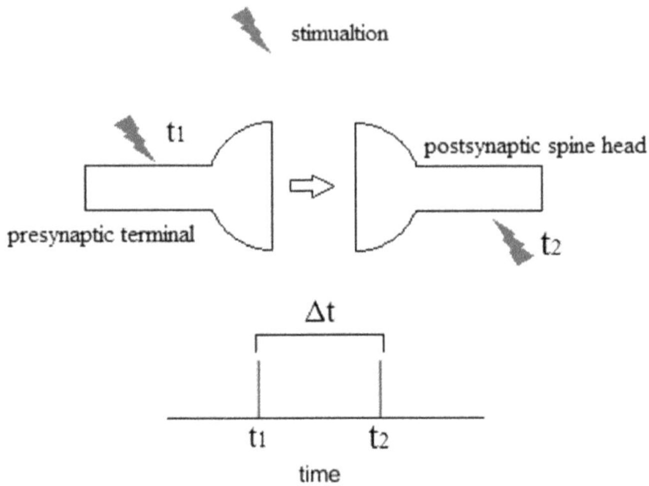


Fig. 2.5. Pairing stimulation. A pairing stimulation contains a presynaptic stimulation at time $t_1$ and a postsynaptic stimulation at $t_2$. The time interval between the pairing stimulation is $t_2 - t_1 = \Delta t$.

alone is sufficient to induce certain types of postsynaptic activities, such as LTP [27, 28]; therefore, a paired pre/postsynaptic stimulation protocol is commonly applied. Paired stimulation at both the presynaptic and postsynaptic neurons are used to create a pairing of the EPSP and the bAP, which leads to large depolarization and $Ca^{2+}$ elevation by N-methyl-D-aspartate receptor (NMDAR) in the postsynaptic spine head [29] (Fig. 2.5). When simulating the pairing HFS protocol, a bAP is introduced at the postsynaptic neuron after each presynaptic stimulus pulse with a small time interval between the two stimuli.

## 2.4. Amyloid and Tau Hypothesis of AD

AD is characterized by the abnormal accumulation of harmful proteins, A$\beta$ protein plaques and neurofibrillary tangles (NFTs) in the nervous system [3]. It is widely believed that these hallmarks are in connection with a massive loss of neurons and synapses in the patients' brains. AD has been identified for more than 100 years, however, to date, the cause of AD is still not fully understood.

Scientists believe that AD is a result of multiple factors and to explain the underlying mechanisms of AD pathophysiology, various hypotheses have been proposed, such as the amyloid hypothesis, the tau hyperphosphorylation hypothesis, the oxidative stress hypothesis, the cholesterol hypothesis and the vascular hypothesis. In this section, we discuss the most dominant hypotheses, the amyloid hypothesis and the tau hypothesis, and relevant pathways involved in them.

### 2.4.1. *Amyloid hypothesis*

The amyloid hypothesis was first proposed in the early 1990s [30]. It suggests that AD is initiated by A$\beta$ oligomers and A$\beta$ plaques in brain tissue. A$\beta$ deposition may directly injure brain neurons and contribute to neuronal toxicity, neuronal death and the subsequent degeneration and cognitive deficiencies in AD patients. During the past 20 years, A$\beta$ deposition has been widely accepted to be the leading cause of AD by most researchers and many experimental research projects are being carried out based on the amyloid hypothesis.

A$\beta$ is produced from a sequence beginning with the cleavage of its precursor, called the amyloid precursor protein (APP), by certain secretase enzymes [31]. The process of APP cleavage is illustrated in Fig. 2.6. APP is a transmembrane protein with a long extracellular N-terminal domain, a transmembrane domain, and a short intracellular C-terminal domain [32]. APP can be cleaved by two different kinds of enzymes, $\alpha$-secretase and $\beta$-secretase, at different sites to produce two different peptides. The $\alpha$-secretase cleaves APP within the critical A$\beta$ region that lies in the part of the extracellular N-terminal and transmembrane domains. The cleavage releases a large soluble amino-terminal fragment, sAPP$\alpha$, and a smaller C-terminal fragment, which will be further cleaved by $\gamma$-secretase. Cleavage by $\alpha$-secretase prevents the generation and release of the A$\beta$ peptide and, thereby, it is called the non-amyloidogenic pathway in APP processing. In the alternative amyloidogenic pathway, APP is cleaved by $\beta$-secretase outside the critical

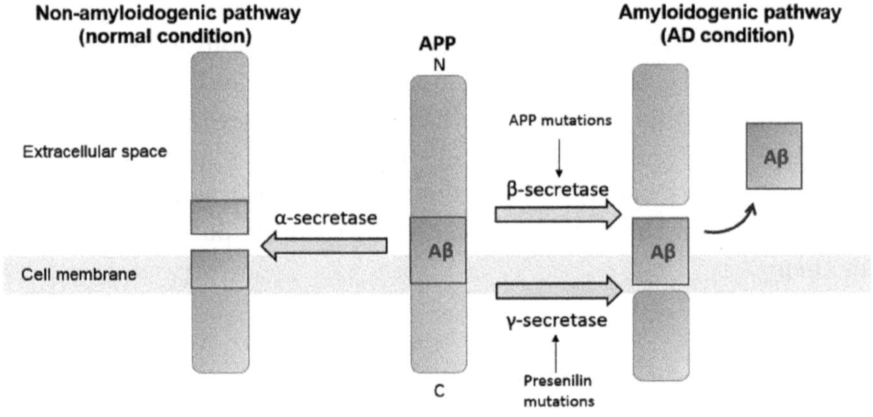


Fig. 2.6. The process of APP cleavage. APP undergoes a series of proteolytic cleavage, mediated by different secretase enzymes. Under normal condition, it is mostly cleaved by $\alpha$-secretase at the A$\beta$ domain (non-amyloidogenic pathway). Under AD condition, APP is cleaved by $\beta$- and $\gamma$-secretase enzymes, which releases the neurotoxic A$\beta$ peptides (amyloidogenic pathway). Either mutations in the APP gene (inhibit cleavage by $\alpha$-secretase) and in the presenilin-1 and presenilin-2 genes (increase cleavage by $\gamma$-secretase) in AD promotes the amyloidogenic pathway and leads to excess production of A$\beta$ peptide.

A$\beta$ region, releasing sAPP$\beta$ and a smaller C-terminal fragment. The $\gamma$-secretase further cleaves the C-terminal fragment releasing the A$\beta$. The majority of A$\beta$ peptides are 40 residues in length (A$\beta_{40}$), while a small proportion ($\approx$10%) are 42 residues in length (A$\beta_{42}$) [33]. Compared with A$\beta_{40}$, A$\beta_{42}$ is more hydrophobic and readily forms amyloid plaques, which are considered to play a causal role in the progression of AD [34]. Therefore, A$\beta_{42}$ is believed to be the most neurotoxic form of the A$\beta$ peptides.

Under normal physiological conditions, A$\beta$ peptides can be removed from brains quickly by its clearance mechanisms. For example, A$\beta$ peptides can be degraded proteolytically by multiple enzymes, especially neprilysin (NEP) and insulin-degrading enzyme (IDE) [35]. The balance between production and clearance of A$\beta$ determines steady levels of A$\beta$. An imbalance will result in the deposition of A$\beta$ and the formation of amyloid plaques. In familial Alzheimer's disease (FAD), mutations of APP, presenilins 1 and 2 genes will lead into increasing production of A$\beta$ and the ratio of $A\beta_{42}/A\beta_{40}$ increases the cleavage activity of $\beta$-secretase and $\gamma$-secretase [31]. In Sporadic Alzheimer's disease (SAD), rather than overproduction of $A\beta$, the formation of amyloid plaques is more likely attributed to the failure of $A\beta$ clearance. The initial cause of this failure is still unclear.

$A\beta$ can self-associate and exists in many different assembly forms, ranging from oligomers to fibrils and amyloid plaques. Initially, fibrillar $A\beta$ that constitutes amyloid plaque was assumed to be toxic. However, recent research suggests that $A\beta$ oligomers are the most toxic form, and may result in nerve damage [36, 37]. $A\beta$ oligomers at pathophysiological levels are considered to contribute to the activation of inflammatory neuronal cascades, oxidative stress, mitochondrial disturbances, $Ca^{2+}$ metabolism deregulation, tau protein phosphorylation and neuronal apoptosis [38]. Moreover, $A\beta$ oligomers can inhibit hippocampal long-term potentiation (LTP) and facilitate hippocampal long-term depression and, thus, cause impairments in synaptic plasticity, learning, and memory.

$A\beta$ is proposed to cause neuronal toxicity via a variety of pathways, and several hypotheses extending from the amyloid hypothesis attempt to further elucidate the alterations resulting from amyloid pathology.

Among them, the $Ca^{2+}$ hypothesis is one of the most important, and feasible hypotheses. We discuss the $Ca^{2+}$ hypothesis in detail in Chapter 6.

### 2.4.2. *Tau hypothesis*

The tau hypothesis proposes that abnormal phosphorylation of tau protein induces the formation of NFTs, which is one of the primary hallmarks of AD besides $A\beta$ plaques [39].

#### 2.4.2.1. *Tau as a component of neuronal cytoskeleton*

Tau protein is highly enriched in neurons and plays important physiological roles under normal conditions. Tau is an essential component of the neuronal cytoskeleton with a molecular weight ranging from 45 kDa to 65 kDa [40]. As one of the most abundant microtubule-associated proteins, tau promotes the formation of axonal microtubules and contributes to their stability, and therefore, drives neurite outgrowth [41]. A combination of protein kinases and phosphatases balances the amount of tau phosphorylation [40]. Failure to maintain this balance is involved in AD, Parkinson's disease, and many other neurodegenerative diseases [42].

In the AD brain, tau presents in a hyperphosphorylated state, which leads to aberrant secondary structures and loss of function [43]. Hyperphosphorylated tau shows a reduced ability to bind to the microtubules and thus, reduce their assembly. Under normal condition, tau is located in axonal microtubules, while in AD, it abnormally translocates to neuropil, where it aggregates and accumulates, disturbing inclusions, cell bodies and dendritic processes. The aggregation of tau in neuropil is one of the prominent cytopathological hallmarks that are found within the AD brain [44].

The aggregation of tau requires several sequential steps from abnormal tau protein to NFTs (Fig. 2.7). Initially, tau protein is synthesized as a single chain polypeptide and then its conformation is altered by post-translational modifications, promoting tau dimerization in an antiparallel manner [45]. Stable tau dimers form tau oligomers, which aggregate to form protomers, subunits of filaments. Two protomers twist around each other with a crossover repeat

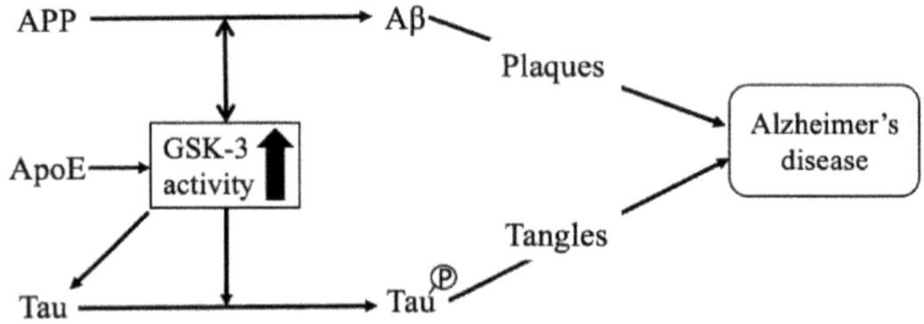


Fig. 2.7. A schematic of the dysregulation of GSK-3 pathway in Alzheimer's disease. Either mutation in APP or PS and AD and the presence of ApoE can lead to increased production of A$\beta$, which consequently increases the activity of GSK-3. GSK-3, in turn, promotes the production of A$\beta$ and the phosphorylation of Tau protein. Ultimately, these alterations result in the formation of amyloid plaques and neurofibrillary tangles, and lead to synaptic damage and cell death in AD.

of 80 nm to form paired helical filaments (PHFs), a characteristic of AD neuronal pathology [46], which ultimately assemble to form NFTs. NFTs have large molecular weights and can be observed microscopically [47].

Hyperphosphorylated tau can also sequester normal tau and other neuronal microtubule-associated proteins (MAPs). This further contributes to disassembled microtubules, disruption of the axonal cytoskeleton, and causes neuronal death [48]. Tau oligomers released from dead neurons enter the extracellular environment, leading to microglial cell activation and consequently trigger further progressive neuronal degeneration.

Tau pathology is suggested to be a result of either elevated protein kinase activity or a reduction in the activity of protein phosphatase. Analysis of phosphorylated tau isolated from AD brains has identified numerous target serine or threonine residues [49]. It has been demonstrated that MAP-kinase, glycogen synthase kinase-type 3 (GSK-3), and/or cyclin-dependent kinase 5 (CDK5) are the main kinases involved in Tau phosphorylation. However, not all Tau phosphorylation events can be attributed to these kinases in AD [49].

#### 2.4.2.2. *GSK-3 as a link between amyloid and tau*

In AD, abnormal A$\beta$ is suggested to promote the activation of GSK-3 (or so-called tau kinase I) and further leads to hyperphosphorylation of tau, which may serve as a link between the amyloid hypothesis and the tau hypothesis [50].

There is evidence that shows in both FAD and SAD, upregulation of GSK-3 can be found to potentially modify Tau protein and other substrates that may contribute to the pathology of neurodegeneration (Fig. 2.7). GSK-3 inhibitors are studied as a treatment strategy for AD by different pharmaceutical companies [51].

GSK-3 is a constitutively active kinase and can phosphorylate a variety of protein substrates, therefore, the regulation of GSK-3 is mainly based on inhibition of its activity through different signaling pathways [52]. Evidence shows that GSK-3 can modify most of all 40 phosphorylation sites in tau taken from the brain of AD patients [53].

A$\beta$ peptide interacts with the insulin and Wnt signaling pathway and consequently, increase the activation state of GSK-3. A$\beta$ peptide acts as an antagonist of insulin to inhibit the activation of PI3 kinase and Akt. Akt is a key serine/threonine protein that is involved in the phosphorylation of GSK-3$\alpha$ (Ser21) or GSK3$\beta$ (Ser9) [54]. Therefore, lack of activated Akt by the A$\beta$ peptide results in an increase in the activity of GSK-3. On the other hand, the A$\beta$ peptide also increases the activity of GSK-3 by disturbing the canonical Wnt pathway. The Wnt pathway is a pathway which leads to an inactivation of GSK-3 and, therefore, a disturbance in the Wnt pathway by the A$\beta$ peptide consequently results in activation of GSK-3 [55]. In addition, the mutation in presenilin 1 (PS1) found in AD is suggested to regulate GSK-3 in both A$\beta$-dependent and A$\beta$-independent manners [56].

ApoE gene has been confirmed to be a genetic risk for SAD. Individuals with ApoE genotype have an increased risk of developing SAD [57]. Evidence is there for a potential link between ApoE and GSK-3 [50, 58]. Among all ApoE isoforms, ApoE4 is suggested to have the highest effect in activating GSK-3 [59]. Besides, ApoE interacts with LDL receptor-related protein 6 (Lrp6), a coreceptor of Wnt frizzled receptor, and thus, inhibit the Wnt signaling pathway [60].

#### 2.4.2.3. *Tau toxicity*

Although the pathology of tau protein is well established, it is still controversial by which mechanism tau exerts its neuronal toxicity. Experimental data suggest that A$\beta$ is the initial trigger, however, tau accumulation may play a central role in neurodegeneration [61]. And a link such as GSK-3 between A$\beta$ and tau potentially drives the neural pathologies and the manifestations of clinical symptoms. Further investigation is needed to study how tau contributes to the progression of neuronal death.

Pseudo-phosphorylated variants of tau (phospho-tau) proteins have experimentally shown toxicity [62]. In *in vitro* experiments, phospho-tau proteins function as mediators of the microtubule stabilization, which might be the underlying mechanism for tau toxicity [63]. In animal models, it suggests that the phospho-tau proteins, not the aggregated ones are toxic [64, 65]. Surprisingly, the aggregated tau, instead, suggest providing some level of neuroprotection [66].

## References

[1] Bradl M. and Lassmann H. (2010). Oligodendrocytes: Biology and pathology. *Acta Neuropathol.*, *119*(1), pp. 37–53.

[2] Araque A. and Navarrete M. (2010). Glial cells in neuronal network function. *Philos. Trans. R. Soc. Lond., B, Biol. Sci.*, *365*(1551), pp. 2375–2381.

[3] Mattson M.P. (2004). Pathways towards and away from Alzheimer's disease. *Nature*, *430*(7000), pp. 631–639.

[4] Fukazawa Y. *et al.* (2003). Hippocampal LTP is accompanied by enhanced F-actin content within the dendritic spine that is essential for late LTP maintenance *in vivo*. *Neuron*, *38*(3), pp. 447–460.

[5] Cingolani L.A. and Goda Y. (2008). Actin in action: The interplay between the actin cytoskeleton and synaptic efficacy. *Nat. Rev. Neurosci.*, *9*(5), pp. 344–356.

[6] Dillon C. and Goda Y. (2005). The actin cytoskeleton: Integrating form and function at the synapse. *Annu. Rev. Neurosci.*, *28*, pp. 25–55.

[7] Sekino Y., Kojima N. and Shirao T. (2007). Role of actin cytoskeleton in dendritic spine morphogenesis. *Neurochem. Int.*, *51*(2), pp. 92–104.

[8] Levitan I.B. and Kaczmarek L.K. (2015). *The Neuron: Cell and Molecular Biology* (Oxford University Press, USA).

[9] Schwartz J.H. (1979). Axonal transport: Components, mechanisms, and specificity. *Annu. Rev. Neurosci.*, *2*(1), pp. 467–504.

[10] Stuart G.J. and Sakmann B. (1994). Active propagation of somatic action potentials into neocortical pyramidal cell dendrites. *Nature*, *367*(6458), pp. 69–72.

[11] Kole M.H. *et al.* (2008). Action potential generation requires a high sodium channel density in the axon initial segment. *Nat. Neurosci.*, *11*(2), pp. 178–186.

[12] Chow C.C. and White J.A. (1996). Spontaneous action potentials due to channel fluctuations. *Biophys. J.*, *71*(6), p. 3013.

[13] Smith M.C. (1967). Theories of the psychological refractory period. *Psychol. Bull.*, *67*(3), p. 202.

[14] Foust A., Popovic M., Zecevic D. and McCormick D.A. (2010). Action potentials initiate in the axon initial segment and propagate through axon collaterals reliably in cerebellar Purkinje neurons. *J. Neurosci.*, *30*(20), pp. 6891–6902.

[15] Tackmann W. and Lehmann H. (1974). Refractory period in human sensory nerve fibres. *Euro. Neurol.*, *12*(5–6), pp. 277–292.

[16] Bressloff P.C. (2014). Waves in neural media. *Lecture Notes on Mathematical Modelling in the Life Sciences.*

[17] Ma J. and Lowe G. (2004). Action potential backpropagation and multiglomerular signaling in the rat vomeronasal system. *J. Neurosci.*, *24*(42), pp. 9341–9352.

[18] Pinheiro P.S. and Mulle C. (2008). Presynaptic glutamate receptors: Physiological functions and mechanisms of action. *Nat. Rev. Neurosci.*, *9*(6), pp. 423–436.

[19] Nakanishi S. (1992). Molecular diversity of glutamate receptors and implications for brain function. *Science*, *258*(5082), pp. 597–603.

[20] Ghosh A. and Greenberg M.E. (1995). Calcium signaling in neurons: Molecular mechanisms and cellular consequences. *Science*, *268*(5208), p. 239.

[21] Bryant H.L. and Segundo J.P. (1976). Spike initiation by transmembrane current: A white-noise analysis. *J. Physiol.*, *260*(2), pp. 279–314.

[22] Scarsi F., Tessadori J., Chiappalone M. and Pasquale V. (2017). Investigating the impact of electrical stimulation temporal distribution on cortical network responses. *BMC Neurosci.*, *18*(1), pp. 1–13.

[23] Raymond C.R. (2007). LTP forms 1, 2 and 3: Different mechanisms for the 'long' in long-term potentiation. *Trends Neurosci.*, *30*(4), pp. 167–175.

[24] Lee H.-K., Barbarosie M., Kameyama K., Bear M.F. and Huganir R.L. (2000). Regulation of distinct AMPA receptor phosphorylation sites during bidirectional synaptic plasticity [10.1038/35016089]. *Nature*, *405*(6789), pp. 955–959.

[25] Sabatini B.L., Oertner T.G. and Svoboda K. (2002). The life cycle of $Ca^{2+}$ ions in dendritic spines. *Neuron*, *33*(3), pp. 439–452.

[26] Palmer L.M. and Stuart G.J. (2009). Membrane potential changes in dendritic spines during action potentials and synaptic input. *J. Neurosci.: Off. J. Soc. Neurosci.*, *29*(21), pp. 6897–6903.

[27] Pike F.G., Meredith R.M., Olding A.W.A. and Paulsen O. (1999). Postsynaptic bursting is essential for 'Hebbian' induction of associative long-term potentiation at excitatory synapses in rat hippocampus. *J. Physiol.*, *518*(Pt 2), pp. 571–576.

[28] Mansvelder H.D. and McGehee D.S. (2000). Long-term potentiation of excitatory inputs to brain reward areas by nicotine. *Neuron*, *27*(2), pp. 349–357.

[29] Caporale N. and Dan Y. (2008). Spike timing-dependent plasticity: A Hebbian learning rule. *Ann. Rev. Neurosci.*, *31*, pp. 25–46.

[30] Selkoe D.J. (1991). The molecular pathology of Alzheimer's disease. *Neuron*, *6*(4), pp. 487–498.

[31] Duyckaerts C., Delatour B. and Potier M.-C. (2009). Classification and basic pathology of Alzheimer disease. *Acta Neuropathol.*, *118*(1), pp. 5–36.

[32] Turner P.R., O'Connor K., Tate W.P. and Abraham W.C. (2003). Roles of amyloid precursor protein and its fragments in regulating neural activity, plasticity and memory. *Prog. Neurobiol.*, *70*, pp. 1–32.

[33] LaFerla F.M., Green K.N. and Oddo S. (2007). Intracellular amyloid-beta in Alzheimer's disease. *Nat. Rev.*, *8*(7), pp. 499–509.

[34] Patterson C. *et al.* (2008). Diagnosis and treatment of dementia: 1. Risk assessment and primary prevention of Alzheimer disease. *Can. Med. Assoc. J.*, *178*(5), pp. 548–556.

[35] Jiang Q. *et al.* (2008). ApoE promotes the proteolytic degradation of Abeta. *Neuron*, *58*(5), pp. 681–693.
[36] Crews L. and Masliah E. (2010). Molecular mechanisms of neurodegeneration in Alzheimer's disease. *Hum. Mol. Genet.*, *19*(R1), pp. R12–R20.
[37] Shankar G.M. and Walsh D.M. (2009). Alzheimer's disease: Synaptic dysfunction and Abeta. *Mol. Neurodegener.*, *4*, p. 48.
[38] Berridge M.J. (2010). Calcium hypothesis of Alzheimer's disease. *Pflug. Arch. Eur. J. Phy.*, *459*(3), pp. 441–449.
[39] Bancher C. *et al.* (1989). Accumulation of abnormally phosphorylated tau precedes the formation of neurofibrillary tangles in Alzheimer's disease. *Brain Res.*, *477*(1–2), pp. 90–99.
[40] Hirokawa N., Shiomura Y. and Okabe S. (1988). Tau proteins: The molecular structure and mode of binding on microtubules. *J. Cell Biol.*, *107*(4), pp. 1449–1459.
[41] Kadavath H. *et al.* (2015). Tau stabilizes microtubules by binding at the interface between tubulin heterodimers. *Proc. Natl. Acad. Sci. USA*, *112*(24), pp. 7501–7506.
[42] Ittner L.M. *et al.* (2008). Parkinsonism and impaired axonal transport in a mouse model of frontotemporal dementia. *Proc. Natl. Acad. Sci. USA*, *105*(41), pp. 15997–16002.
[43] Terry R. (1998). The cytoskeleton in Alzheimer disease. In *Ageing and Dementia* (pp. 141–145). (Springer).
[44] Velasco M.E., Smith M.A., Siedlak S.L., Nunomura A. and Perry G. (1998). Striation is the characteristic neuritic abnormality in Alzheimer disease. *Brain Res.*, *813*(2), pp. 329–333.
[45] Martin L., Latypova X. and Terro, F. (2011). Post-translational modifications of tau protein: Implications for Alzheimer's disease. *Neurochem. Int.*, *58*(4), pp. 458–471.
[46] Bulic B., Pickhardt M., Mandelkow E.-M. and Mandelkow E. (2010). Tau protein and tau aggregation inhibitors. *Neuropharm.*, *59*(4–5), pp. 276–289.
[47] Shelton S.B. and Johnson G.V. (2004). Cyclin-dependent kinase-5 in neurodegeneration. *J. Neurochem.*, *88*(6), pp. 1313–1326.
[48] Iqbal K., Alonso A.D.C. and Grundke-Iqbal I. (2008). Cytosolic abnormally hyperphosphorylated tau but not paired helical filaments sequester normal MAPs and inhibit microtubule assembly. *J. Alzheimer's Dis.*, *14*(4), pp. 365–370.
[49] Arendt T., Holzer M., Fruth R., Brückner M. and Gärtner U. (1998). Phosphorylation of tau, A$\beta$-formation, and apoptosis after *in vivo* inhibition of PP-1 and PP-2A. *Neurobiol. Aging*, *19*(1), pp. 3–13.
[50] Hernández F., de Barreda E.G., Fuster-Matanzo A., Lucas J.J. and Avila J. (2010). GSK3: A possible link between beta amyloid peptide and tau protein. *Exp. Neurol.*, *223*(2), pp. 322–325.
[51] Hernandez F., Nido J.D., Avila J. and Villanueva N. (2009). GSK3 inhibitors and disease. *Mini Rev. Med. Chem.*, *9*(9), pp. 1024–1029.
[52] Jope R.S. and Johnson G.V. (2004). The glamour and gloom of glycogen synthase kinase-3. *Trends Biochem. Sci.*, *29*(2), pp. 95–102.

[53] Hanger D.P., Anderton B.H. and Noble W. (2009). Tau phosphorylation: The therapeutic challenge for neurodegenerative disease. *Trends Mol. Med.*, *15*(3), pp. 112–119.
[54] Townsend M., Mehta T. and Selkoe D.J. (2007). Soluble A$\beta$ inhibits specific signal transduction cascades common to the insulin receptor pathway. *J. Biol. Chem.*, *282*(46), pp. 33305–33312.
[55] Magdesian M.H. *et al.* (2008). Amyloid-$\beta$ binds to the extracellular cysteine-rich domain of Frizzled and inhibits Wnt/$\beta$-catenin signaling. *J. Biol. Chem.*, *283*(14), pp. 9359–9368.
[56] Baki L. *et al.* (2004). PS1 activates PI3K thus inhibiting GSK-3 activity and tau overphosphorylation: Effects of FAD mutations. *EMBO J.*, *23*(13), pp. 2586–2596.
[57] Strittmatter W.J. *et al.* (1993). Apolipoprotein E: High-avidity binding to beta-amyloid and increased frequency of type 4 allele in late-onset familial Alzheimer disease. *Proc. Natl. Acad. Sci. USA*, *90*(5), pp. 1977–1981.
[58] Cselenyi C.S. *et al.* (2008). LRP6 transduces a canonical Wnt signal independently of Axin degradation by inhibiting GSK3's phosphorylation of $\beta$-catenin. *Proc. Natl. Acad. Sci. USA*, *105*(23), pp. 8032–8037.
[59] Cedazo-Minguez A. *et al.* (2003). Apolipoprotein E and $\beta$-amyloid (1–42) regulation of glycogen synthase kinase-3$\beta$. *J. Neurochem.*, *87*(5), pp. 1152–1164.
[60] Caruso A. *et al.* (2006). Inhibition of the canonical Wnt signaling pathway by apolipoprotein E4 in PC12 cells. *J. Neurochem.*, *98*(2), pp. 364–371.
[61] Mokhtar S.H., Bakhuraysah M.M., Cram D.S. and Petratos S. (2013). The Beta-amyloid protein of Alzheimer's disease: Communication breakdown by modifying the neuronal cytoskeleton. *Intern. J. Alzheimer's Dis. 2013*, 910502.
[62] Brandt R., Hundelt M. and Shahani N. (2005). Tau alteration and neuronal degeneration in tauopathies: Mechanisms and models. *Biochimica et Biophysica Acta (BBA) — Molecular Basis of Disease*, *1739*(2–3), pp. 331–354.
[63] Alonso A.d.C., Grundke-Iqbal I., Barra H.S. and Iqbal K. (1997). Abnormal phosphorylation of tau and the mechanism of Alzheimer neurofibrillary degeneration: Sequestration of microtubule-associated proteins 1 and 2 and the disassembly of microtubules by the abnormal tau. *Proc. Natl. Acad. Sci. USA*, *94*(1), pp. 298–303.
[64] Lucas J.J. *et al.* (2001). Decreased nuclear $\beta$-catenin, tau hyperphosphorylation and neurodegeneration in GSK-$\beta$ conditional transgenic mice. *EMBO J.*, *20*(1–2), pp. 27–39.
[65] Jackson G.R. *et al.* (2002). Human wild-type tau interacts with wingless pathway components and produces neurofibrillary pathology in Drosophila. *Neuron*, *34*(4), pp. 509–519.
[66] Lee H.-G. *et al.* (2005). Tau phosphorylation in Alzheimer's disease: Pathogen or protector? *Trends Mol. Med.*, *11*(4), pp. 164–169.

## Chapter 3

# Initiation of Action Potential: Biochemistry, Biophysics, and Relevance to AD

### 3.1. Introduction

As mentioned in Chapters 1 and 2, action potential starts in the region named the axon initial segment (AIS), which is a non-myelinated region found at the beginning of the axons of neurons and is an example of a complex small system that is crucial for signal transmission in neurons. It is only recently that scientists started exploring this region for cues that degenerate the signal transmission. In this chapter, we explore some of the details of proteins associated with this region and how the dynamic interactions among the proteins instigate the conditions within the AIS that initiate action potential.

The main proteins in the AIS are voltage-gated sodium channels ($Na_V$), voltage-gated potassium channels ($K_V$), microtubules, casein kinase 2 (CK2) and ankyrin-G (AnkG). One of the important aspects of the AIS is the maintenance of neuronal polarity so that action potential travels in one-direction along the axon towards the synaptic terminals. Neuronal polarity is the result of fine tuning assemblies of the proteins present in the AIS; the mechanisms involved are amazingly robust, and we intend to elucidate these dynamic arrangements in this chapter in some detail but there are references at the end of the chapter for further study. Alterations to neuronal polarity due to mutation or injury to the AIS-related proteins can cause various neurogenerative diseases, including Alzheimer's disease [1], and the loss of proteins that are essential for the proper functioning of the AIS may be a factor in AD [2]. For example, AnkG, one of the AIS proteins, is found in lower concentrations in AD transgenic mice as compared to wild-type mice. This lower population of AnkG can cause various abnormalities in the AIS mechanisms: for example, alteration in the filtering capability of AIS is responsible for the entry of dendritic proteins into the axons [2]. The reduced concentration of AnkG leads to the loss of $Na_V$ and $K_V$ present in the AIS. The AnkG concentration affects the ion channels' concentration in the AIS because the interactions of AnkG with the ion channels restrict them to the AIS [3]. Restriction of the high density of voltage-gated ion channels in the AIS is the main reason why this region acts as AP initiator zone. We discuss cell and molecular biology of the AIS from a systemic point of view so that we can understand the complexity associated with neuronal signal transmission at the small system level, and its interconnectivity to the larger picture of signal transmission in brain. We discuss in some detail the AIS structure, its components and their function; why the AIS is responsible for the initiation of AP, and the disease conditions, such as AD, due to dysfunctions associated with the AIS proteins. This chapter is a prelude to the next chapter in which we develop a biophysical model of the AIS protein interactions.

## 3.2. Biology of AIS

The AIS, the length of which varies between 10 to 60 $\mu$m [4], has been shown to grow in cultured neurons after three to four days [4]; and the development of the structure of the AIS is an internal process and does not require any extracellular processes [5, 6] (Fig. 3.1). The AIS maintains the neuronal polarity through specific structures, shapes and functions by acting as a diffusion barrier and a cytoplasmic selective filter [3, 4, 7]. The barrier and filtering capabilities of the AIS restrict the entry of dendritic proteins into this region and prevent the transformation of the AIS into dendrites. The main function of the AIS is to act as an AP initiator zone due to its molecular organization, which is heavily enriched by the presence

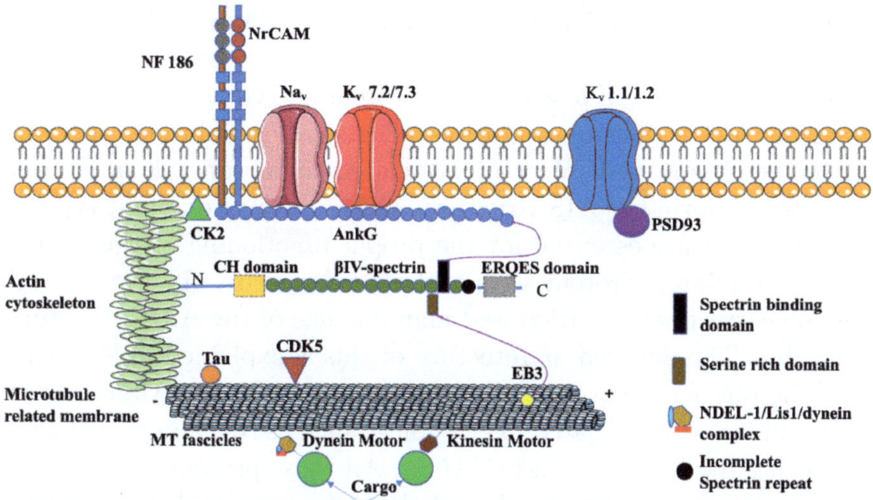


Fig. 3.1. Intracellular structure of the AIS. The plasma membrane of the AIS consists of ion channels (Na$_v$ and K$_v$ channels) and cell adhesion molecules (CAMs), such as NF186 and NrCAM, that are restricted by AnkG. The initial part of AnkG (the membrane binding domain (MBD)) is specific for anchoring the Na$_v$, and K$_v$ channels (K$_v$ 7.2/7.3) and CAMs at the plasma membrane. The proteins that bind to AnkG have a specific amino acid sequence called AnkG binding motifs (ABM), which are absent in K$_v$ 1.1/1.2, and it is the reason for their binding with PSD 93 (purple circle). CK2 is also present at the plasma membrane and this enhances the interaction between AnkG and the Na$_v$ channels. The spectrin–actin membrane situated beneath the plasma membrane contains actin and $\beta$-IV spectrin and they also restrict the entry of proteins into the AIS structure through interactions with AnkG. After the MBD, the presence of a spectrin–binding domain (black rectangle) in AnkG is responsible for the binding between $\beta$-IV spectrin and AnkG. Following the second membrane, the AIS contains a microtubule-related membrane that comprises of the bundles of microtubules known as microtubule fascicles. Microtubule fascicles are also restricted by AnkG due to interactions with an EB3 protein (yellow circle). Moreover, the microtubule fascicles contain tau proteins (orange circle) and dynein and kinesin motors. The kinesin and dynein motors are responsible for the transport of proteins (big green circles) from one end of the AIS to the other while the microtubule fascicles act as the road for these molecular motors. Moreover, CDK5 (red triangle) in the microtubule membrane is responsible for the phosphorylation of NDEL-1, a protein essential for binding with the dynein motors. This figure is produced using Servier Medical Art (https://smart.servier .com).

of the voltage-gated ion channels, especially $Na_V$ and $K_V$ channels [8–10]. The presence of ion channels and other proteins in the AIS structure depends on a scaffolding protein known as AnkG [11, 12].

The AIS structure comprises of plasma membrane, spectrin-actin membrane and microtubule-related membrane [4], and these membranes form the AIS structure, which contains different types of proteins that are essential for the proper functioning of this region. Every individual protein within the structure of AIS is necessary for the correct organization and maintenance of the AIS's structural assembly. The detailed architecture of this is explained in Fig. 3.1.

The plasma membrane is the topmost layer of the AIS structure and contains voltage-gated ion channels, such as $Na_V$ and $K_V$ channels and cell adhesion molecules (CAMs). All these proteins play critical roles in the proper functioning of the AIS, such as AP generation. $Na_V$ channels are highly enriched in the plasma membrane of the AIS [13, 14] (Fig. 3.1). Major isoforms of $Na_V$ channels are $Na_V$ 1.1, 1.2, 1.3 and 1.6 [4, 15], and out of the four isoforms, all are found in the structure of the AIS, except for $Na_V$ 1.3. In the early developmental stage, $Na_V$ 1.2 is present in the AIS but is later replaced by $Na_V$ 1.6 in the adult central nervous system (CNS) [16]. The mechanism involved in this replacement is still unknown. $Na_V$ 1.1 and $Na_V$ 1.6 are present in the proximal (beginning) and distal (end) parts of the AIS, respectively [16, 17]. The proper functioning of the $Na_V$ channel depends on its internal structural organization.

The $Na_V$ channel has four domains (I–IV) and every domain has six transmembrane segments (TS1–TS6) [4, 104] (Fig. 3.2). TS1–TS4 of all the four domains are sensitive towards voltage and are known as voltage sensing segments. However, the TS5–TS6 segments of all domains are known as pore forming segments and are responsible for the entry of $Na^+$ ions during voltage changes in the membrane as sensed by the voltage sensing segments. Of all the voltage sensing segments, the TS4 segment is highly unstable. The TS4 segment gets stabilized within the domain due to ionic pair interactions between the positively charged residue of TS4 and the negatively charged residue of TS2. As soon as the voltage sensing segments sense an incoming voltage input, the ionic pair between TS4 and TS2 breaks

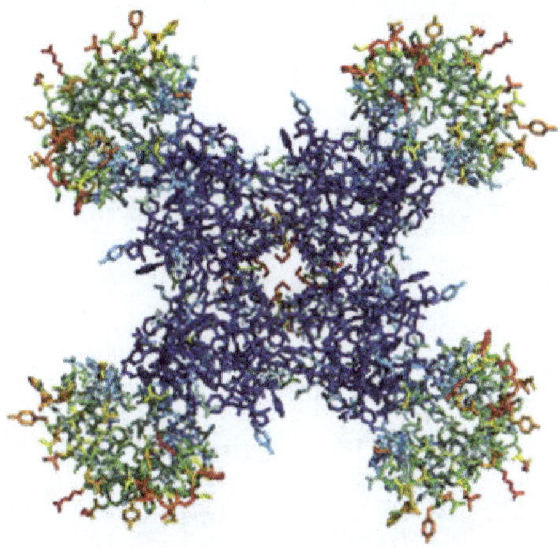

Fig. 3.2. Crystallographic structure of a $Na_v$ channel. The green region at the corners represents the transmembrane segments (TS1–TS4) of the four domains (I–IV) and the blue region refers to the pore-forming segments formed by TS5–TS6 of each domain. In the middle, the gap shows the path for the influx and efflux of $Na^+$ ions during opening and closing of $Na_v$ channels due to changes in the membrane potential. (This figure is retrieved from https://www.guidetop harmacology.org/images/introductions/82_2.jpeg) [104].

and this results in the conformational changes in the $Na_v$ channel due to the outward movement of the TS4 segment [18] (Fig. 3.3). The presence of the $Na_v$ channel in the AIS is highly dependent on AnkG.

Clustering of the $Na_v$ channels in the AIS is carried out by binding with AnkG. The $Na_v$ channel has an ankyrin-binding motif (ABM) present between domains II and III that binds to the membrane-binding domain (MBD) of AnkG [5, 6] (Fig. 3.3). To understand the accumulation of $Na_v$ channels in the AIS, Fache *et al.* [18] conducted an experiment using chimeric CD4 containing a II–III linker of $Na_v$ 1.2. In this experiment chimeric CD4 is supposed to mimic the function of $Na_v$ 1.2 channels due to the presence of the II–III linker of $Na_v$ 1.2 in chimeric CD4. When chimeric CD4 was then transfected into laboratory-extracted neurons, the results clearly showed that chimeric CD4 was restricted to the AIS. This restriction was due to the interaction of chimeric CD4 with the MBD of AnkG [19].

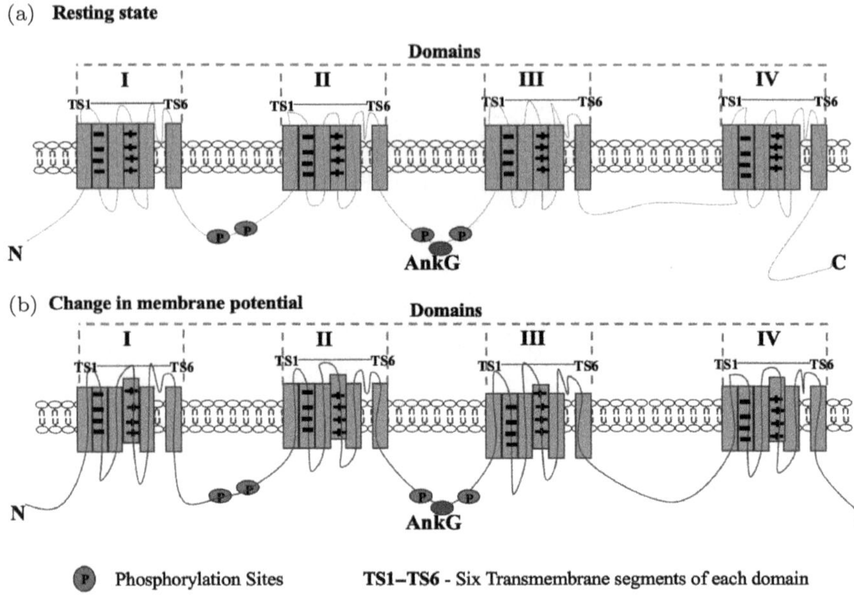


Fig. 3.3. Schematic representation of $Na_v$ channels. The $Na_v$ channel comprises four domains (I–IV) and each domain contains six transmembrane segments (TS1–TS6). Out of the six trans-membrane segments, TS1–TS4 are voltage sensing segments while TS5 and TS6 are pore-forming segments. $Na^+$ ions are transported through the TS5 and TS6 segments when the $Na_v$ channels are open. Various phosphorylation sites are present in $Na_v$ channels and they enhance the interaction of these channels with AnkG after phosphorylation by CK2. The binding of the $Na_v$ channels with AnkG depends on the AIS motif between the II and III domains of the $Na_v$ channels. Image (a) represents the $Na_v$ channel at its resting state (no change in voltage across the membrane), where the TS4 of all four domains is stabilized inside this structure due to the ionic pair interactions between the positively charged residues (lysine or arginine) of TS4 and the negatively charged residues of TS2. Image (b) shows the outward movement of the TS4 segments in all domains. The movement of TS4 segments is due to changes in the membrane voltage. As voltage sensing segments (TS1–TS4) recognize the strong voltage input, the ionic pair between the TS4 and TS2 segments of all domains break and the TS4 segment moves outwards, which leads to conformational changes in the $Na_v$ channels. Conformational changes result in the movement of $Na^+$ ions from outside the membrane to inside the membrane. The influx of $Na^+$ ions results in a change in the membrane voltage from negative to positive; this is called the depolarization of the membrane. This figure is produced using Servier Medical Art (https://smart.servier.com).

Similar to the AIS, $Na_V$ channels are also present in the soma region of neurons. However, the density of $Na_V$ channels in the AIS is much higher than that in the soma [7]: the high number of $Na_V$ channels in the AIS is supported by Kole *et al.* [8], and their experiment suggests that the density of $Na_V$ channels is 50-fold higher than in the soma and they also conclude that the AIS requires a minimum $Na_V$ channel density of 2500 Ps $\mu m^{-2}$ for AP initiation and backpropagation [12]. The high density of $Na_V$ channels in the structure of the AIS supports the AIS skeleton to overcome the load induced during AP initiation. Moreover, due to their high density, $Na_V$ channels in the AIS (1) activate and deactivate much faster than the $Na_V$ channels present in the soma, and (2) the membrane potential changes at a lower voltage as compared to the soma. The density of $Na_V$ channels in the AIS can be influenced by other proteins present, such as the fibroblast growth factor 14 (FGF14) and subunits of $Na_V$ channels [16, 17]. Mutations of FGF14 can result in the loss of $Na_V$ channel density in the AIS. According to *in vitro* studies, $Na_V$ channel mutations result in a reduction in the frequency of $Na_V$ channels and the $Na^+$ current in the AIS [20]. Therefore, the reductions in $Na_V$ channels can be responsible for the high voltage requirement to initiate an AP.

Overall, $Na_V$ channels in the AIS are the key players and any alteration can disturb the initiation of AP due to the large voltage requirement. Mutations in the $Na_V$ channel can develop medical conditions in humans, such as cognitive impairment and poor coordination of hand, speech and eye movements, known as the spinocerebellar ataxia [16]. Moreover, alteration in $Na_V$ channels may be responsible for the breakdown of AnkG when activated by $Ca^{2+}$-dependent cysteine calpain and this process is known as proteolysis [21]. Proteolysis of AnkG can result in the disruption of the AIS structure due to the weak binding of proteins with AnkG.

### 3.2.1. $K_V$ *channels*

The $K_V$ channel is another type of voltage gated ion channel present in the plasma membrane of the AIS. $K_V$ channels are different in structure and formation as well as in functionality compared

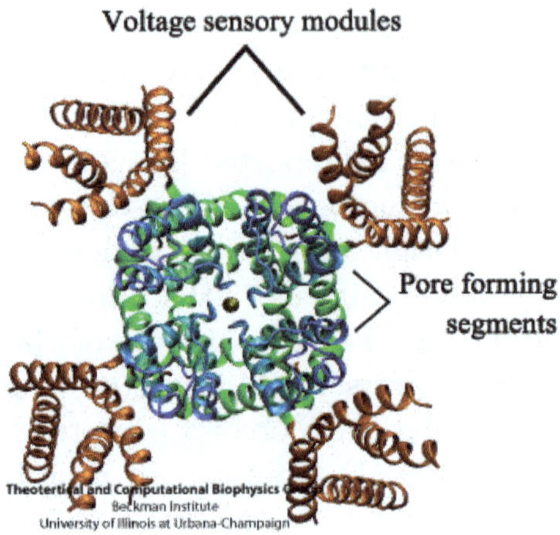


Fig. 3.4. Crystallographic structure of the $K_v$ channel. The brown extensions at the corners represent the transmembrane segments (TS1–TS4) serving as voltage sensing modules. TS5–TS6 are known as pore forming segments (blue and green regions). Due to the change in membrane potential, conformational changes occur within the $K_v$ channel and this results in the opening of the channel (center blue and green region). Opening of the $K_v$ channel helps in changing the membrane voltage from positive to negative due the efflux of $K^+$ ions (This image is owned by the Theoretical and Computational Biophysics Group, NIH Center for Macromolecular Modeling and Bioinformatics, at the Beckman Institute, University of Illinois at Urbana-Champaign, and is retrieved from https ://www.ks.uiuc.edu/Research/kvchannel/).

to $Na_v$ channels. $K_v$ channels contain only one domain with six transmembrane segments (TS1–TS6) (Figs. 3.4 and 3.5). Out of the six segments, TS1–TS4 are known as voltage sensory modules and the remaining two segments (TS5 and TS6) are pore-forming modules. $K_v$ channels consist of a tetramerization domain (T1) at the N-terminal and this is responsible for the binding of $K_v\,\beta$ subunits and other proteins, such as kinesin motors. $K_v\,\beta$ subunits increase the binding affinity of $K_v$ channels towards the AIS. There are several isoforms of $K_v$ channels, such as KCNQ2/3, $K_v$ 1.1/1.2/1.4, and $K_v$ 2.2, all of which are completely different from each other [22]. $K_v$ 1.1 and $K_v$ 1.2 can be found at the end segment of the AIS and share space with $Na_v$ 1.6 [7].

The accumulation of different $K_v$ channel subunits into the AIS is different from each other. $K_v$ 2/3 or $K_v$ 7.2/7.3 is restricted in the

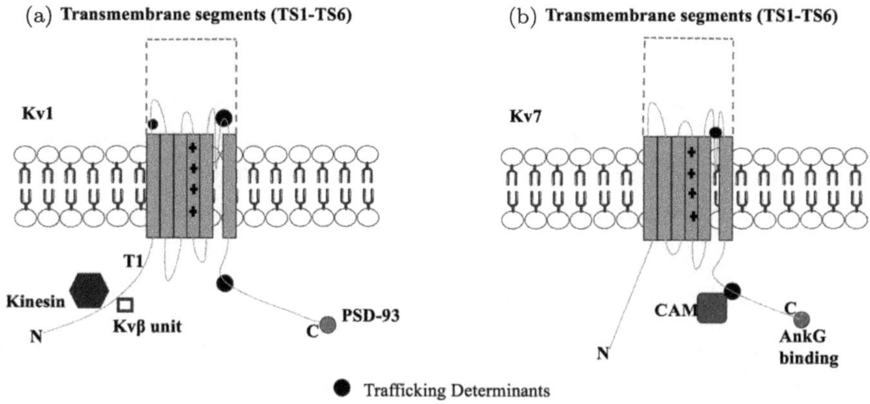


Fig. 3.5. Schematic representation of the $K_v$ channel. Both channels, $K_v$ 1 (a) and $K_v$ 7 (b), have almost similar structures of a single domain divided into six transmembrane segments (TS1–TS6). The TS4 segment contains positive residues that act as a voltage sensing module and, together with TS1–TS3, T5 and T6 comprise of the pore forming segments. Several amino acids in both types of $K_v$ channels act as the trafficking determinants (TDs) (black circle) and are responsible for the maintenance, expression and trafficking of the $K_v$ channels [22]. The TDs at $K_v$ channels also regulate the interaction of $K_v$ channels with other proteins. The N-terminal of $K_v$ 1 channels consists of a tetramerization domain (T1) and this serves as a binding site for the kinesin motors and $\beta$ subunits of the $K_v$ 1 channel. The C-terminal of the $K_v$ 1 channel has a binding motif that is responsible for the restriction of the $K_v$ 1 channel into the AIS through PSD 93 (orange circle). $K_v$ 7 channels have TDs (black circle) near the C-terminal and are responsible for the binding of CAMs. Tethering the $K_v$ 7 channels into the AIS depends on ABM, which is present at the end of the C-terminal. This figure is produced using Servier Medical Art (http://smart.servier.com).

AIS by AnkG because of the presence of ABM, similar to the $Na_v$ channels [7, 16]. However, $K_v$ 1.1 and 1.2 do not bind with AnkG due to the absence of ABM and are anchored into the structure of the AIS through postsynaptic density protein 93 (PSD93) or chapsin 11 [6, 17]. PSD93 is a part of the PDZ domain family that contains the membrane-associated guanylated kinase family of scaffolding proteins (MAGUK) [16]. $K_v$ channel trafficking in the AIS is enhanced by its phosphorylation by cyclin-dependent kinase 5 (CDK5).

In the AIS, the $K_v$ channels modulate AP firing as well as its shape, initiation and spike patterning by restricting neuronal excitability and repolarizing the membrane by allowing an efflux of $K^+$ ions out of the membrane [23, 24]. Mutation in any of the $K_v$ channels can lead to alterations in these functions. Moreover, mutation in particular $K_v$ channel subunits (KCNQ 2/3 or $K_v$ 7.2/7.3) can cause epilepsy in new-born human babies [25] due to a defect in $K_v$ channel trafficking into the AIS, which can reduce the size of the $K^+$ current.

### 3.2.2. *Cell adhesion molecules*

Cell adhesion molecules (CAMs) of the L1 family are also enriched in the AIS. Neurofascin 186 (NF186) and neuronal cell adhesion molecules (NrCAM) are examples of CAMs available in the plasma membrane of the AIS. Both NF186 and NrCAM have an AnkG binding motif and this motif helps in the interaction of CAMs with AnkG (Fig. 3.1) [4]. Mutation of NF186 results in the inhibition of gephyrin, which weakens the GABAergic synapses by inhibiting the clustering of $GABA_A$ receptors ($GABA_AR$). Overall, CAMs maintain the GABAergic synapse between the axon terminal and the AIS, which is crucial for AP initiation. *In-vivo* studies reveal that NF-186 is important for stabilizing the AIS, and the recruitment of NrCAM into this region [26, 27].

### 3.2.3. *Spectrin–actin membrane*

The second membrane of the AIS structure lies beneath the plasma membrane and is composed of $\beta$IV spectrin and actin (Fig. 3.1). The main function of this membrane is to provide mechanical support to the AIS [28]. The spectrin family of cytoskeleton proteins was first noticed in erythrocytes as a rod shape, and it is a tetramer with two units each of $\alpha$ and $\beta$ subunits [29]. It is an antiparallel heterodimer made up of $\alpha$ and $\beta$ subunits with a length of ~100 nm that interact in a head to head way resulting in the formation of a tetramer ~200 nm long [30, 31]. In the erythrocyte membrane, the subunits of spectrin have 106 residues and they form a helix bundle [32]. The $\alpha$ subunit of spectrin has 20 repeats of spectrin and an SH3 domain at the ninth segment [33]. Two EF-hand motifs are present between the 20th spectrin repeat and the C-terminal [4, 32]. The $\beta$ subunit of spectrin has 17 spectrin repeats, an actin-binding domain with protein 4.1, adducin and phosphatidyl-inositol 4, 5-biphosphate ($PIP_2$) at the N-terminal and a pleckstrin homology domain (PH) towards the C-terminal [33]. All $\beta$ spectrin binds with AnkG on spectrin repeat 15.

In mammals, the five genes of $\beta$ spectrin are present but $\beta$IV-spectrin is found in the AIS, which has been localized in the mouse at chromosome 19q13.13 [4, 29]. After alternative splicing, the $\beta$IV-spectrin generates six isoforms, from $\beta$IV-spectrin $\Sigma1$–$\Sigma6$. $\beta$IV-spectrin $\Sigma1$ is the longest isoform and has 36 exons and 2559 amino acids. In the brain, the size of $\beta$IV-spectrin $\Sigma1$ is $\sim$9.0 kb and this has been detected on day 19 of the developmental phase; the concentration of the $\beta$IV-spectrin $\Sigma1$ continues to increase until adulthood [32]. The $\beta$IV-spectrin $\Sigma6$ isoform does not have any actin-binding domain at the N-terminal, and a total of nine repeats of spectrin are present in the full size of 4.7 kb [34].

The recruitment of $\beta$IV spectrin is highly dependent on AnkG [35]. The ERQES domain (a specific domain) with positively and negatively charged residues is present after the 17th repeat and before the PH domain. This proline-rich domain is not present in other $\beta$ spectrin isoforms and is highly charged because of the presence of glutamate, glutamine and arginine at 23.3%, 8.8%, and 26.6%, respectively [34]. The $\beta$IV-spectrin and AnkG play a crucial role in the correct localization of membrane proteins in the AIS. According to various studies, the $\beta$IV-spectrin participates partly in the localization of AnkG and, together, they help the AIS act as a barrier and decrease the concentration of somato-dendritic proteins, such as GluR1 and telecephalin, before they enter the AIS [36]. An optical tweezer experiment showed that $\beta$IV-spectrin regulated the lateral mobility of L1 CAMs, and this function of $\beta$IV-spectrin indirectly depends on the interaction of L1 CAMs with AnkG [37].

According to genetic studies on *C. elegans*, $\beta$-spectrin maintains the function of nerve axons to overcome the stress that occurs during movement, so a deficiency of $\beta$-spectrin can lead to paralysis [38].

### 3.2.4. *Ability of actin and AIS to act as barriers*

The actin-based structure provides passage to axonal proteins and restricts the entry of dendritic proteins into the axon [39]. In cultured cortical neurons, the existence of actin can be noticed at three days-*in-vitro* (DIV 3) by 58% and 100% by DIV 7 [40]. The actin patches available at the axons are higher in number than in the dendrites and, in the AIS, they are present 13 $\mu$m away from the cell body [40]. SEM imaging reveals the structure of actin by showing the presence of actin filaments and synaptopodin. Synaptopodin is a protein, known as the actin-interacting regulatory protein, that has multiple binding sites for $\alpha$-actinin [24]. $\alpha$-actinin comes from the family of actin-binding proteins that crosslinks and bundles actin filaments. $\alpha$-actinin creates short branched packages of actin filaments [24]. All actin filaments are present towards the positive end of the cell body [40] and they result in the formation of a diffusion barrier and this prevents the entry of unnecessary protein into the axon.

Clustering of transmembrane proteins with the spectrin–actin-based cytoskeleton via AnkG is essential for the creation of the diffusion barrier in the AIS. The diffusion barrier develops when all the proteins are available in this region. According to the picket fence model, the formation of the diffusion barrier reduces the mobility of membrane proteins in the structure of AIS because of the crowding of the proteins. In contrast, Albrecht *et al.* [39] conducted a study to propose a new model called the "actin fence model". To test this model, single glycosylphosphatidylinositol-anchored GFP (GPI-GFP) trajectories were measured within the AIS of primary rat hippocampal neurons using high-density single-particle tracking (SPT). SPT analysis was undertaken at different time points. Reduction in the mobility of GPI-GFP in the AIS was noticed after DIV 3. The outcomes of this study revealed that GPI-GFP was restricted in this region because of the actin cytoskeleton and there was also a reduction observed in the mobility of membrane proteins between adjacent actin rings [41]. In contrast, Nakada *et al.* stated that the formation of barrier takes place at DIV 10; however, treatment with jasplakinolide (an actin promoting agent) enhanced the polymerization of f-actin resulting in the creation of a barrier at DIV 5 [42].

### 3.2.5. *Microtubule-related membrane*

The microtubule-related membrane is essential for the transportation of various proteins that act as a pathway for the molecular transporters or molecular motors. Moreover, various post-translation modifications (PTMs), such as phosphorylation acetylation, also contribute to the correct functioning of the microtubules.

Microtubules (MTs) are hollow in shape with a 25 nm diameter [43, 44] (Figs. 3.1 and 3.6). In the AIS, MTs are present in bundles known as fascicles. Fascicle formation may occur because of the interaction of MT with AnkG through an end binding protein (EB) [4]. MT fascicles are collections of MTs that align parallel to each other and then cross-link into a bundle. There are three to seven fascicles in total, and the number of MTs present in fascicles range from 2 to 25 [4, 45].

MTs consist of $\alpha\beta$-tubulin, tau, an end-binding protein, spastin and motor proteins. $\alpha\beta$-tubulin, a heterodimer made up of straight protofilaments [45, 46]. Tubulin dimers maintain the MT dynamics. MT requires the histone deacetylase 6 (HDAC6) enzyme to deacetylase tubulin and tau [47]. HDAC6 is involved in axon development, and the inhibition of HDAC6 can cause the loss of AIS assembly [46, 47]. Microtubule-associated proteins, such as tau, end-binding protein (EB), spastin and motor proteins, maintain the dynamics of the MTs [44, 45]. The EB protein is found in the three forms EBs (1–3), and they are known as plus-end tracking proteins. EB1 and EB3 have unique roles and play a part in axonal transport. EB3 acts as a link between AnkG and MT through the EB homology domain [6, 48]. The properties of MTs are modified with the help of CK2. A few experimental analyses have provided evidence showing that inhibition of CK2 increases the rate of acetylation of tubulin by three times and impairs the function of the MTs [49]. Post-translational modifications (PTMs) of tyrosination, detyrosination,

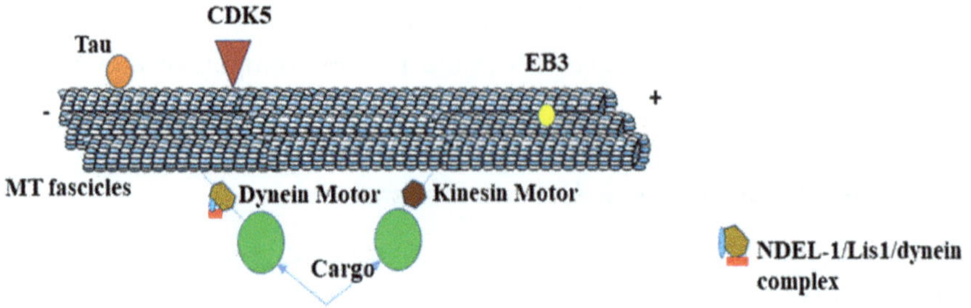


Fig. 3.6. Schematic representation of a microtubule in the AIS. Microtubules are in the AIS in the form of bundles called fascicles. Each fascicle contains 2–25 microtubules parallel to each other. Microtubules act as pathways for the transportation of proteins through the molecular motors, kinesin (brown hexagon) and dynein (gold hexagon). Kinesin motors move towards the plus end of the microtubule and dynein motors move towards the minus end. Transportation of protein cargo (green circle) towards the axon terminal and soma is accomplished by kinesin and dynein motors, respectively. The dynein motors are accompanied by NDEL1 (red outlined rectangle) that are responsible for the activation of the dynein motors. A microtubule consists of tau (orange circle) and the hyperphosphorylation of tau can disrupt the function of other proteins present in this region. The activity of the microtubules is maintained by CDK-5 kinase (blue triangle). Microtubule binding with AnkG relies on the EB3 protein (yellow circle) and this binding is important for the correct localization of the microtubules within the AIS.

polyglutamylation and polyglycylation are examples of modifications that take place in MTs [6]. MAP dysfunction leads to various diseases, such as AD, amyotrophic lateral sclerosis, multiple sclerosis, and chronic traumatic encephalopathy, in athletes suffering concussion [44, 45].

The presence of MTs in the form of fascicles and their involvement in cargo transport make MTs a noteworthy member of the AIS [4, 50]. MTs also play a vital role in the structure of the AIS by maintaining its functions and assembly.

### 3.2.6. *MT-based protein transport*

MTs participate in protein trafficking with the help the kinesin and the dynein superfamily [6, 51, 52]. An alteration in any of these motors results in damage to the assembly of the AIS and forces the axons to acquire the molecular characteristics of dendrites [49]. The motor protein binds with the motors and moves along with MTs. Adenosine triphosphate (ATP) is the energy source for the mobility of these motors [53, 54].

### 3.2.7. *Kinesin motors*

Kinesin motors move along MTs towards their plus end [6]. The kinesin family has 45 types of kinesin proteins [55]. Kinesin-1 is the main motor for axonal transport (Fig 3.6). It contains a heavy chain dimer at the N-terminal and two kinesin light chains at the C-terminal. KIF5A, KIF5B, KIF5C are three isoforms of kinesin-1 [6]. The N-terminal of kinesin has a stalk domain (responsible for dimerization through the coiled-coil region), and the C-terminal is a binding site for the protein that needs to be transported [55]. KIF5 is known for the transport of the vesicle associated membrane protein (VAMP2) and the amyloid precursor protein (APP) to the axon, and *N*-methyl-D-aspartate (NR2B) is transported by, KIF17 (another protein from the kinesin family), to the dendrites [52]. Song *et al.* [54] conducted several experiments using chimeric KIF17, with a KIF5B tail, and chimeric KIF5B, with a KIF17 tail, to understand the presence cytoplasmic selective filter transport in AIS. Outcomes indicate that chimeric KIF 17 took VAMP2 as cargo, and chimeric KIF5B took NR2B as cargo; hence, the results showed no evidence of VAMP2 in the axons while, on the other hand, NR2B was found in both axons and dendrites. This indicates that KIF5B can transport cargoes in both axons and dendrites and the presence of a cytoplasmic filter in the AIS has been found between DIV 3 and 5 [56]. Transport through kinesin motors is important for neuronal polarity because kinesin family proteins are responsible for the precise supply of the components in both axons and dendrites [54].

### 3.2.8. *Dynein motors*

Another motor that also co-localizes with kinesin is called dynein, which moves along the microtubules but towards the minus end [57, 58] (Fig 3.6). The dynein motor is known for the transportation of Golgi bodies from the axons and the rerouting of soma-dendritic protein cargos to the soma. Activation of the dynein motor depends on the nuclear distribution element-like 1 (NDEL1) [59, 60]. The presence of CDK-5 was also noticed in the AIS and this promotes the phosphorylation of NDEL1 [60]. The phosphorylated form of NDEL 1 binds with the dynein motor. Alterations to CDK-5 impair the formation NDEL-dynein binding, thus affecting the assembly of the AIS structure to changes in protein transport [44, 60]. NDEL1 enrichment in the AIS is due to its binding with AnkG, while the subsequent depletion of NDEL1 results in the inactivation of the dynein motor and this allows the entry of dendritic proteins into the proximal axon. Binding of NDEL1 and AnkG occurs at the C-terminal of NDEL1. The activation of dynein forces the cargo containing dendritic proteins to move back to the soma. Strong binding between dynein motors and MTs prevent the exclusion of dynein motors from the AIS structure [60].

### 3.2.9. *Importance of AnkG in AIS*

The ankyrins are a group of scaffolding proteins. In vertebrates, ankyrin is present in three isoforms: AnkR, AnkB, and AnkG, that have similar molecular structures [6, 61]. AnkR is known as the first isoform of ankyrin and is crucial for the maintenance of the shape and functioning of erythrocytes. AnkB is present in tissues, including the brain, thymus, and heart and skeletal muscles. In the heart, AnkB plays a critical role in the organization of the transport and ion channels, as the correct localization of these components is necessary for correct $Ca^+$ signaling and heart function. The third, and most important, ankyrin, AnkG, can be found in almost all tissues of the brain, epithelium, kidney, and muscles. In the brain, AnkG is available in three different isoform protein sizes, i.e., 190 kDa, 270 kDa and 480 kDa [61]. AnkG is crucial for AIS assembly and the maintenance of neuronal polarity [62]. AnkG is the first protein that clusters at the proximal axon; however, the mechanism behind this is still unknown. The presence of AnkG in the proximal axon may

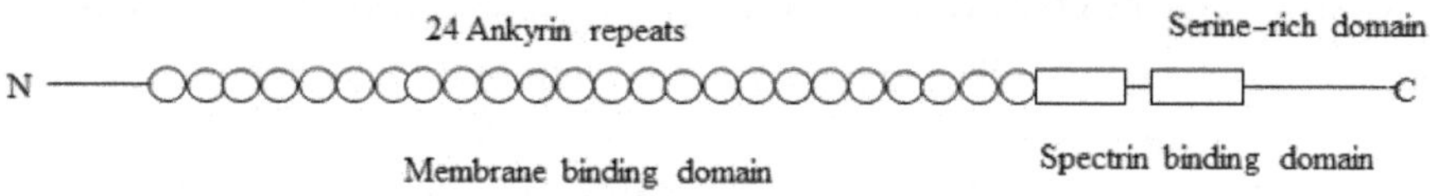


Fig. 3.7. Schematic representation of the structure of AnkG.

be due to teamwork between the multiple domains present at AnkG [63].

AnkG is the chief organizer of the AIS because the accumulation and expression of almost all AIS proteins depends on AnkG. AnkG comprises MBD at the N-terminal made up of 24 ankyrin repeats, a spectrin-binding domain, a serine-rich domain (only in the 270 and 480 kDa isoforms) with a long tail and a C-terminal domain (Fig. 3.7). The MBD of AnkG is responsible for the recruitment of ion channels ($Na_V$, $K_V$), molecular motors and CAMs (NF186, NrCAM) into the AIS [16, 62, 64–66]. The interaction of AnkG with the second membrane, the spectrin–actin membrane, is accomplished by the spectrin-binding domain with the help of a ZU5 motif of 160 amino acids. Other motifs, DAR999 and S2417, are also essential for the interaction of spectrin with the small and large isoforms of AnkG [61]. A regulatory domain, also known as the death domain, is an unstructured domain with 300 amino acids, which regulates the interactions of all proteins with AnkG-binding domains [65].

One of the most important functions of AnkG is to maintain neuronal polarity. Neuronal polarity in the AIS is maintained by the correct localization of AIS proteins and depends highly on AnkG. AnkG interacts with the ion channels, spectrin isoforms and actin, and stabilizes them within the structure [61]. The tight interaction of proteins with AnkG within the AIS favors AP initiation. Injury or mutations can cause weak binding between AIS-related proteins and AnkG. The weaker binding of proteins with AnkG increases the probability of the removal of protein from the structure [67]. Every membrane protein within this region is directly or indirectly associated with AnkG. AnkG and its binding members are known for targeting the GABAergic inter-neuronal presynaptic inputs that are necessary for controlling the AP firing frequency and the excitability of neurons in the brain region [68]. GABAergic interneurons are composed of $GABA_AR$ clusters, and the organization of GABAergic interneurons is dependent on the binding of $GABA_AR$ with a scaffolding protein, called gephyrin, which is indirectly associated with NF186 [61].

Any alteration in AnkG results in the disassembly of the AIS structure and a disruption in neuronal polarity [69] (Fig. 3.8). A mutation in AnkG has a direct impact on the expression of membrane proteins and their functions. The absence of AnkG can cause disturbances in the filtering or transport capability of the AIS where unnecessary proteins or dendritic proteins enter the axon region. The entry of these dendritic proteins forces the axon to adapt dendritic molecular characteristics and this means the axon will perform the same functions as the dendrites. The loss of actin, spectrin and CAMs by 90%, 90%, and 98%, respectively, has been noticed in AnkG mutated mice [16]. A study of Purkinje cells with an AnkG mutation demonstrated the importance of AnkG in AIS by observing that they contained no voltage gated ion channels ($Na_v$ and $K_v$ channels) due to the absence of AnkG. The lack of ion channels suggested the absence of interactions between the voltage gated ion channels and AnkG in AIS. The absence of voltage gated ion channels can also lead to the non-initiation of AP and their generation at the AIS [66].

## 3.3. History and Mechanisms Related to AP

The history of AP is fascinating. Luigi Galvani (1786) became the first researcher to introduce the relationship between electricity and movement in living organisms [70]. He noticed movements in frog legs after using an electric scalpel. His associate, Alessandro Volta, continued his work and invented the galvanic cell. Another scientist, Carlo Matteucci, also conducted the experiments on frog muscles by considering them to be electric conductors and, on the basis of his work, German physician Emil du Bois-Reymond discovered AP [70].

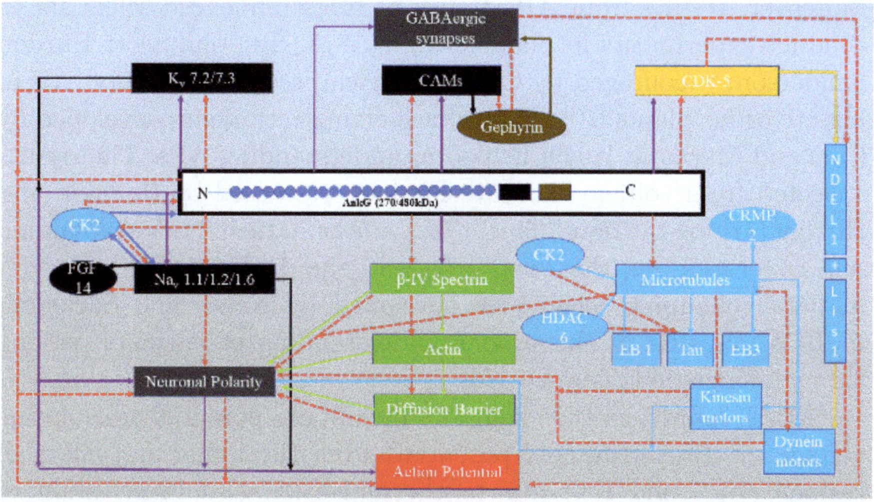


Fig. 3.8. Networks representing the role of AnkG in the AIS. All the purple arrows represent the relationships between AnkG and other proteins, and the relationships between AnkG and the specific function of the AIS (diffusion barrier, action potential or neuronal polarity). In the AIS, AnkG maintains the neuronal polarity (light gray). Neuronal polarity is maintained by the diffusion barrier ability of the AIS. For this region to act as a diffusion barrier, it is only possible when all the necessary proteins are recruited correctly. Recruitment of proteins in the AIS is dependent on AnkG. Moreover, the correct localization of proteins in the AIS promotes the initiation of AP (red rectangle). As shown, AnkG recruits the $K_v$ and $Na_v$ channels and CAMs (black rectangle) and they are responsible for the depolarization and repolarization phase of the AP process, respectively. CAMs bind with the gephyrin protein, which plays a crucial part in the GABAergic synapses (gold sphere). In addition, AnkG also recruits $\beta$-IV spectrin and this is associated with actin; together, they act as a diffusion barrier and this helps in maintaining neuronal polarity. AnkG also has a relationship with microtubules (blue rectangles) connected through the EB3 protein. Microtubules act as the pathway for the molecular motors (kinesin and dynein) (blue rectangles) and are responsible for the transport of proteins through the axons. The transport ability of molecular motors also plays a part in maintaining neuronal polarity by transporting the correct proteins in and out of the axons. The activation of the dynein motors depends on AnkG because NDEL-1 is an activator, and this binds to the dynein motors after interaction with AnkG. The red dotted line indicates the alteration in proteins due to the mutation in AnkG and this has a direct negative impact on AP initiation in the AIS.

Moreover, results from AP studies in frogs influenced scientists to conduct experiments in plants and, in 1873, Sir John Scott Burdon Sanderson, encouraged by Charles Darwin, recorded the first AP in insectivorous plants [70]. The voltage clamp techniques developed by Cole and Marmont [1949] helped in understanding APs. The results collected from voltage clamp techniques were used by Hodgkin and Huxley [71] for the development of a mathematical model explaining AP generation in the axon of giant squid. Moreover, upgrades in patch clamping techniques developed by Neher and Sackman, in 1976, assisted in the proper understanding of currents passing through single ion channels [70].

AP is defined as the change in membrane potential (membrane voltage or transmembrane potential) from the influx and efflux of cations ($Na^+$ and $K^+$) due to the activation and inactivation of voltage gated ion channels (Fig. 3.9). AP is one of the most crucial mechanisms in the human brain and is essential for communication between neurons. The AIS acts as an AP initiator zone due to its molecular organization and, especially, the presence of high densities of voltage gated ion channels ($Na_v$ and $K_v$ channels) [71]. The structural organization of this region supports to overcome the electricity induced by the soma during AP and prepares for the following AP [21]. The AIS is heavily populated with all the key players associated with AP, such as $Na_v$, $K_v$ and $Na^+/K^+$ pumps. Proper functioning of all these proteins is necessary for the independent steps involved in AP. $Na_v$ channels are responsible for the depolarization phase, and the repolarization (positive to negative) phase depends on the $K_v$ channels. Moreover, the hyperpolarization phase is regulated by $Na^+/K^+$ pumps [72]. Depolarization is defined as the movement of the internal membrane voltage from negative to positive and repolarization is the movement/change of the membrane voltage from positive to negative. Hyperpolarization is the phase where the membrane is highly negatively charged (beyond its resting value) [72, 73]. $Na_v$ and $K_v$ channels work using passive transport because the movement of ions takes place from a higher to a lower gradient. Moreover, the $Na^+/K^+$ pumps relies on active transport and uses adenosine triphosphate (ATP) as an energy source. The detailed mechanisms of AP within neurons are explained in Fig. 3.9.

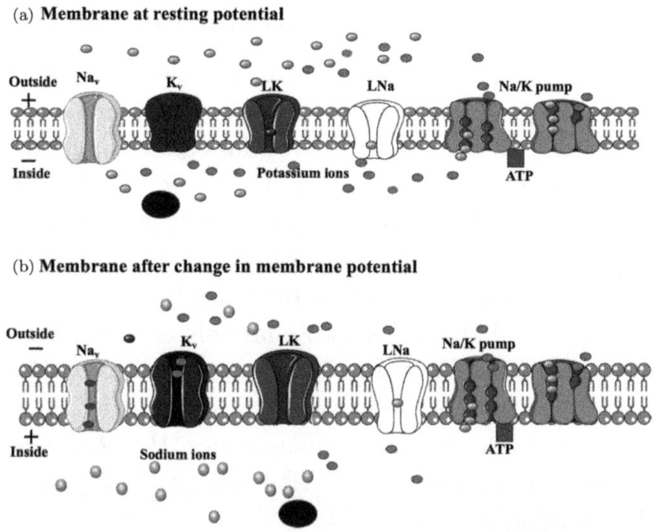


Fig. 3.9. Schematic representation of change in neuronal membrane due to AP. The AIS membrane comprises a voltage-gated ion channel ($Na_v$, $K_v$), $Na^+/K^+$ pumps and leak channels (LK and LNa). Image (a) represents the membrane in the resting state where the influx and efflux of cations ($Na^+$ and $K^+$ ions) are almost zero. At the resting state, the $Na_v$ and $K_v$ channels remain shut. The membrane voltage in the resting state in most neurons lies near 70 mV. During the resting stage, the net charge inside the membrane is negative, while it remains positive outside the membrane. The interior of the membrane has a higher $K^+$ ion concentration than $Na^+$ ions. However, the exterior of the membrane contains a higher $Na^+$ ion concentration than $K^+$ ions. Image (b) demonstrates the disturbance in the movement of ions due to stimuli (electric current, light, pressure (mechanical and osmotic)). Stimuli introduce conformational changes into the ion channels, and these result in the opening of the $Na_v$ and $K_v$ channels. As the $Na_v$ channel opens, an influx of $Na^+$ ions occurs and the concentration of $Na^+$ ions inside the membrane increases. This change in the ionic concentration inside the membrane tends to move the internal voltage towards a positive value ($-70$ mV to $+40$ mV) and the external voltage becomes negative; this is known as depolarization. When the $K_v$ channel opens, efflux of $K^+$ ions take place with the voltage inside the membrane moving again towards its resting value ($+40$ mV to $-70$ mV); this is called repolarization. Due to the slower movement of $K_v$ ions the channel membrane potential reaches beyond $-70$ mV. To generate another AP, the membrane potential should reach a resting value ($-70$ mV) and this step is accomplished by a $Na^+/K^+$ pump. The $Na^+/K^+$ pump takes three $Na^+$ ions from outside the membrane and two $K^+$ ions from inside the membrane. The black sphere in the figure represents the other proteins present inside the membrane. They cannot pass through due to the absence of a structure similar to ion channels and their larger sizes as compared to the ions. This figure is produced using Servier Medical Art (http://smart.servier.com).

## 3.4. Hodgkin and Huxley's Approximation for AP

Hodgkin and Huxley [71] introduced a model explaining mathematically how AP is produced within the axon of giant squid. Hodgkin and Huxley's (HH) model gave differential equations defining capacitance, current and membrane potential during AP [74]. Their model provides for the involvement of ion channels in action potentials. Voltage-gated ion channels, such as the $Na_v$, $K_v$ and leak (L) channels, were introduced into the model. Leak channels maintain the resting potential and the other two channels are responsible for the influx and efflux of $Na^+$ ions and $K^+$ ions, respectively. Hodgkin and Huxley mentioned in their research that $Na_v$ and $K_v$ channels have different gates embedded in their membranes. Hodgkin and Huxley's model explains the production of AP through the equations and by using a circuit diagram (Fig. 3.10).

Hodgkin and Huxley used an electrical circuit with two major components, membrane capacitance and ions, flowing through channels. Current can be expressed as

$$I_c = C_m * \frac{dV_m}{dt}, \tag{3.1}$$

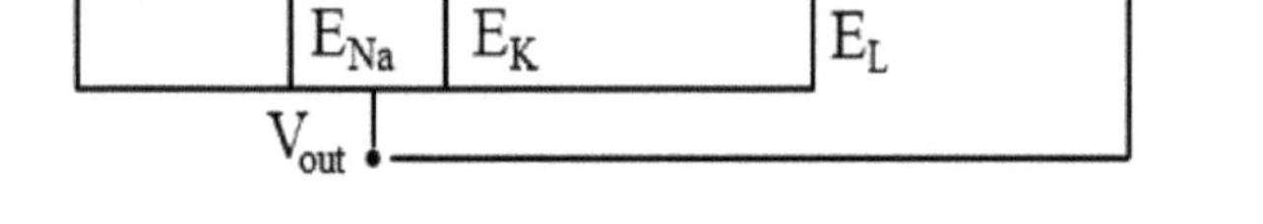


Fig. 3.10. Circuit diagram used in the HH model. $I_c$, $I_{Na}$, $I_K$ and $I_L$ are capacitive currents through the $Na_v$ channel, $K_v$ channel and leak channels, respectively. $G_{Na}$, $G_K$, $G_L$ represent the conductance of $Na^+$, $K^+$ and leak ions, respectively. $G_{Na}$ and $G_K$ are voltage dependent and $G_L$ is a constant in the HH model. $E_{Na}$, $E_K$, $E_L$ are the equilibrium potentials of the $Na^+$, $K^+$ and leak ions and $C_m$ represents the membrane capacitance. Stim, in the circuit, is the external current applied to the electrode. $V_{in}$ and $V_{out}$ are the voltages coming in and out, respectively.

where $I_c$ is the capacitive current and $C_m$ is the membrane capacitance ($\mu F/cm^2$) and this is defined as the rate of change in the charge of the membrane voltage ($V_m(t)$). The behavior of the electric circuit used in the HH model can, therefore, be written as

$$C_m * \frac{dV_m}{dt} + I_{ionic} = I_{ext}. \tag{3.2}$$

Equation (3.2) demonstrates the change in membrane potential due to the passing of ions through the membrane. $I_{ext}$ is the current from the external source (intracellular electrode) applied to the giant squid axon in the model. In the HH model, the total ionic current ($I_{ionic}$) is the sum of the three types of current in the membrane: the sodium ($I_{Na}$), potassium ($I_K$), and leak ($I_L$). The ionic current ($I$) is assumed through Ohm's law and can be written in terms of conductance ($G$), the inverse of the resistance ($G = 1/R) : I = G*V$, and units of $V$ is (mV).

So, considering all the equations above, $I_{ionic}$ can be written as

$$I_{ionic} = G_{Na}(V_m - E_{Na}) + G_K(V_m - E_K) + G_L(V_m - E_L), \tag{3.3}$$

where, $G_{Na}$, $G_K$ and $G_L$ represent the conductance of the potassium, sodium and leak channels. The individual ionic currents can be positive or negative. This depends on whether the membrane voltage is below or above the equilibrium potential. The unit of $G$ is m.mho/$cm^2$. In Eq. (3.3), $E_{Na}$, $E_K$ and $E_L$ are the equilibrium potentials for each ion ($Na^+, K^+, L^+$).

The movement of ions from the ion channels is based on the behavior of gates present in the respective channels, and the conductance of channels relies on these gates. Equation (3.3) indicates the relationship between conductance and the gates within the channels and currents due to the movement of ions through their respective channels. The $K_v$ channel comprises four identical gates denoted as $n$, and the $Na_v$ channel also comprises four gates but three are identical ($m$) and one is different ($h$). The conductance of the $Na_v$ and $K_v$

channels can be written as

$$G_{\mathrm{K}} = n^4 G_{\mathrm{Kmax}}, \tag{3.4a}$$

$$G_{\mathrm{Na}} = m^3 h G_{\mathrm{Namax}}. \tag{3.4b}$$

In Eqs. (3.4a) and (3.4b), $n$ refers to the activation gate and $n^4$ represents the four $n$ gates of the $\mathrm{K_v}$ channel. $\mathrm{G_{kmax}}$ is the maximum conductance when all the gates are open. Equations (3.4a) and (3.4b) indicate that the $\mathrm{Na_v}$ channel conductance is equal to the product of all the gates, and $\mathrm{G_{Namax}}$ is the maximum conductance when all the activation gates are open.

We can insert Eqs. (3.4a), and (3.4b) into Eq. (3.3):

$$I_{\mathrm{ionic}} = m^3 h G_{\mathrm{Namax}}(V_{\mathrm{m}} - E_{\mathrm{Na}}) + n^4 G_{\mathrm{Kmax}}(V_{\mathrm{m}} - E_{\mathrm{K}}) + G_{\mathrm{L}}(V_{\mathrm{m}} - E_{\mathrm{L}}). \tag{3.5}$$

Equation (3.5) gives the total ionic current passing through the membrane during AP.

## 3.5. AIS Dysfunction: AD and Neurological Conditions

Several hypotheses have been introduced to describe the cause of AD over recent decades. The current three main hypotheses are the amyloid deposition hypothesis, the cholinergic hypothesis and the tau hypothesis [75–79]. An explanation for all physiological changes that occur during the development of AD is still unknown.

As described in Chapter 2, the accumulation of extracellular neurotic plaques containing amyloid-$\beta$(A$\beta$), and neurofibrillary tangles (NFTs), consisting mainly of the hyperphosphorylated forms of the microtubule-associated protein, tau, are the two main markers for AD [81]. Most AD patients have SAD, but three genes have been associated with FAD: amyloid-$\beta$ protein precursor (A$\beta$PP), presenilin-1 (PS1) and presenilin-2 (PS2). FAD is clinically similar to SAD but has an earlier onset [82, 83]. A$\beta$ and tau are the significant components of the neurofibrillary tangles (NFT), which are very well known in AD pathogenesis [47].

The cholinergic hypothesis was the first hypothesis to exist to explain AD pathogenesis; it focused mainly on the loss of cholinergic activity in the brains of AD patients. Experimental studies have revealed the role of acetylcholine in learning and memory. Studies on young subjects by blocking cholinergic activity with scopolamine results in the impairment of memory and this defect was reversed by using the cholinergic agonist physostigmine [76]. By implementing this theory, early clinical trials were performed using other acetylcholinesterase inhibitors (AChEIs). Currently, there are three AChEIs (donepezil, rivastigmine and galantamine) that are used for the treatment of mild to moderate AD. These AChEIs are approved by the Food and Drug Administration (Canada). The results of clinical trials show that AChEIs provide a small improvement in AD symptoms. However, even after two years, no visible improvements were noticed in subjects taking AChEIs. Not all AD patients showed positive results after treatment with AChEIs. These negative factors all contribute to the failure of cholinergic activity [76].

According to the tau hypothesis, the hyperphosphorylation of tau plays a critical role in AD pathogenesis. Under pathological conditions, tau self-aggregates into paired helical filaments (PHFs) and these PHFs later turn into NFTs [80]. NFTs will impact MT function and disturb its dynamics. MT dynamics are very important for the correct protein transport by the kinesin and dynein motors. The disassembly of MTs leads to cell death [77, 78].

Amyloid $\beta$ protein (A$\beta$P) deposits in the brain and is the primary causative agent of AD pathology. According to studies in healthy or young brains, A$\beta$ is removed by $\beta$ or $\gamma$-secretase but, in aged brains, the A$\beta$ degradation rate declines and this leads to its deposition. After the deposition of A$\beta$, NFTs, cell loss, and dementia follow. This hypothesis is known as the "amyloid cascade hypothesis". Mutation of the amyloid precursor protein (APP) results in its clustering near $\beta$ or $\gamma$-secretase sites. These are responsible for the high production of A$\beta$. $\gamma$-secretase cleaves to 90% of APP, and the remaining APP is cleaved by $\beta$-secretase, followed by the C-tail in APP being cleaved by $\gamma$-secretase. The main product of this pathway is APP$\alpha$, C83, p3 and A$\beta$ (minor product) [75, 76].

The structure of the AIS is responsible for the correct molecular trafficking to maintain neuronal polarity and the initiation of AP [84]. The region maintains neuronal polarity with the help of different types of proteins clustered in this region as explained previously. Key proteins in the AIS, such as AnkG, voltage-gated ion channels, MTs and kinesin and dynein motors, can cause severe neurological and medical conditions after mutation. Several studies have observed the association of AIS proteins in diseases like epilepsy, bipolar disorder, schizophrenia and many more. Alteration in the voltage-gated ion channels in this region can disrupt the proper signaling between neurons [17, 67, 85]. Studies indicate the degeneration of axons in patients with multiple sclerosis (MS) and the improper localization of $Na_V$ channels may be due to the altered axonal transport that occurs as a result of this disease. Table 3.1 demonstrates the diseases or disorders that are associated with the AIS.

Inadequate functioning of $Na_V$ channels can cause various epileptic phenotypes, such as generalized epilepsy with febrile seizure plus (GEFS+), benign familial neonatal-infantile seizures (BFNS), early infantile epileptic encephalopathy 6, 11, and 13 (EIEE 6, 11 and 13), Ohtahara syndrome, intractable childhood epilepsy with generalized tonic–clonic seizures (ICEGTC) and Dravet syndrome (DS) [86–90]. The $K_V$ channel is equally important as the $Na_V$ channel because it repolarizes the membrane during AP and any mutation in this channel results in the development of various disorders, such as episodic ataxia (EA1), BFNS, benign childhood epilepsy with centrotemporal spikes (BECTS), temporal lobe epilepsy (TLE) and autism [91–93]. AnkG, as the master organizer, may be the common target in pathogenesis of several diseases because of its necessary interaction with the intracellular proteins of the AIS. Bipolar disorder, AD, schizophrenia and Angel-man syndrome are some of the disorders related to AnkG mutation [94–99]. One study observed long-term loss of memory in a rat model after reductions in the length of the AIS [100]. Together, all AIS proteins are very important and any changes in the structure of the AIS have a negative impact on neurological function.

Table 3.1. List of neurological conditions associated with various proteins present in AIS.

| AIS protein | Disorder | Source |
|---|---|---|
| $Na_v$ | GEFS+ — Epilepsy syndrome with febrile convulsions | [86, 88] |
| | BFNS — Recurrent seizures in newborn babies | [87] |
| | EIEE (6, 11, 13) — Tonic spasms occur within three months of life. Can be 100 times per day for up to 10 seconds each | [87] |
| | ICEGTC — Children in the first year of life with recalcitrant seizures and cognitive decline | [89] |
| | DS — Severe chronic seizures activated by fever in the first year of life | [90] |
| $K_v$ | EA1 — Severe discoordination with or without muscle movement | [93] |
| | BFNS — Recurrent seizures in newborn babies | [90] |
| | BECTS — Idiopathic epilepsy with seizures generally occurring in sleep in children from 2–13 years of age | [91] |
| | TLE — Seizures in the temporal lobe; it is a location or focal form of epilepsy | [92] |
| | Autism — Disorder with difficulty in speech, non-verbal communication and repetitive behavior | [93] |
| AnkG | Bipolar disorder — Disorder with manic depression | [101] |
| | AD — Neurodegenerative disease and the main cause of dementia | [2] |
| | Schizophrenia — Disorder with abnormal behavior, not understanding reality and unrealistic thinking and difficulty in speech | [94] |
| | Angel man syndrome — Genetic disorder with specific facial appearance, seizures and speaking problems | [97] |

### 3.5.1. *Possible roles of the AIS in AD*

Mutation or injury of AIS can cause various neurological medical conditions, such as epilepsy, schizophrenia, bipolar disorder, AD, and many more. In this section, we make a possible connection between the AIS and AD pathogenesis.

In an AD mice model, the area around A$\beta$ plaques showed synaptic losses, axonal swelling and mutation in the neuronal network

[98]. A$\beta$ plaques could damage the near or around region of AIS by targeting AnkG and $\beta$IV spectrin [98, 99]. Moreover, a study on AD transgenic mice showed a decrease in the density and length of this region [99]. Alterations in the density and length may be the result of the calpain-mediated proteolysis of AnkG and $\beta$IV spectrin [91, 98]. The proteolysis of AnkG reduces its concentration and dismantles the structure of the AIS by loosening the proteins anchored by AnkG. Dismantling of this structure results in the disruption of neuronal polarity by abolishing the functions of important proteins present in this region.

MTs are one of the main components of the AIS and are found in an impaired state after the inhibition of histone deacetylase 6 (HDAC6) (Fig. 3.11) [46, 47]. A study on HDAC6 stated that inhibition by inducing extracellular soluble A$\beta$ or HDAC6 inhibitors (TSA, valproic acid, sodium butyrate) into AIS has shown that inhibition could lead to the excessive acetylation of tau and tubulin, and this deteriorates MT function. This may be because the acetylated tau loses its binding capability to MTs and this increases the mobility of the EB3 protein. However, stabilized MTs have EB3 proteins with low mobility [47, 103]. Another possibility is that the acetylated tau may be attached to other AIS proteins impairing their functions in the AIS [103] (Fig. 3.11).

The abnormal hyperphosphorylation of tau has also been noticed in the AIS, leading to the impairment of MT dynamics and this is possibly responsible for the distortion of tau into the somatodendritic compartment via mutations in the transport motor proteins (kinesin and dynein) in MTs [103] (Fig. 3.11). An experimental investigation of the rtg4510 mouse strain revealed that hyperphosphorylation of tau reduces AP firing frequency via disrupting MT dynamics by preventing binding of the EB3 protein to MTs. This study also displayed the relocation of the AIS after abnormal hyperphosphorylation and this relocation may increase the AP threshold [101, 102] (Fig. 3.11). Other than tau, hyperphosphorylation of FGF14 also impacts on the density of the AIS by inhibiting the activity of GSK-3. Consequently, the inhibition of this kinase family member reduces the expression of the FGF14-$Na_v$ complex [103]. Sun

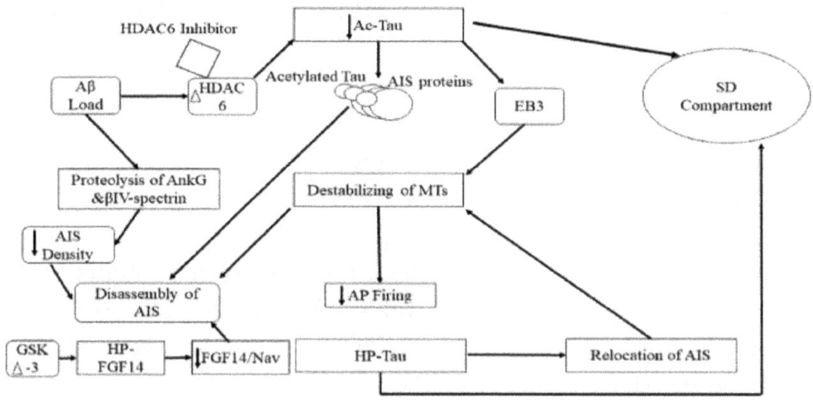


Fig. 3.11. Possible connections of various dysfunctions in the AIS with AD pathogenesis. The introduction of A$\beta$ plaques into the AIS or the inhibition of HDAC-6 can alter the function of HDAC-6 proteins. HDAC-6 prevents the excessive post-translational modification of tau. Mutations of HDAC-6 could destabilize the function of MTs by altering the function of the EB3 protein (EB3 connects MT with AnkG). Hyperacetylation or hyperphosphorylation of tau causes it to bind to other proteins associated with the structure of AIS and alters their functions. The hyperphosphorylation or hyperacetylation of tau can relocate the AIS or they can enter the somato-dendritic compartment after destabilizing the MTs, which is responsible for the correct protein transportation. Destabilized MTs reduce AP firing frequency by altering the AIS. The loading of A$\beta$ onto this structure could be responsible for the proteolysis of AnkG and $\beta$-IV spectrin. The breakdown of AnkG and $\beta$-IV spectrin, dismantles the assembly of AIS by reducing AIS density. Inhibition or mutation of the GSK-3 protein can cause hyperphosphorylation of FGF14, and this lowers its concentration in the $Na_v$ channels and FGF14 in the AIS and dismantles its structure. All alterations to AIS proteins' function due to mutations or inhibition could play a role in AD pathogenesis especially proteolysis of AnkG and the hyperphosphorylation or hyperacetylation of tau.

*et al.* [2] observed the effect of miRNA-342-5p on the hippocampal neurons of AD transgenic and wild-type (WT) mice. Western blot analysis showed a significant decrease in AnkG protein levels in APP/PS1 hippocampal neurons at 2, 3, 5 DIV [2]. Another study by the same research group was undertaken on the cytoplasmic filtering capability of the AIS. They employed dextrans of different sizes, 10 kDa and 70 kDa, in WT and AD mice. The diffusion of small dextrans (10 kDa) into axons was noticed in both types of mice, but the 70 kDa dextran did not pass into the axons in WT mice. However, in AD mice 70 kDa dextran was found in the axons. This indicates a defect in the cytoplasmic filtering capability of the AIS in AD mice. Defects in the filtering mechanism could be due to the low levels of AnkG in AD mice [2].

On the other hand, SH-SY5Y cells showed no change in AnkG-related mRNA levels but displayed low AnkG levels. These results indicate that the direct targeting of AnkG mRNA 3′ UTR by miRNA-342-5p represses the translation process of proteins but not mRNA degradation. The low level of AnkG proteins could affect protein filtering and trafficking (Fig. 3.11) in the AIS and may contribute to AD pathogenesis [2].

## References

[1] William T. and Bleiler L. (2013). Alzheimer's disease facts and figures. *Alzheimer's and Dement.*, *9*(2), pp. 208–245.

[2] Sun X., Yu W., Mingxue G. and Zhang Y. (2014). MiR-342-5p decreases ankyrin G levels in Alzheimer's disease transgenic mouse models. *Cell Rep.*, *6*(2), pp. 264–270.

[3] Leterrier C. (2015). Nanoscale architecture of the axon initial segment reveals an organized and robust scaffold. *Cell Rep.*, *13*(12), pp. 2781–2793.

[4] Jones S.L. and Tatyana M.S. (2016). Axon initial segment cytoskeleton: Architecture, development, and role in neuron polarity. Neural Plasticity.

[5] Boiko T. (2003). Functional specialization of the axon initial segment by isoform-specific sodium channel targeting. TL-23. *J. Neuroscience*, *23*(6), pp. 2306–2313.

[6] Leterrier C. and Bénédicte D. (2014). No pasaran! Role of the axon initial segment in the regulation of protein transport and the maintenance of axonal identity. *Seminars Cell Developmental Biol.*, *27*, pp. 44–51.

[7] Rasband M. (2010). The axon initial segment and the maintenance of neuronal polarity. *Nature Rev. Neurosci.*, *11*(8), pp. 552–562.

[8] Lemaillet G., Walker B. and Lambert S. (2003). Identification of a conserved ankyrin-binding motif in the family of sodium channel $\alpha$ subunits. *J. Biol. Chem.*, *278*(30), pp. 27333–27339.

[9] Rasband M. (2009). Converging on the origins of axonal ion channel clustering. *PLoS Genetics*, *5*(1), pp. 1–2.

[10] Guglielmi L. (2015). Update on the implication of potassium channels in autism: $K^+$ channelautism spectrum disorder. *Frontiers Cellular Neurosci.*, *9*(March), pp. 1–14.

[11] Satake T., Yamashita K., Hayashi K., Miyatake S., Tamura-Nakano M., Doi H., Furuta Y., Shioi G., Miura E., Takeo Y.H., Yoshida K., Yahikozawa H., Matsumoto N., Yuzaki M. and Suzuki A. (2017). MTCL1 plays an essential role in maintaining Purkinje neuron axon initial segment. *EMBO J.*, *36*(9), pp. 1227–1242.

[12] Kole M.H.P., Ilschner S.U., Kampa B.M., Williams S.R., Ruben P.C. and Stuart G.C. (2008). Action potential generation requires a high sodium channel density in the axon initial segment. *Nature Neurosci., 11*(2), pp. 178–186.

[13] Baranauskas G., David Y. and Fleidervish I.A. (2013). Spatial mismatch between the $Na^+$ flux and spike initiation in axon initial segment. *Proc. Natl. Acade. Sci. USA, 110*(10), pp. 4051–4056.

[14] Takeo K., Ailani D., Hiramoto M. and Hiromi Y. (2009). Intra-axonal patterning: Intrinsic compartmentalization of the axonal membrane in drosophila neurons. *Neuron., 64*(2), pp. 188–199.

[15] Akin E.J., Solé L., Johnson B., Beheiry M.E., Masson J.B., Krapf D. and Tamkun M.M. (2016). Single-molecule imaging of $Na_v$ 1.6 on the surface of hippocampal neurons reveals somatic nanoclusters. *Biophys. J. Sep., 20, 111*(6), pp. 1235–1247.

[16] Yoshimura T. and Rasband M.N. (2014). Axon initial segments: Diverse and dynamic neuronal compartments. *Curr. Opinion Neurobio., 27*, pp. 96–102.

[17] Ogawa Y. and Rasband M.N. (2008). The functional organization and assembly of the axon initial segment. *Curr. Opinion Neurobiol., 18*(3), pp. 307–313.

[18] Fache M.P., Moussif A., Fernandes F., Giraud P., Garrido J.J. and Dargent B. (2004). Endocytotic elimination and domain-selective tethering constitute a potential mechanism of protein segregation at the axonal initial segment. *J. Cell Biology, 166*(4), pp. 571–578.

[19] Xu X. and Shrager P. (2005). Dependence of axon initial segment formation on $Na^+$ channel expression. *J. Neuroscience Res., 79*(4), pp. 428–441.

[20] Kole M.H.P. and Stuart G.S. (2012). Signal processing in the axon initial segment. *Neuron., 73*(2), pp. 235–247.

[21] Rowan M.J., Tranquil E. and Christie J.M. (2014). Distinct $K_v$ channel subtypes contribute to differences in spike signaling properties in the axon initial segment and presynaptic boutons of cerebellar interneurons. *J. Neurosci., May 34*(19), pp. 6611–6623.

[22] Sánchez-Ponce D., Blázquez-Llorca L., DeFelipe J., Garrido J.J. and Muñoz A. (2012). Colocalization of $\alpha$-actinin and synaptopodin in the pyramidal cell axon initial segment. *Cereb. Cortex., 22*(7), pp. 1648–1661.

[23] Jensen C.S., Watanabe S., Ingrid S.J., Klaphaak J., Yamane A., Schmitt N., Olesen S.P., Trimmer J.S., Rasmussen H.B. and Misonou H. (2017). Trafficking of $K_v2.1$ channels to the axon initial segment by a novel non-conventional secretory pathway. *J. Neurosci., 37*(48), pp. 3510–3516.

[24] Kriebel M., Metzger J., Trinks S., Chugh D., Harvey R.J., Harvey K. and Volkmer H. (2011). The cell adhesion molecule neurofascin stabilizes axo-axonic GABAergic terminals at the axon initial segment. *J. Biol. Chem., 286*(27), pp. 24385–24393.

[25] Zonta B., Desmazieres A., Rinaldi A., Tait S., Sherman D.L., Nolan M.F. and Brophy P.J. (2011). A critical role for Neurofascin in regulating

action potential initiation through maintenance of the axon initial segment. *Neuron.*, *69*(5), pp. 945–956.

[26] Zhang C. and Rasband M.N. (2016). Cytoskeletal control of axon domain assembly and function. *Curr. Opinion Neurobiol.*, *39*, 116–121.

[27] Tse W.T., Tang J., Jin O., Korsgren C., John K.M., Kung A.L., Gwynn B., Peters L.L. and Lux S.E. (2001). A new spectrin, BIV, has a major truncated isoform that associates with promyelocytic leukemia protein nuclear bodies and the nuclear matrix. *J. Biol. Chem.*, *276*(26), pp. 23974–23985.

[28] Goodman S.R., Zimmer W.E., Clark M.B., Zagon I.S., Barker J.E. and Bloom M.L. (1995). Brain spectrin: of mice and men. *Brain Res. Bull.*, *36*(6), pp. 593–606.

[29] Viel A. and Branton D. (1996). Spectrin: On the path from structure to function. *Curr. Opinion Cell Biol.*, *8*(1), pp. 49–55.

[30] Berghs S., Aggujaro D., Dirkx R., Jr, Maksimova E., Stabach P., Hermel J.M., Zhang J.P., Philbrick W., Slepnev V., Ort T. and Solimena M. (2000). BetaIV spectrin, a new spectrin localized at axon initial segments and nodes of ranvier in the central and peripheral nervous system. *J. Cell Biol.*, *151*(5), pp. 985–1002.

[31] Saka S.K., Honigmann A., Eggeling C., Hell S.W., Lang T. and Rizzoli S.O. (2014). Organization of the plasma membrane. *Nature Commun.*, *5*(May), pp. 1–14.

[32] Komada M. and Soriano P. (2002). $\beta$IV-spectrin regulates sodium channel clustering through ankyrin-G at axon initial segments and nodes of ranvier. *J. Cell Biol.*, *156*(2), pp. 337–348.

[33] Yang Y., Ogawa Y., Hedstrom K.L. and Rasband M.N. (2007). $\beta$IV spectrin is recruited to axon initial segments and nodes of ranvier by Ankyrin G. *J. Cell Biol.*, *176*(4), pp. 509–19.

[34] Kazunari N., Akiyama H., Komada M. and Kamiguchi H. (2007). $\beta$IV-Spectrin forms a diffusion barrier against L1CAM at the axon initial segment. *Mole. Cellular Neurosci.*, *34*(3), pp. 422–430.

[35] Bettina W., Forscher P. and Mellman I. (1999). A diffusion barrier maintains distribution of membrane proteins in polarized neurons. *Nature*, *397*(6721), pp. 698–701.

[36] Baines A.J. (2010). Évolution du squelette membranaire dépendant de la spectrine. *Trans. Clin. Biol.*, *17*(3), pp. 95–103.

[37] Balasanyan V., Watanabe K., Dempsey W.P., Lewis T.L., Jr, Trinh L.A. and Arnold D.B. (2017). Structure and function of an actin-based filter in the proximal axon. *Cell Rep.*, *21*(10), pp. 2696–2705.

[38] Watanabe K., Al-Bassam S., Miyazaki Y., Wandless T.J., Webster P. and Arnold D.B. (2012). Networks of polarized actin filaments in the axon initial segment provide a mechanism for sorting axonal and dendritic proteins. *Cell Rep.*, *2*(6), pp. 1546–1553.

[39] Albrecht D., Winterflood C.M., Sadeghi M., Tschager T., Noé F. and Ewers H. (2016). Nanoscopic compartmentalization of membrane protein motion at the axon initial segment. *J. Cell Biol.*, *215*(1), pp. 37–46.

[40] Nakada C., Ritchie K., Oba Y., Nakamura M., Hotta Y., Iino R., Kasai R.S., Yamaguchi K., Fujiwara T. and Kusumi A. (2003). Accumulation of anchored proteins forms membrane diffusion barriers during neuronal polarization. *Nature Cell Biol.*, *5*(7), pp. 626–632.
[41] Hoogenraad C.C. and Bradke F. (2009). Control of neuronal polarity and plasticity — A renaissance for microtubules? *Trends Cell Biol, 19*(12), pp. 669–676.
[42] Kapitein L.C. and Hoogenraad C.C. (2011). Which way to go? Cytoskeletal organization and polarized transport in neurons. *Mole. Cellular Neurosci., 46*(1), pp. 9–20.
[43] Hung P.J., Song C., Deek J., Miller H.P., Li Y., Choi M.C. Wilson L., Feinstein S.C. and Safinya C.R. (2016). Tau mediates microtubule bundle architectures mimicking fascicles of microtubules found in the axon initial segment. *Nature Commun., 7*, pp. 1–9.
[44] Tapia M., Wandosell F. and Garrido J.J. (2010). Impaired function of Hdac6 slows down axonal growth and interferes with axon initial segment development. *PLoS One., 5*(9).
[45] Tsushima H., Emanuele M., Polenghi A., Esposito A., Vassalli M., Barberis A., Difato F. and Chieregatti E. (2015). HDAC6 and RhoA are novel players in abeta-driven disruption of neuronal polarity. *Nature Commun., 6*.
[46] Leterrier C., Vacher H., Fache M.P., d'Ortoli S.A., Castets F., Autillo-Touati A. and Dargent B. (2011). End-binding proteins EB3 and EB1 link microtubules to ankyrin G in the axon initial segment. *Proc. Acad. Sci., 108*(21), pp. 8826–8831.
[47] Sanchez-Ponce D., Muñoz A. and Garrido J.J. (2011). Casein kinase 2 and microtubules control axon initial segment formation. *Molecular Cellular Neurosci., 46*(1), pp. 222–234.
[48] Nakata T. and Hirokawa N. (2003). Microtubules provide directional cues for polarized axonal transport through interaction with kinesin motor head. *J. Cell Biol., 162*(6), pp. 1045–1055.
[49] Gumy L.F. and Hoogenraad C.C. (2018). Local mechanisms regulating selective cargo entry and long-range trafficking in axons. *Curr. Opinion Neurobio., 51*, pp. 23–28.
[50] Shaohua X. and Jan L.Y. (2009). A gate keeper for axonal transport. *Cell, 136*(6), pp. 996–998.
[51] Encalada S.E. and Goldstein L.S. (2014). Biophysical challenges to axonal transport: Motor-cargo deficiencies and neurodegeneration. *Ann. Rev. Biophys., 43*(1), pp. 141–169.
[52] Jenkins B., Decker H., Bentley M., Luisi J. and Banker G. (2012). A novel split kinesin assay identifies motor proteins that interact with distinct vesicle populations. *J. Cell Biol., 198*(4), pp. 749–761.
[53] Barry J., Gu Y., Jukkola P., O'Neill B., Gu H., Mohler P.J., Rajamani K.T. and Gu C. (2014). Ankyrin-G directly binds to kinesin-1 to transport voltage-gated $Na^+$ channels into axons. *Develop. Cell, 28*(2), pp. 117–131.

[54] Song A.H., Wang D., Chen G., Li Y., Luo J., Duan S. and Poo M.M. (2009). A selective filter for cytoplasmic transport at the axon initial segment. *Cell, 136*(6), pp. 1148–1160.
[55] Kuijpers M., van de Willige D., Freal A., Chazeau A., Franker M.A., Hofenk J., Rodrigues R.J.C., Kapitein L.C., Akhmanova A., Jaarsma D. and Hoogenraad C.C. (2016). Dynein regulator NDEL1 controls polarized cargo transport at the axon initial segment. *Neuron, 89*(3), pp. 461–471.
[56] Vallee R.B., Williams J.C., Varma D. and Barnhart L.E. (2004). Dynein: An ancient motor protein involved in multiple modes of transport. *J. Neurobiol., 58*(2), pp. 189–200.
[57] Egan M.J., Kaeling T. and Samara L.P. (2012). Lis1 is an initiation factor for dynein-driven organelle transport. *J. Cell Biol., 197*(7), pp. 971–982.
[58] Klinman E., Tokito M. and Holzbaur E.L.F. (2017). CDK5-dependent activation of dynein in the axon initial segment regulates polarized cargo transport in neurons. *Traffic, 18*(12), pp. 808–24.
[59] Smith K.R. and Penzes P. (2018). Ankyrins: Roles in synaptic biology and pathology. *Mol. Cellular Neurosci., 91*, pp. 131–139.
[60] Cunha S.R. and Mohler P.J. (2009). Ankyrin protein networks in membrane formation and stabilization. *J. Cell. Mol. Med., 13*(11–12), pp. 4364–4376.
[61] He M., Jenkins. and Bennett V. (2012) Cysteine 70 of ankyrin-G is S-palmitoylated and is required for function of ankyrin-G in membrane domain assembly. *J. Bio. Chem., 287*(52), p. 43995.
[62] North C. (1998). Restriction of 480/270-KD ankyrin. *J. Cell Biol., 142*(6), pp. 1571–1581.
[63] Nelson A.D. and Jenkins P.M. (2017). Axonal membranes and their domains: Assembly and function of the axon initial segment and node of ranvier. *Frontiers Cellular Neurosci., 11*(May), pp. 1–17.
[64] Zhou D., Lambert S., Malen P.L., Carpenter S., Boland L.M. and Bennett V. (1998). AnkyrinG is required for clustering of voltage-gated Na channels at axon initial segments and for normal action potential firing. *J. Cell Biol., 143*(5), pp. 1295–1304.
[65] Leterrier C., Brachet A., Fache M.P. and Dargent B. (2010). Voltage-gated sodium channel organization in neurons: Protein interactions and trafficking pathways. *Neurosci. Lett., 486*(2), pp. 92–100.
[66] Ango F., di Cristo G., Higashiyama Bennett V., Wu P. and Huang Z.J. (2004). Ankyrin-based subcellular gradient of neurofascin: An immunoglobulin family protein, directs GABAergic innervation at purkinje axon initial segment. *Cell., 119*(2), pp. 257–272.
[67] Le Bras B., Fréal A., Czarnecki A., Legendre P., Bullier E., Komada M., Brophy P.J., Davenne M. and Couraud F. (2014). *In vivo* assembly of the axon initial segment in motor neurons. *Brain structure & Function, 219*(4), pp. 1433–1450.
[68] Coombs J.S., Curtis D.R. and Eccles J.C. (1957). The Interpretation of spike potentials of motoneurons. *J. Physiol., 139*(2), pp. 232–249.
[69] Platkiewicz J. and Brette R. (2010). A threshold equation for action potential initiation. *PLoS Computational Biology, 6*(7), e1000850.

[70] Schmidt-Hieber C., Jonas P. and Bischofberger J. (2008). Action potential initiation and propagation in hippocampal mossy fibre axons. *Journal of Physiol., 7*, pp. 1849–1857.

[71] Hodgkin A.L. and Huxley A.F. (1952). A quantitative description of membrane current and its application to conduction and excitation in nerve. *J. Physiol., 117*(4), pp. 500–544.

[72] Checler F. and Turner A.J. (2012). Journal of neurochemistry special issue on Alzheimer's disease: 'Amyloid cascade hypothesis-20 years on. *J. Neurochem., 120*, pp. 2011–2012.

[73] Craig L.A., Hong N.S. and McDonald R.J. (2011). Revisiting the cholinergic hypothesis in the development of Alzheimer's disease. *Neurosci. Biobehav. Rev., 35*(6), pp. 1397–1409.

[74] Hardy J. and Higgins G. (1992). Alzheimer's Disease: The amyloid cascade hypothesis. *Science, 256*(5054), pp. 184–185.

[75] Kametani F. and Hasegawa M. (2018). Reconsideration of amyloid hypothesis and tau hypothesis in Alzheimer's disease. *Frontiers Neurosci., 12*, p. 25.

[76] Karran E., Mercken M. and Strooper B.D. (2011). The amyloid cascade hypothesis for Alzheimer's disease: An appraisal for the development of therapeutics. *Nature Rev. Drug Discovery, 10*(9), pp. 698–712.

[77] Maccioni R.B., Farías G., Morales I. and Navarrete L. (2010). The revitalized tau hypothesis on Alzheimer's disease. *Arch. Medical Res., 41*(3), pp. 226–231.

[78] Goedert M. and Spillantini M.G. (2006). Reviews a century of Alzheimers' disease. *Science, 314*(November), pp. 777–781.

[79] Hardy J. and Selkoe D.J. (2002). The amyloid hypothesis of Alzheimer's disease: Progress and problems on the road to the amyloid hypothesis of Alzheim progress and problems on the road to. *Science, New Series, 297*(5580), pp. 353–356.

[80] Pimplikar S.W. (2009). Reassessing the amyloid cascade hypothesis of Alzheimer's disease. *Inte. J. Biochem. Cell Biol., 41*(6), pp. 1261–1268.

[81] Brachet A., Leterrier C., Irondelle M., Fache M.P., Racine V., Sibarita J.B., Choquet D. and Dargent B. (2010). Ankyrin G restricts ion channel diffusion at the axonal initial segment before the establishment of the diffusion barrier. *J. Cell Biol., 191*(2), pp. 383–395.

[82] Kress, G.J. and Mennerick, S. (2009). Action potential initiation and propagation: Upstream influences on neurotransmission. *Neuroscience, 158*(1), pp. 211–222.

[83] Fujiwara T., Sugawara T., Mazaki-Miyazaki E. and Takahashi Y. (2003). Mutations of sodium channel α-subunit type 1 (SCN1A) in intractable childhood epilepsies with frequent generalized tonic–clonic seizures. *Brain, 1*(6), pp. 6–7.

[84] Huang W., Liu M., Yan S.F. and Yan N. (2017). Structure-based assessment of disease-related mutations in human voltage-gated sodium channels. *Protein Cell, 8*(6), pp. 401–438.

[85] George A.L. (2005). Inherited disorders of voltage-gated sodium channels. *J. Clinical Invest., 115*(8).

[86] Nabbout R., Chemaly N., Chipaux M., Barcia G., Bouis C., Dubouch C., Leunen D., Jambaqué I., Dulac O., Dellatolas G. and Chiron C. (2013). Encephalopathy in children with Dravet syndrome is not a pure consequence of epilepsy. *Orphanet J. Rare Dise, 8*, p. 176.

[87] Saeed M., Azam M., Shabbir N. and Qamar S.A. (2014). Is 'benign childhood epilepsy with centrotemporal spikes' always benign? *Iranian J. Child Neurol., 8*(3), pp. 38–43.

[88] Blair R.D.G. (2012). Temporal lobe epilepsy semiology. *Epilepsy Res. Treatment*, pp. 1–10.

[89] Zuberi S.M., Eunson L.H., Spauschus A., De Silva R., Tolmie J., Wood N. W., McWilliam R.C., Stephenson J.B., Kullmann D.M. and Hanna M.G. (2010). A Novel mutation in the human voltage-gated potassium channel gene (Kv1.1) associates with episodic ataxia type 1 and sometimes with partial epilepsy. *Brain, 133*(5), p. 1569.

[90] Bi C., Wu J., Jiang T., Liu Q., Cai W., Yu P., Cai T., Zhao M., Jiang Y.H. and Sun Z.S. (2012). Mutations of ANK3 identified by exome sequencing are associated with autism susceptibility. *Human Mutation, 33*(12), pp. 1635–1638.

[91] Buffington S.A. and Rasband M.N. (2011). The axon initial segment in nervous system disease and injury. *European J. Neurosci., 34*(10), pp. 1609–1619.

[92] Harty R.C., Kim T.H., Thomas E.A., Cardamone L., Jones N.C., Petrou S. and Wimmer V.C. (2013). Axon initial segment structural plasticity in animal models of genetic and acquired epilepsy. *Epilepsy Res., 105*(3), pp. 272–279.

[93] Kaphzan H., Buffington S.A., Jung J.I., Rasband M.N. and Klann E. (2011). Alterations in intrinsic membrane properties and the axon initial segment in a mouse model of Angelman syndrome. *J. Neurosci., 31*(48), 17637–17648.

[94] Peltola M.A., Kuja-Panula J., Liuhanen J., Võikar V., Piepponen P., Hiekkalinna T., Taira T., Lauri S.E., Suvisaari J., Kulesskaya N., Paunio T. and Rauvala H. (2016). AMIGO-Kv2.1 potassium channel complex is associated with schizophrenia-related phenotypes. *Schizophrenia Bull., 42*(1), pp. 191–201.

[95] Rueckert E.H., Barker D., Ruderfer D., Bergen S.E., O'Dushlaine C., Luce C.J., Sheridan S.D., Theriault K.M., Chambert K., Moran J., Purcell S.M., Madison J.M., Haggarty S.J. and Sklar P. (2013). Cis-acting regulation of brain-specific ANK3 gene expression by a genetic variant associated with bipolar disorder. *Mol. Psychiatry, 18*(8), pp. 922–929.

[96] Evans M.D., Dumitrescu A.S., Kruijssen D., Taylor S.E. and Grubb M.S. (2015). Rapid modulation of axon initial segment length influences repetitive spike firing. *Cell Rep., 13*(6), pp. 1233–1245.

[97] Zhu S., Cordner Z.A., Xiong J., Chiu C.T., Artola A., Zuo Y., Nelson A.D., Kim T.Y., Zaika N., Woolums B.M., Hess E.J., Wang X., Chuang

D.M., Pletnikov M.M., Jenkins P.M., Tamashiro K.L. and Ross C.A. (2017). Genetic disruption of ankyrin-G in adult mouse forebrain causes cortical synapse alteration and behavior reminiscent of bipolar disorder. *Proc. Natl. Acade. Sci., 114*(39), pp. 10479–10484.

[98] Marin M.A., Ziburkus J., Jankowsky J. and Rasband M.N. (2016). Amyloid-$\beta$ plaques disrupt axon initial segments. *Exp. Neurol., 281*, pp. 93–98.

[99] Sohn P.D., Tracy T.E., Son H.I., Zhou Y., Leite R.E., Miller B.L., Seeley W.W., Grinberg L.T. and Gan L. (2016). Acetylated tau destabilizes the cytoskeleton in the axon initial segment and is mislocalized to the somatodendritic compartment. *Mol. Neurodegeneration, 11*(1), pp. 1–13.

[100] Kadavath H., Hofele F.V., Biernat J., Kumar S., Tepper K., Urlaub H., Mandelkow E. and Zweckstetter M. (2015). Tau stabilizes microtubules by binding at the interface between tubulin heterodimers. *Proc. Natl. Acade. Sci., 112*(24), pp. 7501–7506.

[101] Gong C.X. and Iqbal K. (2008). Hyperphosphorylation of microtubule-associated protein tau: A promising therapeutic target for Alzheimer disease. *Curr. Med. Chem., 15*(23), pp. 2321–2328.

[102] Hatch R.J., Wei Y., Xia D. and Götz J. (2017). Hyperphosphorylated tau causes reduced hippocampal CA1 excitability by relocating the axon initial segment. *Acta Neuropathologica, 133*(5), pp. 717–730.

[103] Hsu W.J., Wildburger N.C., Haidacher S.J., Nenov M.N., Folorunso O., Singh A.K., Chesson B.C., Franklin W.F., Cortez I., Sadygov R.G., Dineley K.T., Rudra J.S., Taglialatela G., Lichti C.F., Denner L., and Laezza F. (2017). PPARgamma agonists rescue increased phosphorylation of FGF14 at S226 in the Tg2576 mouse model of Alzheimer's disease. *Exp. Neurol., 295*, pp. 1–17.

[104] Catterall W.A., Goldin A.L., and Waxman S.G. (2019). Voltage-gated sodium channels (version 2019.4) in the IUPHAR/BPS Guide to Pharmacology Database. *IUPHAR/BPS Guide to Pharmacology CITE, 2019*(4). Available from: https://doi.org/10.2218/gtopdb/F82/2019.4.

# Chapter 4

# Computational Biophysics of Protein Interactions in the AIS

## 4.1. Introduction

Communication between neurons through action potentials (APs) is necessary for the nervous system to perform its functions as discussed in the previous chapters. An AP is characterized by the changes in membrane potential or membrane voltage due to the efflux and influx of cations. Several studies have confirmed that AP is initiated at the axon initial segment [1–4], which is a non-myelinated region (Fig. 4.1), with a length of 10–60 $\mu$m, at the beginning of an axon [1, 5–10]. We have discussed the internal structure of the AIS region extensively in Chapter 3, but we focus on more specific mechanisms in this chapter in order to develop a biophysical model of protein interactions [11].

As mentioned before, the main proteins in the AIS are voltage-gated sodium channels ($Na_V$), voltage-gated potassium channels ($K_V$), microtubules (MTs), casein kinase 2 (CK2) and ankyrin-G (AnkG). AnkG is an essential protein in the AIS because the recruitment of other proteins ($K_V$, $Na_V$, NF186, and NrCAM) into the AIS depends on it [7, 12]. To evaluate the importance of AIS in neurological processes and to show its relationships with various neurological diseases, it is necessary to understand the structural organization of the AIS region (Fig. 4.1).

CK2, a serine/threonine-specific kinase, is acidophilic in nature and mediates phosphorylation of various proteins in the AIS [23].

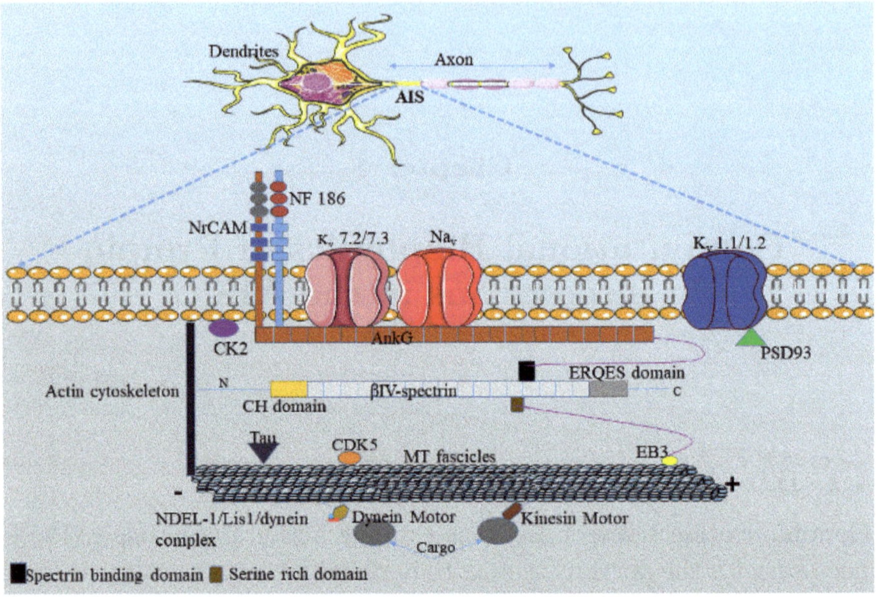


Fig. 4.1. A schematic diagram of AIS structure. The AIS region consists of three membranes: the plasma membrane, the spectrin–actin membrane and the microtubule related membrane (see also Fig. 3.1). Each membrane carries out different functions that favor AP initiation at the AIS. AnkG acts as the master organizer and all the proteins in the AIS are directly or indirectly associated with it [5, 12–15]. The initial part of AnkG is called the membrane binding domain (MBD), to which the anchoring of $Na_v$, and $K_v$ channels ($K_v$ 7.2/7.3) occurs [5, 6]. At the AIS, the proteins binding with AnkG have a specific amino acid sequence, known as the AIS motif, except for the $K_v$1.1/1.2 channels [16]. Instead of AnkG, the recruitment of $K_v$1.1/1.2 channels into the AIS depends on the post-synaptic density protein (PSD 93) [17, 18]. The MBD is followed by a spectrin binding domain in the AnkG structure, and this is responsible for the binding between $\beta$-IV spectrin and AnkG [7]. The microtubule fascicles in the AIS are also restricted by AnkG due to the interactions with an end-binding 3 protein (EB3) [7, 19]. According to the experimental evidence, the high-density presence of $Na_v$ channels in the AIS is correlated to the AP initiation [1, 7]; and it depends not only on AnkG but also on CK2-facilitated phosphorylation [20]. The experimental investigations have also shown that CK2 plays a dynamic role in the AIS region, especially in the accumulation of $Na_v$ channels into the AIS [20–22].

CK2 is constitutively active and does not require any second messenger or phosphorylation event to be activated [23, 24]. During the developmental stage, the early presence of CK2 in neurons has been observed, even before axon formation, and the experiments have shown the shortening of axons by 30% after CK2 inhibition [25]. CK2 plays a very critical role in AnkG-$Na_v$ binding in order to recruit $Na_v$ channels into the AIS facilitating the phosphorylation of $Na_v$ channels to increase their affinity towards AnkG [26, 27]. $Na_v$ channels comprise of a special type of amino acid sequence called the AIS motif, where the binding with AnkG takes place. The AIS motif contains four serine residues at the different positions, S1112, S1123, S1124, and S1126 (Fig. 4.2) [16]. In addition to that, the AIS motif also contains the acidic residues: glutamate (E) and aspartate (D) residues are present at 1111 and 1113 positions, respectively (Fig. 4.2). All these residues have different functions in the AIS motif: the serine residues are responsible for the phosphorylation of $Na_v$ channels through CK2, and acidic residues (glutamate and aspartate) increase the affinity of the phosphorylation process [23]. All these serine sites are phosphorylated by CK2 and this increases the binding affinity of the $Na_v$ channels towards AnkG in order to restrict them at the AIS [20, 23, 24].

Any alteration in neuronal transmission can cause various neurological conditions, such as Alzheimer's disease (AD), epilepsy, schizophrenia and bipolar disorder [6, 28–32]. AD is the most common neurodegenerative disease in humans; thus, it is the most common cause of dementia and, as yet, there is no known cure for AD [33–39]. The experimental studies to decipher the role of AIS in AD pathogenesis show the loss of various AIS related proteins, including AnkG and $Na_v$ channels, in the AD brain [40, 41]. The loss of these proteins may be due to a disturbance in protein trafficking because of the altered function of molecular motors, such as kinesin and dynein [40].

Biological systems are complex as their features can be explained using different methods, and this makes them very difficult to study and to predict their behaviors [42]. However, computational models can be used to understand the plausible molecular mechanisms

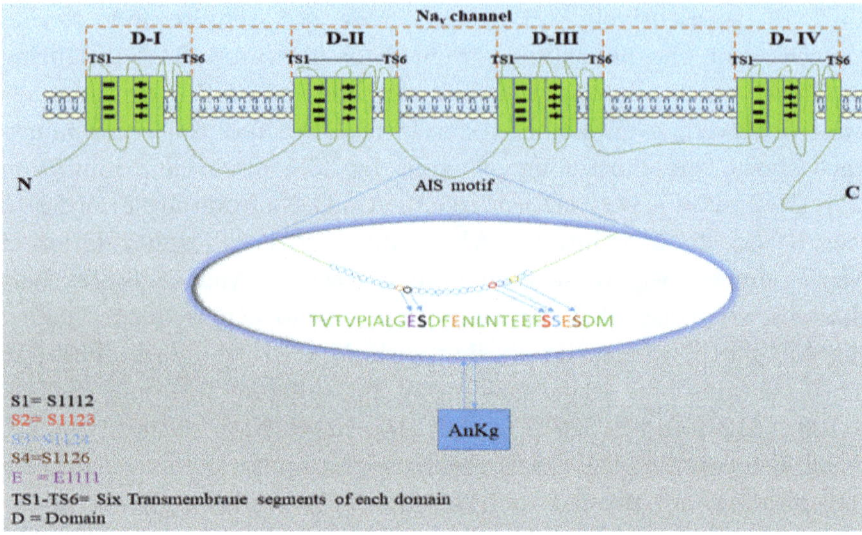


Fig. 4.2. A schematic representation of the AIS motif within Na$_v$ channels. The Na$_v$ channels comprise of four domains (I–IV), and each domain contains six transmembrane segments (TS1–TS6). The AIS motif (enlarged area) is located between the II and III domains of the Na$_v$ channels, and is made up of different 27 amino acids, with three main ones: serine (S) at the positions 1112, 1123, 1124, and 1126; glutamate (E) at the positions 1111 and 1115; and aspartate (D) at the position, 1113. Serine is important because the phosphorylation of Na$_v$ channels is accomplished by CK2, and CK2 is specific for serine or threonine. The presence of negatively charged residues, such as glutamate and aspartate, are also significant because they provide the support for the CK2-mediated phosphorylation of serine. Moreover, glutamate and aspartate assist CK2 in the identification of serine as a target for phosphorylation. This figure was produced using Servier Medical Art (https://smart.servier.com).

in an organism at the molecular level if we carefully develop the questions to be investigated. As we mentioned before, AP initiation is associated with high numbers of Na$_v$ channels in the AIS and these channels are phosphorylated by CK2 at their serine sites before their accumulation in the AIS. Further, AnkG recruits the phosphorylated Na$_v$ channels in this region. The dynamics of the phosphorylated channels, CK2, and AnkG would have crucial effects on AP initiation, and any impairment of these interactions would perhaps explain the experimentally observed link of AnkG to AD pathogenesis. Hence, we develop a computational model to investigate the role of serine-specific phosphorylation of $Na_v$ channels by CK2 before the accumulation of $Na_v$ channels into the AIS. In addition, we test the effect of changes in the concentration of CK2 and AnkG on the accumulation of $Na_v$ channels at the AIS. Overall, the main aim of this model is to investigate the importance of AIS related proteins (CK2, $Na_v$ and AnkG) in AP initiation and their association with Alzheimer's disease (AD) pathogenesis.

## 4.2. Development of a Computational Model

It is necessary to make simplifying assumptions in the development of the model so that we can understand the interactions between the proteins within AIS. Based on the currently available data and insights, the following assumptions are made:

(1) The AIS region is considered to be a cylinder with a length of $40\,\mu m$ and a diameter of $1.5\,\mu m$ [4]. A total of 12 species are present in our model which includes four different $Na_v$ channels (NaA, NaB, NaC, $N_{1111}$), four AnkG species (G1, G2, G3, G4) and four CK2 species (C1, C2, C3, C4).

(2) In this model, four types of $Na_v$ channels are considered based on different phosphorylation events within the channels. We assume five phosphorylation conditions: no site phosphorylation, single-site phosphorylation, double-site phosphorylation, triple-site phosphorylation and four-site phosphorylation. The no site phosphorylation means that all serine sites are in the non-phosphorylated state. We do not consider any $Na_v$ species in this model as not being phosphorylated because without any serine phosphorylation, $Na_v$–AnkG interactions are impossible [20]. Similarly, none of the $Na_v$ species are assumed to be in the single-site phosphorylation condition as well, because AnkG does not bind to the $Na_v$ channels with single-site phosphorylation [20]. Four $Na_v$ species in the model are assumed based on double-site phosphorylation, triple-site phosphorylation and four-site phosphorylation. Due to the lack of literature on the concentration of $Na_v$ channels before its restriction into the AIS

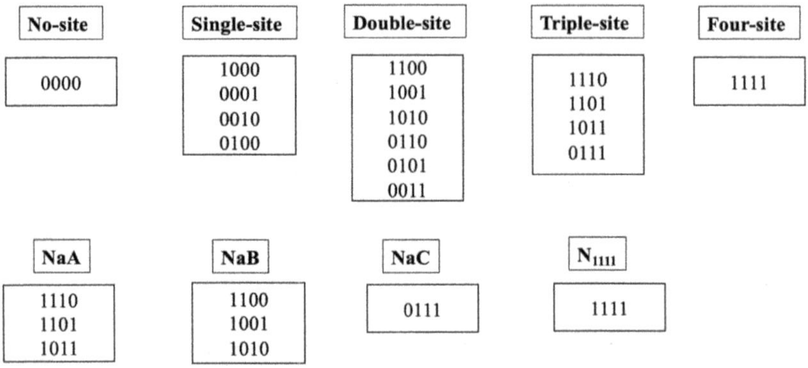


Fig. 4.3. Classification of the $Na_v$ species in this model. None of the $Na_v$ species in the model are in the no phosphorylation or single-site phosphorylation conditions. In the double-site phosphorylation, two serine sites are phosphorylated, and there is a total of six combinations. Out of the six, only the first three combinations are considered for $Na_v$–AnkG binding. These three combinations are considered as one and called NaB in the model. Combinations of the double phosphorylation condition, other than first three (NaB), are ignored on the basis of the absence of a S1 site because the affinity of $Na_v$ channel binding with AnkG is high in the presence of the S1 site [20, 43]. In the triple-site phosphorylation conditions, all combinations favor the interaction of $Na_v$-AnkG. We, therefore, consider these first three combinations because of the presence of S1 and assume them to be another $Na_v$ species (NaA) in the model; the fourth combination of the triple phosphorylation condition is taken as the third $Na_v$ species (NaC) even in the absence of a S1 site because of the phosphorylation of the remaining three serine sites (S2, S3 and S4) can favor $Na_v$–AnkG interactions but with a low binding affinity [20]. The fourth $Na_v$ species (N1111) in the model is based on the presence of all serine sites being in the phosphorylated state.

we estimate that the concentration of each of NaA, NaB, NaC, $N_{1111}$ as 3.32 $^{*}E^{-12}(\mu M)$ (see Fig. 4.3).

(3) The four phosphorylation sites in the AIS motifs (S1112, S1123, S1124 and S1126) are denoted by S1 = S1112, S2 = S1123, S3 = S1124, and S4 = S1126. All these sites within the $Na_V$ channels are responsible for equally strong interactions between the $Na_v$ channels and AnkG [44].
Here are the phosphorylation sequences for all four sites in the $Na_v$ species:

- The phosphorylation of S1 is the first preference for CK2 because the regions near S1 (E, S1, D, F, E) fulfil the criteria

for the minimum consensus sequence (S/T, X, X, D/E/pS/pT) for CK2 phosphorylation. The presence of acidic residues at the $n+1$ and $n+3$ positions makes S1 the most suitable site for phosphorylation by CK2 [23].
- S3 also fulfills the minimum criterion for the consensus sequence (S/T, X, X, D/E/pS/pT). S3 also has acidic residues at the $n+1$ and $n+3$ positions, similar to S1. However, S3 is assumed to be the second preference after S1 because S1 is located near the N-terminal.
- S4 is considered the third phosphorylation site because of the presence of an acidic residue (D) at the $n+1$ position. The fourth, and final, phosphorylation site is S2 (S2, pS, E, pS, D) because the region near S2 fulfills the minimum consensus sequence due to the presence of phosphorylated S or T at the $n+1$ and $n+3$ positions (see Fig. 4.4).

(4) Phosphorylation of the $Na_v$ species (NaA, NaB, NaC, $N_{1111}$) by CK2 (C1, C2, C3, C4) and its interaction with AnkG (G1, G2, G3, G4) in this model are based on the mass action law.

(5) In general, the phosphorylation is modeled using Michaelis–Menten kinetics but due to rapid binding and unbinding of CK2 to the sites, we assume that the binding-unbinding can be modeled as elementary reactions and follows the mass action kinetics within the first few minutes; i.e., the serine related phosphorylation of $Na_v$ channels by CK2 is considered to be a reversible elementary reaction. Because of the importance of the serine residues, the main aim is to understand the binding of CK2 with the $Na_v$ channels. Protein kinases have significant effects on the target protein they bind to. Protein kinases go on and off by auto-phosphorylation by binding with inhibitors or activator proteins or other small molecules. Their activity is highly regulated. However, this is not the case with CK2 due to its pleiotropic nature: CK2 does not require any mediators to activate because CK2 is always active. There is no evidence in the literature for any regulation of CK2 activity. CK2 not only phosphorylates the $Na_v$ and $K_v$ channels but also AnkG. Another reason for considering the mass action law to model

**Consensus Sequence for CK2 mediated phosphorylation in AIS motif of $Na_v$ channels**

TVTVPIALG E **S** D F E NLNTEEFSSESDM

-2 -1 n +1 +2 +3..............................

Fig. 4.4. Consensus sequence for CK2 mediated phosphorylation. The AIS motif of $Na_v$ channels and the requirements for CK2-mediated phosphorylation. S represents serine residues, E refers to glutamate, and D refers to aspartate. D and E assist CK2 to find a suitable serine for phosphorylation [21, 22]. The positions of residues within the motif are represented by $n - 1$, $n$, $n + 1$ and so on. The potential targets of CK2 for phosphorylation are serine/threonine residues in a motif surrounded by any of the acidic residues, such as E and D. CK2-mediated phosphorylation of serine/threonine requires at least one acidic residue between the $n - 4$ to $n + 7$ positions [21, 22]. The most favorable serine for CK2-mediated phosphorylation is when an acidic residue is present at the $n + 3$ position within a motif. In some instances, the phosphorylation by CK2 is enabled when an acidic residue is present at the $n + 1$ position and this is the second most preferred position after $n + 3$ [21, 22]. The minimum consensus sequence for CK2 phosphorylation is S/T, X, X, D/E/pS/pT, where S/T is a serine or threonine in the nth position, followed by any type residue at the $n + 1$ and $n + 2$ positions. At the $n + 3$ position, D/E/pS/pT stands for aspartate (D) or glutamate (E) or phosphorylated serine/threonine (pS/pT), respectively. Phosphorylated serine or threonine mimics the function of acidic residues and makes serine/threonine at the $n$ position suitable for CK2 phosphorylation. The presence of basic residues (lysine, proline, and lysine) at the $n + 1$, $n + 2$ or $n + 3$ positions has a negative impact on CK2-mediated phosphorylation. In this figure, the presence of a negative residue (E) at $n + 3$ in the sequence makes S at the $n$th position suitable for phosphorylation by CK2 [43]. S at the $n$th position is phosphorylated first because it is near the N-terminal when compared to the other serine residues. Similarly, other serine residues in the AIS motif can be identified as targets for CK2 by assuming them at $n$ positions and by checking nearby for the presence of negative residues (D/E/pS/pT).

the phosphorylation of $Na_v$ species (NaA, NaB, NaC, $N_{1111}$) is the abundance of CK2 in brain; therefore, we assume negligible competition of other species for CK2. In the absence of data, the phosphorylation of all four $Na_v$ species is assumed to occur at the same association ($K_{on}$) and dissociations rates ($K_{off}$).

(6) In the brain, more than 300 substrates are available for CK2-mediated phosphorylation [21, 23]. We have no data on the exact number of CK2 molecules required for phosphorylation of the $Na_V$ channels. As CK2 is present in abundance, we assume that the total CK2 concentration is $5.28{*}E^{-8}(\mu M)$. Specifically, for this model, total CK2 concentration is divided into four equal concentrations designated as C1, C2, C3 and C4 in order to phosphorylate NaA, NaB, NaC and $N_{1111}$, respectively.

(7) According to the literature, total number of $Na_V$ channels present in the AIS after binding with AnkG are 100–300 molecules/$\mu$m [1]. On the other hand, according to Srinivasan *et al.* (1988), AnkG numbers are ten times higher than the $Na_V$ channels [45]. Based on this observation, in this model initial AnkG concentration is assumed to be $1.32{*}E^{-13}$ $\mu$M. Similar to CK2 species, for this model, we divide AnkG species into four species named as G1, G2, G3 and G4 with the equal concentration of $3.32{*}E^{-14}$ $\mu$M; G1, G2, G3, G4 bind with the phosphorylated form of NaA, NaB, NaC and $N_{1111}$, respectively.

### 4.2.1. *Equations in the model*

The schematic of the model is shown in Fig. 4.5 and the kinetic parameters in this model are obtained from the literature and through calculations (Table. 4.1). The initial concentrations of all the species in this model are based on the literature (Table. 4.2).

### 4.2.2. *Phosphorylation of $Na_v$ species by CK2*

Equations. (4.1)–(4.20) model the dynamic changes in all $Na_V$ species (NaA, NaB, NaC and $N_{1111}$) after phosphorylation mediated by CK2 (C1, C2, C3, C4) based on ordinary differential equations (ODEs). The phosphorylated forms of the $Na_V$ species are denoted as $pNaA$, $pNaB$, $pNaC$ and $pN_{1111}$. $[NaA]$, $[NaB]$, $[NaC]$, $[N_{1111}]$, $[pNaA]$, $[pNaB]$, $[pNaC]$, $[pN_{1111}]$, $[C1]$, $[C2]$, $[C3]$ and $[C4]$ represent the concentration of the species. The phosphorylation of all $Na_V$ species by CK2 species in Eqs. (4.1), (4.6), (4.11), and (4.16) are achieved at specific association rates ($K_{01}, K_{02}, K_{03}$ and $K_{04}$) and dissociation rates ($K_{10}, K_{20}, K_{30}$ and $K_{40}$).

$$NaA + C1 \underset{K_{10}}{\overset{K_{01}}{\rightleftharpoons}} pNaA, \tag{4.1}$$

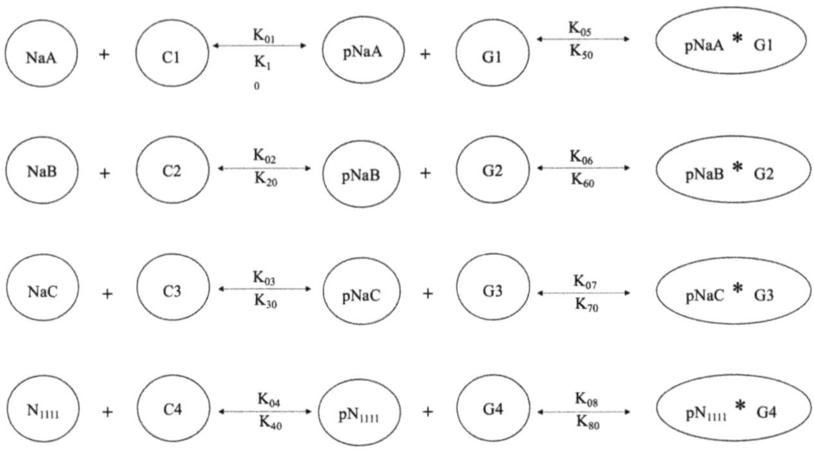


Fig. 4.5. Conceptual model for the accumulation of $Na_v$ channels in AIS. The model contains 12 different species: CK2 (C1, C2, C3, C4), $Na_v$ (NaA, NaB, NaC, N1111), and AnkG (G1, G2, G3, G4); and sixteen rate constants. A total of eight reactions are used to model the phosphorylation of $Na_v$ species through CK2 and the binding of phosphorylated $Na_v$ species to AnkG.

Table 4.1. Initial concentration of all species considered in the model.

| Protein | Initial concentration ($\mu$M) | Source |
|---|---|---|
| NaA | $3.32{*}E^{-12}$ | [9] |
| NaB | $3.32{*}E^{-12}$ | [9] |
| NaC | $3.32{*}E^{-12}$ | [9] |
| $N_{1111}$ | $3.32{*}E^{-12}$ | [9] |
| C1 | $1.32{*}E^{-8}$ | [23] |
| C2 | $1.32{*}E^{-8}$ | [23] |
| C3 | $1.32{*}E^{-8}$ | [23] |
| C4 | $1.32{*}E^{-8}$ | [23] |
| G1 | $3.32{*}E^{-14}$ | [45] |
| G2 | $3.32{*}E^{-14}$ | [45] |
| G3 | $3.32{*}E^{-14}$ | [45] |
| G4 | $3.32{*}E^{-14}$ | [45] |

Table 4.2. Rate constants used in this model with their biological meaning.

| Rate constants | Biological meaning | Values | Ref. |
|---|---|---|---|
| $K_{01}$ | Association rate for NaA and C1 | $1{*}E^{12}$ ($\mu M^{-1}s^{-1}$) | [52] |
| $K_{10}$ | Dissociation rate of pNaA into NaA and C12 | $1{*}E^{6}$ ($s^{-1}$) | [52] |
| $K_{02}$ | Association rate for NaB and C2 | $1{*}E^{12}$ ($\mu M^{-1}s^{-1}$) | [52] |
| $K_{20}$ | Dissociation rate of pNaB into NaB and C2 | $1{*}E^{6}$ ($s^{-1}$) | [52] |
| $K_{03}$ | Association rate for NaC and C3 | $1{*}E^{12}$ ($\mu M^{-1}s^{-1}$) | [52] |
| $K_{30}$ | Dissociation rate of pNaC into NaC and C3 | $1{*}E^{6}$ ($s^{-1}$) | [52] |
| $K_{04}$ | Association rate of $N_{1111}$ and C4 | $1{*}E^{12}$ ($\mu M^{-1}s^{-1}$) | [52] |
| $K_{40}$ | Dissociation rate of $pN_{1111}$ into $N_{1111}$ and C4 | $1{*}E^{6}$ ($s^{-1}$) | [52] |
| $K_{05}$ | Association rate of pNaA and G1 | $1.71{*}E^{9}(\mu M^{-1}s^{-1})$ | [20] |
| $K_{50}$ | Dissociation rate of pNaA*G1 into pNaA and G1 | $3.3{*}E^{-3}(s^{-1})$ | [20] |
| $K_{06}$ | Association rate of pNaB and G2 | $8.2 \pm 2.6{*}E^{8}(\mu M^{-1}s^{-1})$ | [20] |
| $K_{60}$ | Dissociation rate of pNaB*G2 into pNaB and G2 | $3.4 \pm 0.2{*}E^{-3}(s^{-1})$ | [20] |
| $K_{07}$ | Association rate of pNaC and G3 | $8.9 \pm 0.2{*}E^{8}(\mu M^{-1}s^{-1})$ | [20] |
| $K_{70}$ | Dissociation rate of pNaC*G3 into pNaC and G3 | $3.3 \pm 0.4[{*}]E^{-3}(s^{-1})$ | [20] |
| $K_{08}$ | Association rate of $pN_{1111}$ and G4 | $4.24{*}E^{9}$ ($\mu M^{-1}s^{-1}$) | [20] |
| $K_{80}$ | Dissociation rate of $pN_{1111}{*}G4$ into $pN_{1111}$ and G4 | $2.1 \pm 0.8{*}E^{-3}(s^{-1})$ | [20] |

$$\frac{d[NaA]}{dt} = -K_{01}[NaA][C1] + K_{10}[pNaA], \tag{4.2}$$

$$\frac{d[C1]}{dt} = -K_{01}[NaA][C1] + K_{10}[pNaA], \tag{4.3}$$

$$\frac{d[pNaA]}{dt} = K_{01}[NaA][C1] - K_{10}[pNaA], \tag{4.4}$$

$$\frac{d[NaA]}{dt} = \frac{d[C1]}{dt} = -\frac{d[pNaA]}{dt}, \tag{4.5}$$

$$NaB + C2 \underset{K_{20}}{\overset{K_{02}}{\Longleftrightarrow}} pNaB, \tag{4.6}$$

$$\frac{d[NaB]}{dt} = -K_{02}[NaB][C2] + K_{20}[pNaB], \tag{4.7}$$

$$\frac{d[C2]}{dt} = -K_{02}[NaB][C2] + K_{20}[pNaB], \tag{4.8}$$

$$\frac{d[pNaB]}{dt} = K_{02}[NaB][C2] - K_{20}[pNaB], \tag{4.9}$$

$$\frac{d[NaB]}{dt} = \frac{d[C2]}{dt} = -\frac{d[pNaB]}{dt}, \tag{4.10}$$

$$NaC + C3 \underset{K_{30}}{\overset{K_{03}}{\Longleftrightarrow}} pNaC, \tag{4.11}$$

$$\frac{d[NaC]}{dt} = -K_{03}[NaC][C3] + K_{30}[pNaC], \tag{4.12}$$

$$\frac{d[C3]}{dt} = -K_{03}[NaC][C3] + K_{30}[pNaC], \tag{4.13}$$

$$\frac{d[pNaC]}{dt} = K_{03}[NaC][C3] - K_{30}[pNaC], \tag{4.14}$$

$$\frac{d[NaC]}{dt} = \frac{d[C3]}{dt} = -\left\{\frac{d[pNaC]}{dt}\right\}, \tag{4.15}$$

$$N_{1111} + C4 \underset{K_{40}}{\overset{K_{04}}{\Longleftrightarrow}} pN_{1111}, \tag{4.16}$$

$$\frac{d[N_{1111}]}{dt} = -K_{04}[N_{1111}][C4] + K_{40}[pN_{1111}], \tag{4.17}$$

$$\frac{d[C4]}{dt} = -K_{04}[N_{1111}][C4] + K_{40}[pN_{1111}], \tag{4.18}$$

$$\frac{d[pN_{1111}]}{dt} = K_{04}[N_{1111}][C4] - K_{40}[pN_{1111}], \quad (4.19)$$

$$\frac{d[N_{1111}]}{dt} = \frac{d[C4]}{dt} = -\left\{\frac{d\,[pN_{1111}]}{dt}\right\}. \quad (4.20)$$

### 4.2.3. *Binding of phosphorylated* $Na_v$ *species with AnkG*

Equations. (4.21)–(4.40) describe the accumulation of $Na_v$ species at the AIS after binding with AnkG. The ODEs for binding between the phosphorylated form of $Na_v$ species ($pNaA$, $pNaB$, $pNaC$ and $pN_{1111}$) and AnkG (G1, G2, G3, G4) are written using the mass action law. $[pNaA]$, $[pNaB]$, $[pNaC]$, $[pN_{1111}]$, $[pNaA{*}G1]$, $[pNaB{*}G2]$, $[pNaC{*}G3]$, $[pN_{1111}{*}G4]$, $[G1]$, $[G2]$, $[G3]$ and $[G4]$ represent the concentrations of the species. The bindings of phosphorylated $Na_v$ species with AnkG species in Eqs. (4.21), (4.26), (4.31) and (4.36) are modeled through the association rates ($K_{05}, K_{06}, K_{07}$ and $K_{08}$) and the dissociation rates ($K_{50}, K_{60}, K_{70}$ and $K_{80}$).

$$pNaA{+}G1 \underset{K_{50}}{\overset{K_{05}}{\Longleftrightarrow}} pNaA{*}G1, \quad (4.21)$$

$$\frac{d[pNaA]}{dt} = -K_{05}[pNaA][G1]{+}K_{05}[pNaA{*}G1], \quad (4.22)$$

$$\frac{d[G1]}{dt} = -K_{05}[pNaA][G1] + K_{50}[pNaA{*}G1], \quad (4.23)$$

$$\frac{d[pNaA{*}G1]}{dt} = K_{05}[pNaA][G1] - K_{05}[pNaA{*}G1], \quad (4.24)$$

$$\frac{d[pNaA]}{dt} = \frac{d[G1]}{dt} = -\left\{\frac{d[pNaA{*}G1]}{dt}\right\}, \quad (4.25)$$

$$pNaB{+}G2 \underset{K_{60}}{\overset{K_{06}}{\Longleftrightarrow}} pNaB{*}G2, \quad (4.26)$$

$$\frac{d[pNaB]}{dt} = -K_{06}[pNaB][G2]{+}K_{60}[pNaB{*}G2], \quad (4.27)$$

$$\frac{d[G2]}{dt} = -K_{06}[pNaB][G2] + K_{50}[pNaB{*}G2], \quad (4.28)$$

$$\frac{d[pNaB{*}G2]}{dt} = K_{06}[pNaB][G2] - K_{60}[pNaB{*}G2], \quad (4.29)$$

$$\frac{d[pNaB]}{dt} = \frac{d[G2]}{dt} = -\left\{\frac{d[pNaB{*}G2]}{dt}\right\}, \quad (4.30)$$

$$pNaC{+}G3 \underset{K_{70}}{\overset{K_{07}}{\Longleftrightarrow}} pNaC{*}G3, \quad (4.31)$$

$$\frac{d[pNaC]}{dt} = -K_{07}[pNaC][G3] + K_{70}[pNaC{*}G3], \quad (4.32)$$

$$\frac{d[G3]}{dt} = -K_{07}[pNaC][G3]{+}K_{70}[pNaC * G3], \quad (4.33)$$

$$\frac{d[pNaC * G3]}{dt} = K_07[pNaC][G3] - K_{70}[pNaC * G3], \quad (4.34)$$

$$\frac{d[pNaC]}{dt} = \frac{d[G3]}{dt} = -\left\{\frac{d[pNaC * G3]}{dt}\right\}, \quad (4.35)$$

$$pN_{1111}{+}G4 \underset{K_{80}}{\overset{K_{08}}{\Longleftrightarrow}} pN_{1111}{*}G4, \quad (4.36)$$

$$\frac{d[pN_{1111}]}{dt} = -K_{08}[pN_{1111}][G4]{+}K_{80}[pN_{1111}{*}G4], \quad (4.37)$$

$$\frac{d[G4]}{dt} = -K_{08}[pN_{1111}][G4]{+}K_{80}[pN_{1111}{*}G4], \quad (4.38)$$

$$\frac{d[pN_{1111}{*}G4]}{dt} = K_{08}[pN_{1111}][G4]{-}K_{80}[pN_{1111}{*}G4], \quad (4.39)$$

$$\frac{d[pN_{1111}]}{dt} = \frac{d[G4]}{dt} = -\left\{\frac{d[pN_{1111}{*}G4]}{dt}\right\}. \quad (4.40)$$

## 4.3. Results from the Model and Interpretations

All the simulations in this study were conducted using Virtual Cell (VCell) software [14].

Using the computational model, we test the role of the serine site in the recruitment of $Na_V$ channels into the AIS in order to understand the nature of binding of all phosphorylated $Na_V$ species (NaA, NaB, NaC and $N_{1111}$) with AnkG species (G1, G2, G3, G4) (Fig. 4.6). The line for $pN_{1111}{*}G4$ in Fig. 4.6 represents the binding of $pN_{1111}$ with G4 and it achieved the highest concentration of the

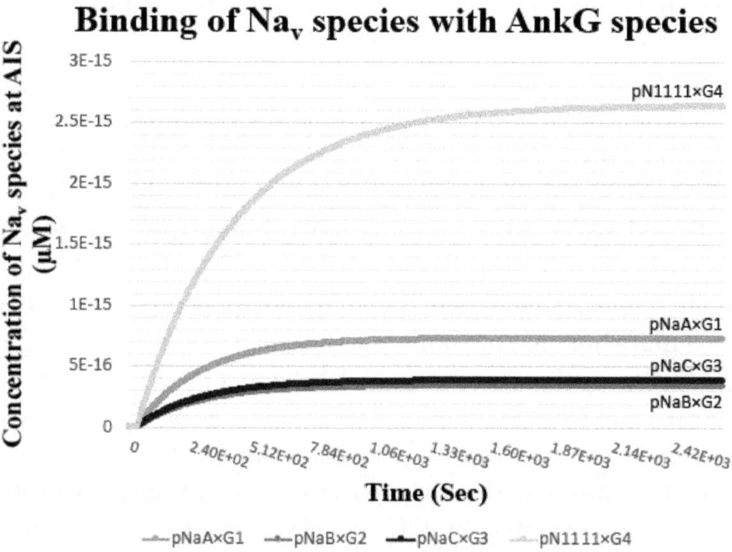


Fig. 4.6. Binding of Na$_v$ species with AnkG (G1, G2, G3, G4) under different phosphorylation conditions. The effect of the presence and absence of serine sites within Na$_v$ channels. The line for pN$_{1111}$∗G4 represents the highest concentration of pN$_{1111}$ due to the presence of all four serine sites for CK2-mediated phosphorylation. The line for pNaA∗G1 denotes the effect of the absence of two serine residues on the pNaA concentration. The concentrations of the remaining two phosphorylated Na$_v$ species (pNaC and pNaB) are found to be at a similar level. However, their concentrations are lower than the concentrations of pN$_{1111}$ and pNaA.

pN$_{1111}$ recruited by G4 (approx. $2.64{*}E^{-15}\mu$M $\sim$ 1565 molecules within the AIS volume) among all Na$_v$ species. These results are supported by the study in which 100–300 Na$_v$ molecules/$\mu$m of the AIS are documented [8]. The higher concentration of pN$_{1111}$ after binding to G4 indicates that in the presence of all serine sites, the binding with AnkG takes place with a strong binding affinity, which corroborated with the experimental evidence [20, 43]. This strong binding affinity is responsible for the high density of Na$_v$ channels at the AIS. Na$_v$ species other than pN$_{1111}$ shows a dramatic decrease in their protein concentrations after binding with AnkG. After binding with G1, the pNaA species had the second highest concentration ($7.2{*}E^{-16}$ $\mu$M $\sim$ 433 molecules). As mentioned in the assumptions, the fact that pNaA consists of three serine sites including S1, could be the reason for this significant reduction of pNaA as compared to the $pN_{1111}$. Remaining $Na_v$ species (pNaB and pNaC) after their respective binding with G2 and G3 show almost similar concentration but less than $pN_{1111}$ and pNaA ($3.8{*}E^{-16}\,\mu M \sim 205$ molecules and $3.8{*}E^{-16}\,\mu M \sim 228$ molecules), respectively. Similar concentrations of pNaB and pNaC after binding with AnkG species, demonstrate the importance of S1 site in the AIS motif. Similar results are given by Bréchet *et al.* [20] where the concentration of the $Na_v$ channels after S1 mutation is the same as the $Na_v$ channels with a double site mutation in the presence of S1 [20]. The S1 site within the AIS motif of a $Na_v$ channel is surrounded by acidic residues, such as glutamate and aspartate. The presence of these acidic residues increases the importance of the S1 site by increasing its probability of being phosphorylated by CK2 as the first preference. This could be the reason for pNaB to show similar behavior compared to the pNaC species (S1 absence).

Therefore, the availability of all serine sites for CK2-mediated phosphorylation is necessary for the regulation of the $Na_v$ channel population at the AIS in order to maintain the conditions suitable for AP initiation. In addition, any alteration in the phosphorylation conditions can cause a disturbance in the recruitment of $Na_v$ into the AIS by preventing $Na_v$ – AnkG binding. According to the literature, any alteration in the $Na_v$ channel densities at the AIS can increase the threshold voltage required for AP initiation. This disturbance in the AIS system can shift the AP initiation zone from AIS to a region such as the nodes of Ranvier (NOR) with a high number of $Na_v$ channels after AIS [2–4, 8]. However, the NOR, as an AP initiator zone, cannot hold the voltage stress during AP due to its position away from the soma [1, 8].

In order to test the significance of initial concentration of CK2 on the $Na_v$ channel recruitment at AIS, we phosphorylate $N_{1111}$ species at ten different CK2 (C2) initial concentrations and obeserve its impact on the binding of $pN_{1111}$ with AnkG (G4) (Fig. 4.7). The initial concentration of C2 is taken within a range of $6.6{*}E^{-9}\,\mu M$–$1.98{*}E^{-8}\,\mu M$. According to the results, at low initial

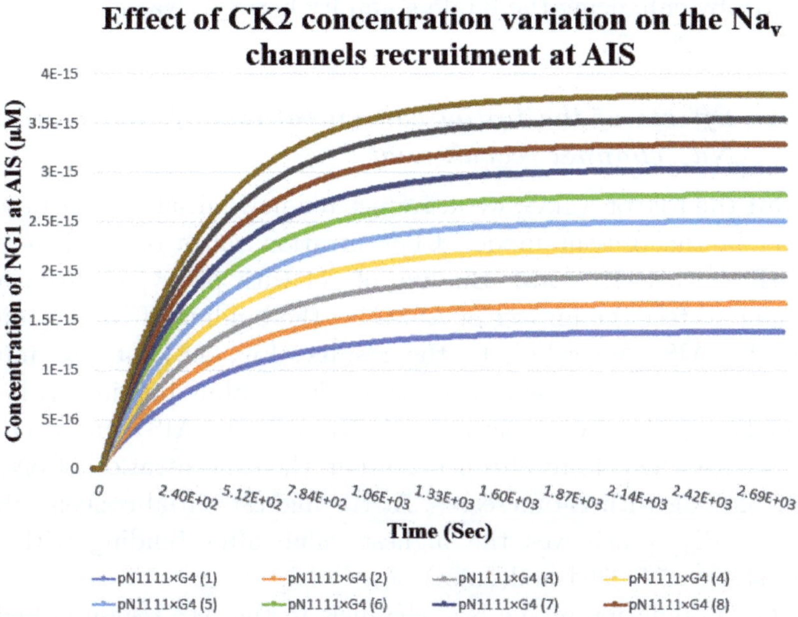


Fig. 4.7. Levels of $pN_{1111}$ at different initial CK2 concentrations. This figure demonstrates the effect of variation in initial CK2 concentration on the $Na_v$ – AnkG binding. Binding between $pN_{1111}$ and G4 is tested at ten different initial CK2 concentration. In the figure, $pN_{1111}{*}G4(1)$ represents the binding of $pN_{1111}$ with G4 at low initial CK2 concentration (1) and $pN_{1111}$ achieves lowest concentration. However, as we increase the initial CK2 concentration, levels of $pN_{1111}$ – G4 binding product also increase.

C2 concentration ($6.6 * E^{-9}$ μM), lesser $pN_{1111}$ molecules bind with G4 ($1.38{*}E^{-15}\mu M$ ∼ 833 molecules). As we increase the initial C2 concentration, the number of $pN_{1111}$ molecules also increased after binding with G4. These results clearly indicate the dependency of binding affinity between $Na_v$ channels and AnkG on the CK2 concentration. Similar results are reported by various studies which supported the importance of CK2 [20, 47]. Hsu *et al.* [47] show that the inhibition of CK2 using 4,5,6,7-tetrabromobenzotriazole (TBB) significantly altered the binding of $Na_v$ channels with AnkG by eliminating their CK2-mediated phosphorylation. In addition to that, CK2 inhibition shortened the axons by 30% [25]. Our model and the experiments reported so far show that CK2 plays a critical role in the AIS by enhancing the binding affinity between $Na_v$ channels and AnkG [20, 25, 47].

### 4.3.1. *Effects of the initial concentrations of AnkG on $Na_v$ channel recruitment*

To test the role of AnkG, we introduce ten new initial concentrations of G4 in the present model. Concentration range of G4 is set to $1.66*E^{-14}$–$4.98*E^{-14}$ $\mu$M. Binding of G4 with $pN_{1111}$ is taken into account to test the impact of AnkG on the binding of $Na_v$ channels into the AIS. According to the results (Fig. 4.8), at low initial concentration of G4, $pN_{1111}$ levels are lower after binding with G4 ($1.36*E^{-15}$ $\mu$M $\sim$ 806 molecules of $pN_{1111}$ in the AIS). However, as we increase the G4 initial concentration, the concentration of $pN_{1111}$ after binding with G4 increases. At the highest initial concentration of G4, $pN_{1111}$ achieves the highest value after binding with G4 ($3.96*E^{-14}$ $\mu M/2394$ molecules).

The population of the $Na_v$ channels in the AIS region is higher than those in the soma region [48]. The high number of $Na_v$ channels in the AIS region is supported by Kole *et al.* [8]; their experimental results suggest that the density of $Na_v$ channels in this region is 50-fold higher than that in the soma [8]. The high number of $Na_v$ channels in the structure of the AIS supports the AIS skeleton in overcoming the load induced during AP initiation. Moreover, due to their high number, $Na_v$ channels in the AIS region activate and deactivate much faster than the $Na_v$ channels present in the soma. Also, the voltage requirement to change the membrane potential in AIS is low as compared to the soma [18]. However, the AnkG's functions are not limited by $Na_v$ channel recruitment because, according to previous studies, AnkG is responsible for the recruitment of ion channels ($Na_v, K_v$), molecular motors (kinesin and dynein) and CAMs (NF186, NrCAM) into the AIS [5, 12–15]. Moreover, the function of AnkG is also related to protein trafficking within the neurons because AnkG is connected with microtubules through the EB3 protein. While the kinesin and dynein motors travel through microtubules in order to facilitate the proper trafficking of the proteins, AnkG related mutations can cause disturbances in

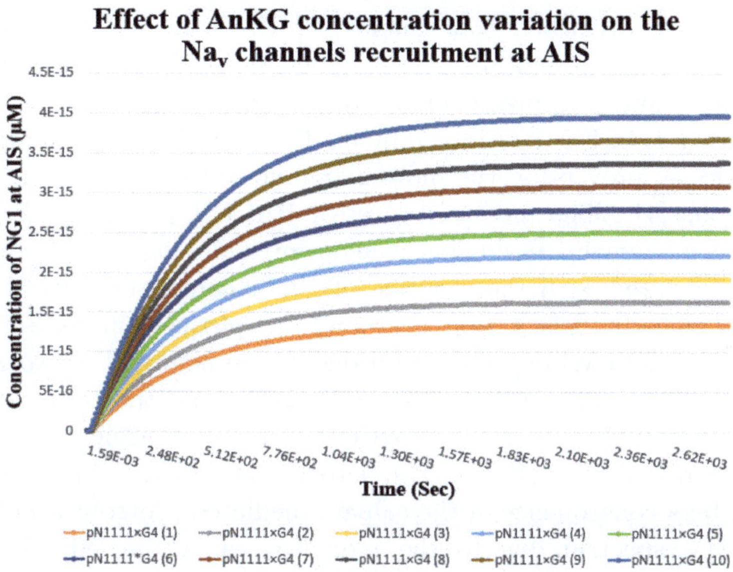


Fig. 4.8. Effect of different AnkG concentration on the Na$_v$ channel recruitment at AIS. Picture above demonstrates the effect of variation in initial AnkG concentration on the Na$_v$–AnkG binding. Binding between $pN_{1111}$–G4 is tested at ten different AnkG initial concentration. In figure, $pN_{1111}$*G4(1) represents the binding of $pN_{1111}$ with G4 at lowest initial AnkG concentration (1) and $pN_{1111}$*G4 binding achieved lowest concentration. However, as we increase the initial AnkG concentration, levels of $pN_{1111}$–G4 binding product also increase.

the transportation of the proteins within neurons [40]. Moreover, dendritic proteins, such as integrin-$\beta$1 and MAP-2, have been found in the axons after AnkG mutation indicating an alteration in protein trafficking [12].

### 4.3.2. *Possible role of the AIS in AD pathogenesis*

The AIS mutations are not only restricted to the reduced ability of a neuron to generate action potential but also are associated with the developing of seizures and epilepsy. Several experimental investigations have suggested the role of the AIS in AD pathogenesis. Experimental evidence shows that the patients with sporadic AD have the risk of developing seizures especially after onset of dementia. On the other hand, the studies using AD brain samples found the presence of disturbed $Na_V$ channels [41]. Disturbed level of $Na_V$ channels have potentially been associated with elevated BACE-1 ($\beta$-site amyloid precursor protein (APP) cleavage enzyme-1). Increased BACE-1 possibly inhibits the trafficking of channels to the AIS by preventing the molecular motors from performing their functions [41]. The increased level of BACE-1 precipitated in AD pathogenesis by increasing pathogenic amyloid $\beta$ peptides production and associated with the cleavage of $Na_V$ channel subunits [41, 49]. In AD mice models, the area around A$\beta$ plaques showed synaptic losses, axonal swelling and mutations in the neuronal network [50]. A$\beta$ plaques could cause damage near or around the region of AIS by targeting AnkG and $\beta$IV spectrin and decreasing the density and length of the AIS [50, 51]. Disturbance in the AIS functions could be a consequence of the calpain-mediated proteolysis of AnkG and $\beta$IV spectrin due to the induction of A$\beta$ [6, 50]. The low AnkG concentration can dismantle the structure of AIS by loosening the proteins anchored by AnkG. Dismantling the structure results in the disruption of neuronal polarity by abolishing the functions of the important proteins present in this region [6]. The results shown by Sun *et al.* [40] support the relationship between the AIS and AD by observing the low level of AnkG in AD transgenic mice [40]. Low level of AnkG could results in the alteration of $Na_V$ channel at the AIS.

The importance of the serine sites (S1, S2, S3 and S4) in the accumulation of $Na_V$ channels into the AIS in order to create suitable conditions for AP initiation has been shown through this modelling study which corroborates with the experimental findings. The results suggest that, in the presence of all the serine sites in $Na_V$ channels for CK2-mediated phosphorylation, the binding between the $Na_V$ channels and AnkG take place with a strong binding affinity. We show the significance of the initial concentrations of CK2 and AnkG on the recruitment of $Na_V$ channels. From the simulation results, we observe that the low initial levels of CK2 and AnkG reduce the population of the $Na_V$ channels at the AIS. At low levels of CK2, the $Na_V$ channels are not fully phosphorylated, which weaken the binding between the $Na_V$ channels and AnkG. Moreover, at low AnkG concentration, the AIS structural assembly dismantles which could create difficulties in the proper trafficking of proteins into the AIS such as $Na_V$ channels. Low levels of $Na_V$ channels increase the voltage requirement to initiate AP. Overall, these results indicate the potentially strong association of AIS proteins in AD pathogenesis. Future experimental investigation into the activity levels of AIS related CK2 in AD brains may shed light into the role of CK2 in AD.

## References

[1] Kole M.H.P. and Stuart G.J. (2012). Signal processing in the axon initial segment. *Neuron, 73*, pp. 235–247.

[2] Lemaillet G., Barbara W. and Stephen L. (2003). Identification of a conserved ankyrin-binding motif in the family of sodium channel $\alpha$ subunits. *J. Bio. Chem., 278*(30), pp. 27333–27339.

[3] Rasband M. (2009). Converging on the origins of axonal ion channel clustering. *PLoS Genetics, 5*(1), pp. 1–2.

[4] Gulledge A.T. and Jaime J. (2016). Neuron morphology influences axon initial. *eNeuro, 3*(1), pp. 1–24.

[5] Nelson A.D. and Jenkins P.M. (2017). Axonal membranes and their domains: assembly and function of the axon initial segment and node of ranvier. *Front. Cell. Neurosci., 11*, pp. 1–17.

[6] Buffington S.A. and Rasband M.N. (2011). The axon initial segment in nervous system disease and injury. *Eur. J. Neurosci., 34*, pp. 1609–1619.

[7] Jones S.L. and Svitkina T.M. (2016). Axon initial segment cytoskeleton: architecture, development, and role in neuron polarity. *Neural Plast., 2016*, pp. 1–19.

[8] Kole M.H.P., Ilschner S.U., Kampa B.M., Williams S.R., Ruben P.C. and Stuart G.J. (2008). Action potential generation requires a high sodium channel density in the axon initial segment. *Nat. Neurosci., 11*, pp. 178–186.

[9] Leterrier C., Potie J., Caillol G., Debarnot C., Rueda B.F. and Dargent B. (2015). Nanoscale architecture of the axon initial segment reveals an organized and robust scaffold. *Cell Rep., 13*, pp. 2781–2793.

[10] Fan X. and Markram H. (2019). A brief history of simulation neuroscience. *Front. Neuroinform., 13*, p. 32.

[11] Bhardwaj P., Kulasiri D. and Samarasinghe S. (2021). Modeling protein-protein interactions in axon initial segment to understand their potential impact on action potential initiation. *Neural Regen. Res., 16*(4), pp. 700–706.

[12] Hedstrom K.L., Ogawa Y. and Rasband M.N. (2008). Ankyrin(G) is required for maintenance of the axon initial segment and neuronal polarity. *J. Cell Biol., 183*, pp. 635–640.

[13] Zhang X. and Bennett V. (1998). Restriction of 480/270-kD Ankyrin(G) to Axon Proximal Segments Requires Multiple Ankyrin(G)-specific Domains. *J. Cell Biol.*, *142*(6), pp. 1571–1581.

[14] Mohler S. (2009). Ankyrin protein networks in membrane formation and stabilization. *J. Cell. Mol. Med., 13* 4364–4376.

[15] Zhou D., Lambert S., Malen P.L., Carpenter S., Boland L.M. and Bennett V. (1998). Ankyrin(G) is required for clustering of voltage-gated Na channels at axon initial segments and for normal action potential firing. *J. Cell Biol., 143*, pp. 1295–1304.

[16] Schafer D.P., Jha S., Liu F., Akella T., McCullough L.D. and Rasband M.N. (2009). Disruption of the axon initial segment cytoskeleton Is a new mechanism for neuronal injury. *J. Neurosci., 29*, pp. 13242–13254.

[17] Leterrier C. and Dargent B. (2014). No Pasaran! Role of the axon initial segment in the regulation of protein transport and the maintenance of axonal identity. *Semin. Cell Dev. Biol.*, *27*, pp. 44–51.

[18] Yoshimura T. and Rasband M.N. (2014). Axon initial segments: Diverse and dynamic neuronal compartments. *Curr. Opin. Neurobiol.*, *27*, pp. 96–102.

[19] Nakata T. and Hirokawa N. (2003). Microtubules provide directional cues for polarized axonal transport through interaction with kinesin motor head. *J. Cell Biol.*, *162*, pp. 1045–1055.

[20] Bréchet A., Fache M.P., Brachet A., Ferracci G., Baude A., Irondelle M., Pereira S., Leterrier C. and Dargent B. (2008). Protein kinase CK2 contributes to the organization of sodium channels in axonal membranes by regulating their interactions with ankyrin G. *J. Cell Biol., 183*, pp. 1101–1114.

[21] Nishi H., Shaytan A. and Panchenko A.R. (2014). Physicochemical mechanisms of protein regulation by phosphorylation. *Front. Genet.*, *5*, pp. 1–10.

[22] Xu M. and Edward C.C. (2015). An ankyrin-G N-terminal gate and protein kinase CK2 dually regulate binding of voltage-gated sodium and KCNQ2/3 potassium channels. *J. Biol. Chem.*, *290*(27), pp. 16619–16632.

[23] Meggio F. and Pinna L.A. (2003). One-thousand-and-one substrates of protein kinase CK2. *J. FASEB.*, *17*, pp. 349–368.

[24] Bian Y., Ye M., Wang C., Cheng K., Song C., Dong M., Pan Y., Qin H. and Zou H. (2013). Global screening of CK2 kinase substrates by an integrated phosphoproteomics workflow. *Sci. Rep.*, *3*, pp. 1–7.

[25] Ponce D., Muñoz A. and Garrido J.J. (2011). Casein kinase 2 and microtubules control axon initial segment formation. *Mol. Cell. Neurosci., 46*, pp. 222–234.

[26] Rasband M.N. (2008). $Na^+$ channels get anchored...with a little help. *J. Cell Biol.*, *183*, pp. 975–977.

[27] Yamada R. and Kuba H. (2016). Structural and functional plasticity at the axon initial segment. *Front. Cell. Neurosci.*, *10*, pp. 1–7.

[28] Bi C., Wu J., Jiang T., Liu Q., Cai W., Yu P., Cai T., Zhao M., Jiang Y.H. and Sun Z.S. (2012). Mutations of ANK3 identified by exome sequencing are associated with autism susceptibility. *Hum. Mutat.*, *33*, pp. 1635–1638.

[29] Harty R.C., Kim T.H., Thomas E.A., Cardamone L., Jones N.C., Petrou S. and Wimmer V.C. (2013). Axon initial segment structural plasticity in animal models of genetic and acquired epilepsy. *Epilepsy Res., 105*, pp. 272–279.

[30] Kaphzan H., Buffington S.A., Jung J.I., Rasband M.N. and Klann E. (2011). Alterations in intrinsic membrane properties and the axon initial segment in a mouse model of angelman syndrome. *J. Neurosci.*, *31*, pp. 17637–17648.

[31] Peltola M.A., Kuja-Panula J., Liuhanen J., Võikar V., Piepponen P., Hiekkalinna T., Taira T., Lauri S.E., Suvisaari J., Kulesskaya N., Paunio T. and Rauvala H. (2016). AMIGO-Kv2.1 potassium channel complex is associated with schizophrenia-related phenotypes. *Schizophr. Bull.*, *42*, pp. 191–201.

[32] Rueckert E.H., Barker D., Ruderfer D., Bergen S.E., O'dushlaine C., Luce C.J., Sheridan S.D., Theriault K.M., Chambert K., Moran J., Purcell S.M., Madison J.M., Haggarty S.J. and Sklar P. (2013). *Cis*-acting regulation of brain-specific ANK3 gene expression by a genetic variant associated with bipolar disorder. *Mol. Psychiatry*, *18*, pp. 922–929.

[33] William T. and Laura B. (2013). Alzheimer's disease facts and figures. *Alzheimer's Dement., 9*, pp. 208–245.

[34] Checler F. and Turner A.J. (2012). Journal of Neurochemistry special issue on Alzheimer's disease: Amyloid cascade hypothesis — 20 years on. *J. Neurochem.*, *120*, pp. 2011–2012.

[35] Craig L.A., Hong N.S. and McDonald R.J. (2011). Revisiting the cholinergic hypothesis in the development of Alzheimer's disease. *Neurosci. Biobehav. Rev.*, *35*, pp. 1397–1409.

[36] Hardy J. and Higgins G. (1992). Alzheimer's disease: The amyloid cascade hypothesis. *Science*, *256*, pp. 184–185.

[37] Kametani F. and Hasegawa M. (2018). Reconsideration of amyloid hypothesis and tau hypothesis in Alzheimer's disease. *Front. Neurosci.*, *12*(25).

[38] Karran E., Mercken M. and Strooper B.D. (2011). The amyloid cascade hypothesis for Alzheimer's disease: An appraisal for the development of therapeutics. *Nat. Rev. Drug Discov., 10*, pp. 698–712.

[39] Maccioni R.B., Farías G., Morales I. and Navarrete L. (2010). The revitalized tau hypothesis on Alzheimer's disease. *Arch. Med. Res., 41*, pp. 226–231.

[40] Sun X., Wu Y., Gu M. and Zhang Y. (2014). MiR-342-5p decreases ankyrin G levels in Alzheimer's disease transgenic mouse models. *Cell Rep., 6*, pp. 264–270.

[41] Kim D.Y., Carey B.W., Wang H., Laura A.M., Binshtok A.M., Wertz M.H., Pettingell W.H., He P., Virginia M., Woolf C. and Kovacs D.M. (2007). BACE1 Regulates Voltage-gated sodium channels and Neuronal activity. *Nature Cell Biol.*, *9*(7), pp. 755–764.

[42] Baley C., Lu H., Glinski G., Wheeler D., Hamilton P., Hendriksen M. and Smith B. (2002). Using computer models to identify optimal conditions for flip-chip assembly and reliability. *Circuit World*, *28*(1), pp. 14–20.

[43] Fache M.P., Moussif A., Fernandes F., Giraud P., Garrido J. and Dargent B. (2004). Endocytotic elimination and domain-selective tethering constitute a

potential mechanism of protein segregation at the axonal initial segment. *J. Cell Biol.*, *166*(4), pp. 571–578.

[44] Cantrell A.R., Scheuer T. and Catterall W.A. (2018). Voltage-dependent neuromodulation of $Na^+$ channels by D1-like dopamine receptors in rat hippocampal neurons. *J. Neurosci.*, *19*, pp. 5301–5310.

[45] Srinivasan Y., Elmer L. and Bennet V. (1988). Ankyrin and spectrin associated with voltage dependent sodium channels in brain. *Nature*, *333*(6169), pp. 177–180.

[46] Moraru I.I., Schaff J.C., Slepchenko B.M., Blinov M.L., Morgan F., Lakshminarayana A., Gao F., Li Y. and Loew L.M. (2008). Virtual cell modelling and simulation software environment. *IET Sys. Biol.*, *2*(5), pp. 352–362.

[47] Hsu W.C.J., Wildburger N.C., Haidacher S.J., Nenov M.N., Folorunso O., Singh A.K., Chesson B.C., Franklin W.F., Cortez I., Sadygov R.G., Dineley K.T., Rudra J.S., Taglialatela G., Lichti C.F., Denner L. and Laezza F. (2017). PPARgamma agonists rescue increased phosphorylation of FGF14 at S226 in the Tg2576 mouse model of Alzheimer's disease. *Exp. Neurol.*, *295*, pp. 1–17.

[48] Rasband M.N. (2010). The axon initial segment and the maintenance of neuronal polarity. *Nat. Rev. Neurosci.*, *11*, pp. 552–562.

[49] Kovacs D.M., Gersbacher M.T. and Kim D. (2010). Alzheimer's secretases regulate voltage-gated sodium channels. *Neurosci Lett.*, *486*(2), pp. 68–72.

[50] Marin M.A., Ziburkus J., Jankowsky J. and Rasband M.N. (2016). Amyloid-$\beta$ plaques disrupt axon initial segments. *Exp. Neurol.*, *281*, pp. 93–98.

[51] Sohn P.D., Tracy T.E., Son H.I., Zhou Y., Leite R.P., Miller B.L., Seeley W.W., Grinberg L.T. and Gan L. (2016). Acetylated tau destabilizes the cytoskeleton in the axon initial segment and is mislocalized to the somatodendritic compartment. *Mol. Neurodegener.*, *11*, pp. 1–13.

[52] Ubersax J.A. and Ferrell J.E. (2007). Mechanisms of specificity in protein phosphorylation. *Nat. Rev. Mol. Cell Biol.*, *8*, pp. 530–541.

# Chapter 5

# Computational Modeling of Presynaptic Mechanisms

## 5.1. Introduction

As we discussed in the previous chapters, synapses have been shown to undergo dynamical modulations in postsynaptic responses following stimulation by specific patterns of AP at the presynaptic terminal. As the patterns of AP change rapidly in the presynaptic terminals, there should be plasticity in the processes in these regions which adapt quickly and make changes that last only a short time, hence the name short-term plasticity. Short-term plasticity is one of the main forms of plasticities displayed in central neurons [1] and it indicates the modulation that lasts from tens of milliseconds to a few minutes. In this chapter, we delve into some mechanisms of short-term plasticity from a biophysical perspective. We develop a mathematical model that proposes a hypothesis for mechanisms that can explain experimental data so far accrued [2].

The short-term plasticity originates from presynaptic mechanisms through modulating the presynaptic release probability of neurotransmitters and the site availability for release [3]. The plastic nature of synaptic transmission provides a theoretical basis for information transfer and memory formation [4, 5]. There are two typical forms of short-term synaptic plasticity in central synapses: short-term facilitation and depression [6]. In synapses displaying short-term facilitation, postsynaptic response by the second of two closely spaced presynaptic APs is larger than by the first one, and thus, synapses show an increase in synaptic efficacy [3]. On the contrary, short-term depression refers to a lower postsynaptic response by the second of two closely spaced presynaptic APs than by the first one, it leads to a reduction in synaptic strength. Mostly, the synaptic strength is a result of a combination of short-term facilitation and depression, depending on the timing of presynaptic stimulation [6]. However, the functional importance of short-term plasticity in the synapse is not well understood.

Facilitation can enhance the information coding and transmission across a synapse by promoting the neurotransmitter release by presynaptic stimulation [7]. The mechanism underlying the facilitation in Schaffer collateral (SC) remains a mystery and is different from the $Ca^{2+}$ buffer saturation mechanism, which accounts for the facilitation in several critical brain areas [8, 9]. Jackman *et al.* [9] show that synaptotagmin 7 (Syt7), a calcium sensor located in the presynaptic terminal, has a significant role in the facilitation in SC. Syt7 has a number of functions including asynchronous release [10], replenishing the release site [11], and secretion of large molecules [12, 13]. When Syt7 is knocked out in several central areas, including SC, a significant reduction in the presynaptic facilitation is observed. Hence, the functional importance of short-term plasticity in the synapse would be further be revealed by investigating the roles of Syt7.

Given the complexity and variability of the synaptic transmission and plasticity [14–16], it is difficult to understand the potential roles of Syt7, as well as Syt7-dependent short-term plasticity, because of the lack of understanding of the mechanisms linking the short-term plasticity to the synaptic functions. But there are serious attempts to understand these mechanisms over the last few decades. Three pools of synaptic vesicles in the presynaptic terminal are observed to exist [17] which are usually categorized as (1) a readily releasable pool (RRP) which rapidly releases, (2) a recycling pool which releases slower, and (3) a reserve pool which releases the slowest (reviewed in [18, 19]). Quantal release hypothesis, where a "quanta" of neurotransmitters, corresponding to the number of neurotransmitters contained in a single vesicle, evokes a miniature excitatory postsynaptic current (mEPSC) and the EPSC evoked by a single AP is in multiples of mEPSC [20]. We depict the three pools hypothesis in Fig. 5.1.

We now understand the factors that regulate the probability of vesicle release: the release requires 10–100 $\mu$M local $Ca^{2+}$ concentration around $Ca^{2+}$ sensors which conduct the release; and the release can be enhanced by build-up of the residue $Ca^{2+}$ concentration ($Ca_{res}$), which is the remaining free $Ca^{2+}$ after the buffering and transporting out of the terminal [3, 22].

We still do not understand the mechanism underlying the transitions of vesicles among the three pools of synaptic vesicles. A number of hypotheses are proposed as to the nature of vesicle release at central synapses, and there are contentious issues related to the single vesicular release [15, 23, 24], and the multivesicular release [25–27]. Furthermore, a vesicle can be released in either of the two modes in the hippocampus: (1) kiss and run, where a small portion of the neurotransmitters contained in a vesicle is released, and (2) full fusion, where all the containing neurotransmitters are released (reviewed in [3, 4]). But, how a vesicle switches between the two modes is unclear.

Mathematical modeling based on the physics of the problem is one of the ways to obtain insights into unknown mechanisms based on plausible reasoning. A general model based on current understanding of the mechanistic relationships of synapses is developed in this chapter. In this general model, there is a subset of parameters that are common to both wild-type (WT) and Syt7 knock-out (KO) neurons in SC [9]. The values of the set of parameters excluding the common parameter set depend on the WT and Syt7 KO experimental data. These two separate sets of parameter values are estimated to capture the behavioral difference between wild-type (WT) and Syt7 knock-out (KO) neurons in SC based on the limited experimental data. The general model and the two sets of parameters enable us to capture the behavioral differences, if any, between WT and Syt7 KO through the computational experiments. Since Syt7 drives short-term facilitation, a behavioral difference caused by Syt7 KO indicates

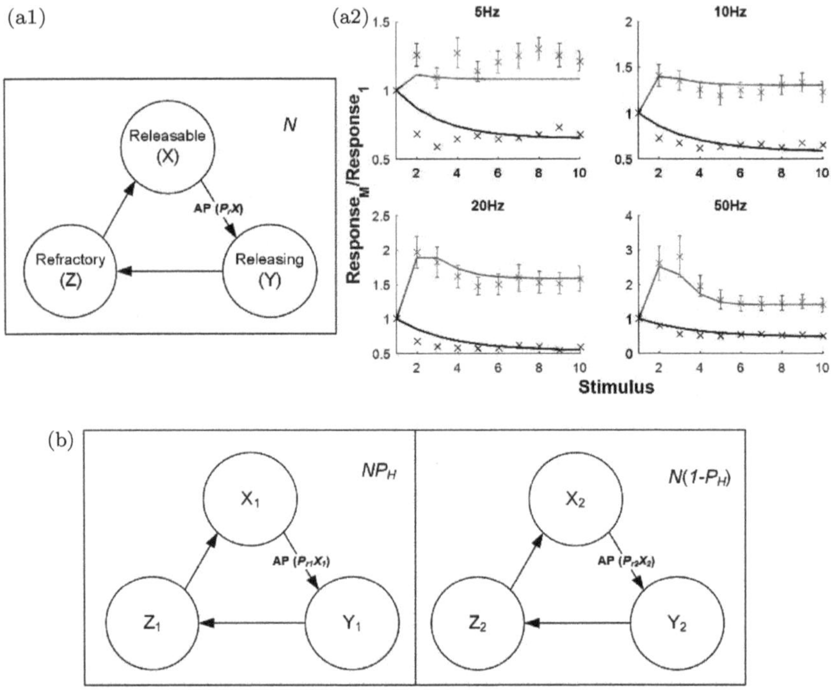


Fig. 5.1. Models of presynaptic releasing. Gray crosses with error bars are wild-type (WT) data and black crosses are knock-out (KO) data. Black and gray lines are corresponding simulation results. Data was retrieved from extended data of Fig. 5.4 in [9]. (a1) The uniform one-population model (1P model) with $N$ release sites and uniform $P_r$ (see Section 5.2 for details). Release sites have three possible states: releasable sites, releasing sites, and refractory sites, using $X$, $Y$ and $Z$ to denote the average fractions of sites at these three states, respectively. AP causes releasable release sites (or $P_r \cdot X$ in terms of the fraction) to release and enter into a brief releasing state. After releasing, the sites become refractory and undergo a very slow recovery process to become releasable. $\mathrm{Ca_{res}}$ is determined to accelerate this recovery process [11, 21]. The mean amplitude of EPSC is calculated by $N \cdot P_r \cdot X \cdot Q$, where $Q$ is the mean amplitude of mEPSC induced by a quanta of neurotransmitters. (a2) After parameter estimation (see Section 5.2 for the procedure), the 1P model captures the trends of WT data. However, a significant mismatch is revealed for the first few stimuli in Syt7 KO data. (b) In the non-uniform two-population model (2P model), release sites are divided into two binomial models that one model with $N \cdot P_H$ release sites and a very high uniform $P_r$, $P_{\mathrm{r1}}$, and the other model with $N(1\text{-}P_H)$ release sites and a low uniform $P_r, P_{\mathrm{r2}}$.

potential functional importance of Syt7 and Syt7-driven short-term plasticity.

Previous models in the literature [28, 29] can only partially capture the trends in experimental data; they fit either WT or Syt7 KO, but not both. By using our model, we understand that (1) release sites are governed by non-uniform release probabilities in SC as shown in the experiments [30–33]: there are at least two sets of release sites; the majority have low release probability (low sites) and the minority have high release probability (high sites) while previous models are formulated based on one set of release sites. In agreement, another computational study also reported heterogeneity in presynaptic release probabilities in the giant synapses in the calyx of Held [33] indicating that the heterogeneity of presynaptic release probabilities across different types of synapses may exist; (2) the basal transmission is contributed by the high sites, and short-term facilitation is contributed by the low sites; and (3) Syt7 KO causes significant alterations to the high sites during the low-frequency stimulation (LFS) and to the low sites during the high-frequency stimulation (HFS). This result suggests that Syt7 may select release sites depending on the release probability and stimulation frequency.

We propose that Syt7 selects release sites to undergo short-term plasticity and modulates the releasing profile of a synapse in SC. In addition to short-term plasticity, SC synapses also undergoes long-term plasticity when subjected to LFS and HFS stimulations. We show a hypothesized system where short-term plasticity can function as a high-pass filter and Syt7 helps to train the target frequency of the filter through modulations on the selected release sites.

## 5.2. Conceptual and Mathematical Modeling

### 5.2.1. *Presynaptic $Ca^{2+}$ dynamics*

The rise of the presynaptic $Ca^{2+}$ level is modeled as the normalization by the amount of $Ca^{2+}$ influx triggered by presynaptic APs. The use-dependent changes of $Ca^{2+}$ influx are ignored so that the $Ca^{2+}$ influx is fixed across APs [3, 9]. Each AP elevates presynaptic $Ca^{2+}$ level by a fixed amount, $\Delta$Ca, and the resulting $Ca^{2+}$ rise decays with a time constant $\tau_{ca}$. However, $Ca^{2+}$ decay may not be governed by a single exponential function: the decay may be biphasic [34]. Therefore, we use a Hill function to represent the decay rate according to the $Ca^{2+}$ level. The ordinary differentiation equation (ODE) governing the $Ca^{2+}$ level (Ca) is given by

$$\frac{d\text{Ca}}{dt} = -\frac{\text{Ca}}{\tau_{ca}}\left(\frac{1}{1+K_{ca}/\text{Ca})}\right)+\Delta_{ca}\delta(t-t_{ap}), \tag{5.1}$$

where $K_{ca}$ is the dissociation constant for $Ca^{2+}$ decay, $\delta(t - t_{ap})$ is the Kronecker delta function, $\Delta_{ca}$ is the normalized rise of $Ca^{2+}$, which is fixed to 1 in this model, after each AP stimulation, and $t_{ap}$ is the time that an AP arrives at the presynaptic terminal.

### 5.2.2. *Short-term plasticity*

#### 5.2.2.1. *Short-term depression*

Depression is mainly caused by the reduction in the availability of release sites in response to closely spaced presynaptic APs when the inter-AP interval is insufficient for the recovery from the depletion due to any releases. The reduction has a number of presynaptic causes, including the depletion of release sites and vesicle pools, the inactivation of release sites, and the inactivation of $Ca^{2+}$ channel [3]. The reduction may also be caused by postsynaptic factors such as postsynaptic receptor saturation or desensitization which decrease mEPSC [35, 36]. The depression undergoes two major phases: a refractory phase which prevents the response of the release site from further APs and followed by a recovery phase which exhibits reduced availability of the release site [23, 37, 38]. We use X, Y, and Z to reflect average fractions of sites at releasable, releasing and refractory states, respectively, and $X+Y+Z = 1$ [29]. The changes of X, Y and Z are modeled as three ODEs [29], as given by Eqs. (5.2), (5.4) and (5.5) in later paragraphs.

In this setup, both releasing and refractory states are unreleasable and the recovery phase of the depression is reflected by the transition of the refractory fraction into the releasable fraction. Hence, the reduction in the availability of release sites as a whole is reflected by the increased fractions of releasing and refractory sites. In addition, the very slow vesicle refilling process by endocytosis is ignored by assuming the three vesicle pools [18, 19], with greater than 500 vesicles [18, 19], are sufficient to replenish the release site without refilling during HFS, at least for a small number of stimuli. The differences in replenishing rates among the pools [19] are ignored because they are less significant for a small number of stimuli.

Once a presynaptic AP arrives at the presynaptic terminal, only releasable sites can release a vesicle with a mean probability of $P_r$. After releasing, releasable sites transfer into the releasing state instantaneously. At the same time, refractory sites are recovering from depression to releasable state with a recovery rate of $k_{\mathrm{rec}}$. Thus, the change of the fraction of releasable sites $X$ with time [28, 29] is

$$\frac{dX}{dt} = -P_r X \cdot \delta(t - t_{ap}) + k_{\mathrm{rec}} \cdot Z. \tag{5.2}$$

The recovery rate $k_{\mathrm{rec}}$ has a slow basal component and a faster $\mathrm{Ca}^{2+}$ controlled component as given by

$$k_{\mathrm{rec}} = k_0 + \frac{k_{\mathrm{max}} - k_0}{1 + (K_r/Ca(t))^{n_r}}, \tag{5.3}$$

where $k_0$ is the basal recovery rate, $k_{\mathrm{max}}$ is the faster recovery rate, $K_r$ is the dissociation constant, Ca is the normalized presynaptic $\mathrm{Ca}^{2+}$ rise following the AP in Eq. (5.1) and $n_r$ is the Hill coefficient. Here, the Hill equation is used to reflect a switch-like behavior of the faster recovery rate having the normalized presynaptic $\mathrm{Ca}^{2+}$ level as the controlling parameter. After the release of vesicles, the sites enter a brief releasing state before becoming refractory, such that

$$\frac{dY}{dt} = P_r X \cdot \delta(t - t_{ap}) - \frac{Y}{\tau_{in}}, \tag{5.4}$$

where $\tau_{in}$ is the time constant for the transition into refractory states.

Finally, the change of the fraction of refractory sites $Z$ with time is determined by the transition of releasing sites to refractory sites and the recovery of refractory sites to releasable sites:

$$\frac{dZ}{dt} = \frac{Y}{\tau_{in}} - k_{\mathrm{rec}} \cdot Z. \tag{5.5}$$

#### 5.2.2.2. *Short-term facilitation*

Facilitation of release is hypothesized to be caused by a number of presynaptic $Ca^{2+}$ related mechanisms, including increased $Ca^{2+}$ influx, $Ca^{2+}$ buffer saturation, and high-affinity $Ca^{2+}$ sensors detecting $Ca_{res}$ level [3, 9]. Therefore, the facilitation is simulated as a $Ca^{2+}$-dependent process. The enhanced $P_r$ is determined by the facilitation and the initial release probability per release site ($P_0$) as given by [22, 33]

$$P_r = \frac{1 - P_0}{1 + (K_f/\mathrm{Ca})^{n_f}} + P_0, \tag{5.6}$$

where $K_f$ is the dissociation constant for $Ca^{2+}$-bound molecules and $n_f$ is the Hill coefficient. $P_r$ is in the range of $P_0$ to 1.

### 5.2.3. *Postsynaptic response: Quantal release and one release site, one vesicle hypothesis*

A "quanta" of neurotransmitters, corresponding to a single vesicle of neurotransmitters, evokes a mEPSC, and the EPSC evoked by a single AP is in multiples of mEPSC [20]. The mean amplitude of the postsynaptic EPSC in a CA1 neuron following an AP is described by a binomial model [31]:

$$\mathrm{EPSC} = N \cdot P \cdot Q, \tag{5.7}$$

where $P$ is the release probability over the population of release sites, and $Q$ is the mean amplitude of their mEPSCs. The underlying assumption for this simple interpretation of postsynaptic EPSC is that $P$ and $Q$ are both uniform across the contributing release sites [31]. $P$ is determined by the mean release probability and the availability of releasable sites, thus, $P = P_r \cdot X$.

#### 5.2.3.1. *Normalization of responses*

Experimentally, the facilitation ratio is measured by dividing the amplitude of the following EPSC by the amplitude of the first EPSC. Therefore, in our study, we define the normalized responses as the amplitudes of corresponding EPSCs evoked by multiple-APs in succession normalized by the amplitude of the EPSC evoked by the first AP as given by

$$\text{Response}_M/\text{Response}_1 = \frac{\text{EPSC}_M}{\text{EPSC}_1} = \frac{N_M \cdot P_{r,M} \cdot X_M \cdot Q_M}{N_1 \cdot P_{r,1} \cdot X_1 \cdot Q_1}, \quad (5.8)$$

where $\text{EPSC}_M$ is the amplitude of the EPSC evoked by the $M$th AP, $N_M$, $P_{r,M}$, $X_M$ and $Q_M$ are the number of the release sites, the mean release probability of release sites, the mean availability of release sites, and mean amplitude of mEPSC at the $M$th AP, respectively. Further assumptions are taken to simplify Eq. (5.8) into Eq. (5.9): (1) the facilitation and depression are both absent ($P_{r,1} = P_0$ and $X_1 = 1$) before the first AP ($\text{Ca}_{\text{res}}$ is at rest); and (2) the number of release sites and their mean amplitude of mEPSC remain unchanged during the multiple APs applied in a short-time window, as given by

$$\text{Response}_M/\text{Response}_1 = \frac{P_{r,M} \cdot X_M}{P_0}. \quad (5.9)$$

The pair–pulse ratio (PPR) is a special case where the second response is normalized:

$$\text{PPR} = \frac{\text{Response}_2}{\text{Response}_1} = \frac{P_{r,2} \cdot X_2}{P_0}. \quad (5.10)$$

5.2.3.2. *Parameter estimation*

Parameters are estimated using Markov chain Monte Carlo (MCMC) method using the MCMC toolbox of Matlab [39]. Two sets of parameters, corresponding to WT and Syt7 KO conditions, respectively, are obtained by minimizing the sum of two squared errors WT and Syt7 KO between normalized model outputs (given by Eq. (5.9)) and the experimental data. The experimental data was obtained from Jackman *et al.* [9], where responses following 10 stimuli, at 4 frequencies (5Hz, 10Hz, 20Hz and 50Hz), are provided and normalized by the first response. Data sets were obtained from multiple animals on each group (WT vs. Syt7 KO), and both peak PPR and $\text{Response}_{10}/\text{Response}_1$ was significantly different for WT and Syt7 KO mice at all synapses over all four frequencies ($P < 0.01$, Student's $t$-test). Since normalized response for

Syt7 (KO) is much smaller than that of WT, a weight is assigned to the squared error of Syt7 KO to balance the contribution of two errors to the total error. The common parameters, which are related to the processes unaltered by Syt7 KO, are conserved between WT and Syt7 KO sets. According to Jackman *et al.* [9], values of $P_0$, $k_0$ and $Ca^{2+}$-related parameters, $K_{ca}$ and $\tau_{\mathrm{Ca}}$, are unaltered between WT and KO conditions. Other parameters have two separate copies, one for WT and the other for Syt7 KO, and are trained separately. During the estimation, each iteration performs two pieces of training, one against WT data followed by the other against Syt7 KO data. Both trainings estimate common parameters, and WT training/Syt7 KO training estimate WT/Syt7 KO copies, respectively. The training is stopped when parameters are converged. This model is programmed in MATLAB (https://www.mathworks.com/products/matlab.html). Source code is available in ModelDB [40] at http://modeldb.yale.edu/239066.

## 5.3. Results

In this section, we investigate the factors regulating short-term plasticity and highlight critical behavior differences between WT and Syt7 KO to support our later discussions.

### 5.3.1. *General model overview*

The general model includes three major processes: (1) presynaptic $Ca^{2+}$ dynamics, (2) short-term plasticity, and (3) postsynaptic response. The presynaptic $Ca^{2+}$ dynamics are simulated with presynaptic APs at different frequencies as inputs. The facilitation and depression for a synapse are $Ca^{2+}$-dependent and they modulate the release of neurotransmitter in response to a presynaptic AP. There are three states for a synapse: releasable, releasing and refractory state. The state transitions are in a cyclic fashion as shown in Fig. 5.1 A1. A synapse is capable of releasing a vesicle after a presynaptic AP only when it is in the releasable state, and the release probability is modulated by the short-term plasticity. The postsynaptic response is evaluated according to the quantal release hypothesis [20]. In SC, the neurons are connected by a larger number of synaptic contacts and each contact contains one release site [16, 41]. The exact releasing profile of these contacts in central neurons is unclear. We assume that there are $N$ contacts/sites in our system excluding silent synapses. All release sites are independent and each one of them has one vesicle ready to be released. Hence, a release site releases one-or-none "quanta" unit per AP. The model selection is based on the consideration of model structure in the literature [28, 29] and information provided in the given data (see Appendix A for explanation).

### 5.3.2. *Non-uniform release probability model*

Previous models were developed according to a binomial model with $N$ release sites and the uniform mean release probability of release sites, $P_r$ (Fig. 5.1(a1), one-population model, abbreviation 1P model). The 1P model partially captures the experimental observations. The depression during the first few stimuli of Syt7 KO predicted by the 1P model is far from the great depression as seen in experimental observations (Fig. 5.1(a2)) [9]. The 1P model does not capture the "gap" between WT and Syt7 KO during the first few stimuli; the 1P model follows either the great facilitation in WT or the great depression in Syt7 KO, but not both.

There are two possibilities for the "gap": (1) the release sites have non-uniform release probabilities [30, 32, 33]. The minority of release sites have a high $P_r$ and is depleted by the first AP, and the majority of release sites have a much lower $P_r$ and produce significantly weaker EPSC following the subsequent APs; and (2) another source of depression exists in addition to the depletion of release sites. For example, use-dependent depression [42] and inactivation of calcium channels [43] change the $Ca^{2+}$ influx. Here, we focus on the first possibility for the following reasons: (1) no significant change of total $Ca^{2+}$ influx between two closely applied stimuli was detected for the given data [9]. Hence, the latter possibility would only occur, if it does occur, at the subset of $Ca^{2+}$ channels that trigger neurotransmitter release; and (2) for the latter possibility, the depression often occurs following the prolonged HFS [3], but the depression as seen in the

Syt7 KO response (Fig. 5.1(a2)) is great at the second stimulus. Hence, the latter possibility contributes less to the great depression in Syt7 KO, at least for the first 10 stimuli.

We develop the simplest non-uniform release probability model (Fig. 5.1(b), two-population model, abbreviation 2P model) based on additional assumptions from the A1 model: (1) there are two sets of release sites, the majority, $N(1 - P_H)$ sites, have a low mean initial release probability ($P_0$) and the rest, $N \cdot P_H$ sites, have a much higher $P_0$. $P_H$ denotes the fraction of high $P_0$ release sites (high sites). $P_r$ has a range between 0 and 1 and is given by the sum of $P_0$ and the facilitation, which is controlled by $\mathrm{Ca_{res}}$ and increases the release probability following AP [14, 15]. The releasable fractions of the two sets are denoted by $X_1$ (high sites) and $X_2$ (low sites) given the initial values of $P_H$ and $1 - P_H$, respectively, at rest; (2) the depression process, the facilitation process and miniature EPSCs (mEPSC with a mean amplitude $Q$) are the same across the two sets; (3) both sets follow the binomial model so that the overall mean EPSC amplitude is $N \cdot Q(P_{r1} \cdot X_1 + P_{r2} \cdot X_2)$; and (4) $P_H$ is conserved between WT and Syt7 KO.

As shown in Fig. 5.2, the 2P model has a good agreement with the "gap" and the general trends over different frequencies. The $P_0$s predicted differed by more than 10-fold; most release sites (83%) have a low $P_0$ of approximately 0.03 and the rest (17%) have a high $P_0$ of approximately 0.55.

Although the estimated parameters of the 2P model (see Table 5.1) may not contain useful quantitative information for the processes governing releasing and plasticity primarily due to the great reduction in the synaptic processes and the uncertainties in the feasible ranges of parameters, the estimated parameters contain useful relative information between WT and Syt7 KO as well as agree well with the general trends of synaptic vesicle releasing. Those bases are adequate and sufficient for our purpose to investigate the dynamic behavioral differences between WT and Syt7 KO in order to understand the functional importance of short-term plasticity.

The differences of Syt7 KO parameters from that of WT are summarized: (1) the maximum fast recovery rate is significantly

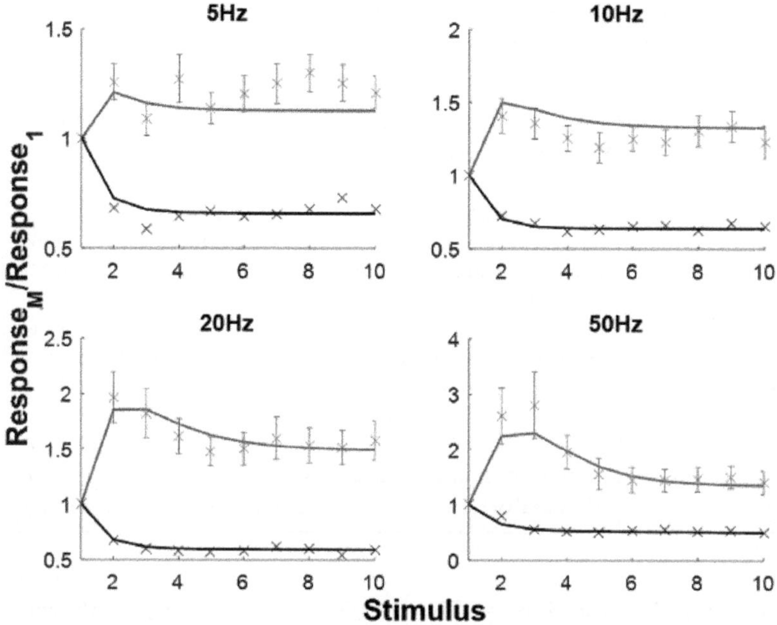


Fig. 5.2. After parameter estimation, the 2P model in Fig. 5.1 captures the trends of both WT and Syt7 KO data.

reduced, but a much smaller $Ca^{2+}$ rise is required to reach the maximum rate (Fig. 5.3(a)). This result indicates that the fast recovery rate has at least two components: one moderate rate responds quickly to $Ca^{2+}$ rise and a Syt7 controlled fast rate which requires a higher $Ca^{2+}$ rise to activate. The higher $Ca^{2+}$ rise may be accumulated by many stimuli. A previous experimental study shows that Syt7 KO causes a significant reduction in the fast replenishment rate of synaptic vesicles following the prolonged HFS while a minor change in the responses following a small number of stimuli [11]. This yields consistency with our simulation results; (2) the facilitation is significantly reduced but is not completely eliminated (Fig. 5.3(b)), which again yields consistency with the experimental observation where a small facilitation is observed when $P_0$ is reduced significantly by decreasing the extracellular $Ca^{2+}$ level [9]; (3) the relationship between $Ca^{2+}$ elevation and $P_r$ is linear for a moderate $Ca^{2+}$ elevation ($\leq$ 2Norm.Ca), and $P_r$ approaches a plateau with the

Table 5.1. Parameter sets for WT and KO conditions.

| Parameters | Biological meaning | WT | Syt7 KO |
|---|---|---|---|
| $\tau_{ca}$ | Decay time constant for $Ca^{2+}$ | 30.1 ms | 30.1 ms |
| $K_{ca}$ | Dissociation constant for $Ca^{2+}$ decay | 1.19 | 1.19 |
| $k_0$ | The basal recovery rate | $0.9\,ms^{-1}$ | $0.9\,ms^{-1}$ |
| $k_{max}$ | Faster recovery rate | $26\,ms^{-1}$ | $8.25\,ms^{-1}$ |
| $K_r$ | Dissociation constant for recovery | 4.05 | 0.65 |
| $n_r$ | Hill coefficient for recovery | 1 | 3.92 |
| $\tau_{in}$ | Time constant for the transition into refractory states | 3 ms | 3 ms |
| $K_f$ | Dissociation constant for $Ca^{2+}$-bound molecules | 2.4 | 19.4 |
| $n_f$ | Hill coefficient for facilitation | 1.15 | 1.5 |
| $P_H$ | Fraction of high $P_0$ release sites (high sites) | 0.17 | 0.17 |

greater $Ca^{2+}$ elevation (>2Norm.Ca); and (4) the difference in $P_r$ between the two sets of release sites reduces while $Ca^{2+}$ level increases in WT, but the difference only reduces by a small amount in Syt7 KO (Fig. 5.3(b)). In other words, the facilitation "activates" the low $P_0$ release sites and Syt7 KO lacks this capability. The consistency of the model outputs against experimental observations assures the believability of the estimated parameters for our latter investigations.

### 5.3.3. *Selective alteration of pair-pulse ratio by KO*

The pair–pulse ratio (PPR) is the ratio of the second response to the first response evoked by two closely spaced APs. PPR reflects the short-term changes of the postsynaptic response as a result of the dynamic modulations by the short-term facilitation and depression. A PPR of 1 denotes the balance between the facilitation and depression, while greater or less than 1 denotes the dominance of the facilitation or the depression, respectively. The conditions for the switching between the facilitation and the depression are not clear. We test the relationship between PPR and several factors including $P_H$, $P_0$ of the high sites and the inter-AP interval. The short-term

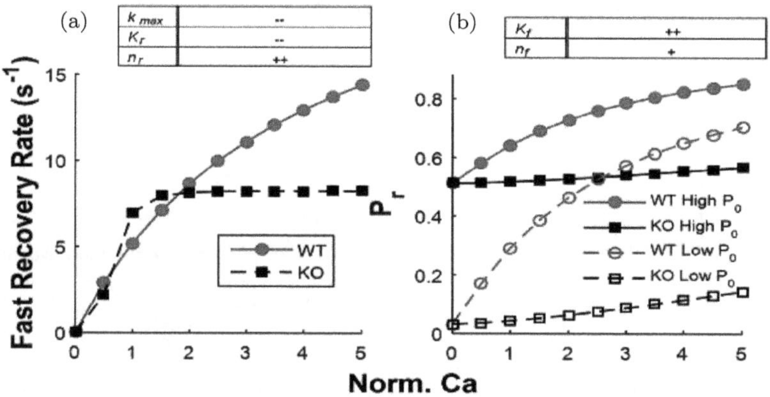


Fig. 5.3. The difference in parameters governing the WT and Syt7 KO responses. Both the recovery and facilitation are controlled by the normalized calcium level (Norm. Ca), and are modeled by Hill equations carrying three parameters: max recovery rate or facilitation, dissociation constant and the Hill coefficient. (a) The change in the fast recovery rates between WT and KO. The table above summarizes the quantitative changes in KO parameters, related to the recovery from depression, from that of WT. −, + and / denote decrease, increase and no change, respectively. Repeats denote a significant change. (b) The change in the facilitation between WT and KO as well as between the two sets of release sites. The max facilitation is 1. The table above summarizes the quantitative changes in KO parameters, related to the facilitation, from that of WT. $P_r = P_0$ when normalized calcium level is 0 (at rest).

change lasts for less than 1s and the degree of the change increases as the inter-AP interval decreases (Fig. 5.4). For an inter-AP interval of 10s, PPR is flat at 1 that indicates a complete recovery from any short-term changes (Fig. 5.4). But for an inter-AP interval of 1s, PPR is slightly above 1 for the low levels of $P_H$ and high $P_0$ (Fig. 5.4) indicating signs of a weak facilitation which produces significant short-term changes only at low $P_r$. These results suggest that the majority of the response for large inter-AP intervals is contributed by the minority of high sites because the facilitation is weak leading to a negligible release probability for the majority of low $P_0$ release sites (low sites). Correspondingly, for small inter-AP intervals, the first response is contributed by the minority of high sites and the second response is contributed by the majority of low sites due to the strong facilitation. Furthermore, the switching between the facilitation and

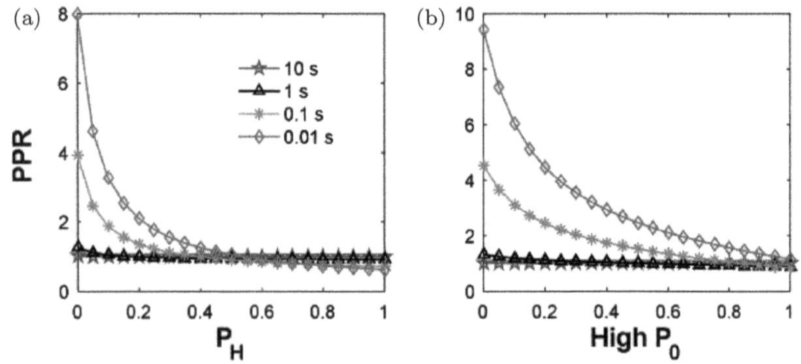


Fig. 5.4. PPR dependence on $P_H$, $P_0$ and the inter-AP interval in WT. (a) PPR as a function of $P_H$ $(0-1)$ and inter-AP interval (10s, 1s, 0.1s, and 0.01s). (b) PPR as a function of high $P_0$ $(0-1)$ and inter-AP interval (10s, 1s, 0.1s, and 0.01s). Other parameters are held at their estimated levels.

the depression depends on $P_H$ and high $P_0$; increasing $P_H$ or high $P_0$ shift the balance towards the depression since more release sites are depleted by the first AP (Fig. 5.4).

Therefore, the balance between the facilitation and the depression requires a proper coordination between $P_H$ and high $P_0$. As for the 0.01s inter-AP interval, the balance requires at least the moderate levels of both $P_H$ and high $P_0$ to cause sufficient depletion by the first AP to offset the strong facilitation at the second AP (Fig. A.2 in Appendix A). But, for large inter-AP intervals, the requirement of the depletion is much less since the facilitation is much weaker (Figs. A.3 and A.4 in Appendix A). Alternatively, the inter-AP interval can be decoded by a proper coordination between $P_H$ and high $P_0$ where the postsynaptic response is constant over stimuli.

The PPR difference is the difference between the PPRs of WT and of Syt7 KO. The difference is very small (close to 0) at the 10s inter-AP interval indicating an insignificant role for Syt7 in largely separated APs, i.e., the basal transmission (Fig. 5.5). The difference increases as the inter-AP interval decreases (Fig. 5.5). For intervals smaller than 1s, the PPR difference follows a decreasing trend as $P_H$ (Fig. 5.5 (a)) or high $P_0$ increase (Fig. 5.5 (b)). These general behaviors of the difference occur because of the decrease in PPR by increasing depression as a result of the relationships between PPR

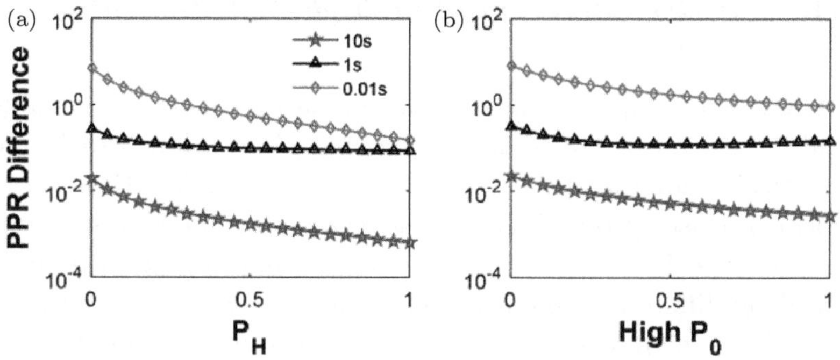


Fig. 5.5. PPR difference between WT and Syt7 KO. PPR difference is calculated by PPR of WT − PPR of KO. (a) PPR difference (log scale) as a function of $P_H$ (0 − 1) and inter-AP interval (10s, 1s, and 0.01s). (b) PPR difference (log scale) as a function of the high $P_0$(0 − 1) and inter-AP interval (10s, 1s, and 0.01s). Other parameters are held at their estimated levels.

and $P_H$, high $P_0$ and inter-AP interval, as discussed previously, with Syt7 playing a minor role in these behaviors. However, at the 1s inter-AP interval, the difference decreases rapidly before reaching a steady level (Fig. 5.5). This result is expected because the high sites dominate the response at large inter-AP intervals. Since the facilitation has insignificant contribution to high sites at large inter-AP intervals (Fig. 5.4), the PPR difference approaches a steady level with the increase of either the portion or $P_0$ of high sites. Once the steady level is reached, the reduction in response by Syt7 KO is mostly contributed by high sites. On the other hand, at the 0.01s inter-AP interval, the set of low sites maintain a considerable contribution to the PPR difference when the portion of high sites increases as indicated by the decreasing line (Fig. 5.5 and Fig. A.5 in Appendix A). These results yield interesting information that the degree of reduction caused by Syt7 KO takes two factors into account: the type of release sites and the inter-AP interval. The majority of the reduction occurs in high sites or low sites for large or small inter-AP intervals, respectively. However, the selection is not simply a function of Syt7 but also originates from the non-uniform releasing structure of the release sites.

### 5.3.4. *Selective alteration of multiple-AP stimulations by KO*

Sustained presynaptic stimulation, depending on the frequency of stimulation, can produce a long-lasting depression which recovers slowly or long-lasting synaptic enhancement [3]. Besides, non-uniform release probability is suggested to affect the synaptic strength especially when long-lasting synaptic enhancement is induced by different stimulation patterns [30]. Syt7 has been shown to contribute significantly to the short-term changes in the second of two closely spaced APs without affecting the first response; the reduction in Syt7 KO is dependent on release site and inter-AP intervals. However, the prolonged stimulations induce greater changes in postsynaptic response, as compared to that of PPR, by accumulating more $Ca_{res}$ for greater facilitation and depleting more release sites before the full recovery. The nature of the relationship between Syt7, the release site and the inter-AP interval still holds for more realistic multiple-AP stimulations environment can suggest a possible role of short-term plasticity in the synapse.

To better reflect the input specificity, we normalize EPSC amplitude of a set of release site by their max EPSC amplitude as given by

$$\begin{aligned}\text{Norm}\cdot\text{Response}_M &= \frac{\text{EPSC}_M}{\text{max EPSC}} = \frac{N\cdot P_H\cdot P_{r1,M}\cdot X_{1,M}\cdot Q}{N\cdot P_H\cdot Q} \\ &= P_{r1,M}\cdot X_{1,M},\end{aligned} \tag{5.11}$$

where, $P_{r1,M}$ and $X_{1,M}$ are the mean release probability and mean availability of high sites at the $M$th AP, respectively. As shown, the normalized response is same to the average release probability of the corresponding set of release sites. This measure reflects the average behavior of a single release site.

Since LFS (100 APs applied in 1HZ) and HFS (100 APs applied in 100Hz) are common protocols for triggering long-term synaptic plasticity in SC, the behavior of 2P model in response to the two stimulations are tested. During first 10 APs of LFS, Syt7 KO causes a significant reduction of the normalized response in only the high

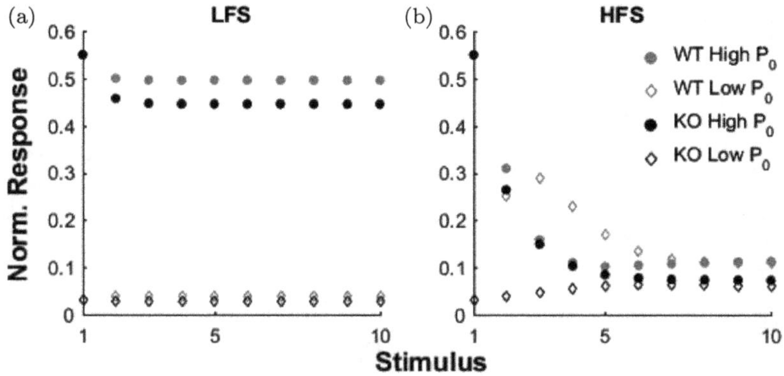


Fig. 5.6. Normalized response during the first 10 stimuli of LFS and HFS. The normalized response of two different sets of release sites in both WT and Syt7 KO during (a) LFS and (b) HFS.

sites (Fig. 5.6 (a)). During the first 10 APs of HFS, both sites show reductions (Fig. 5.6 (b)). But, the reduction in the high sites is minor before the 5th stimuli and becomes significant afterwards, when the fast recovery by Syt7 is switched on. On the other hand, the reduction is always significant for the low sites, especially during first 6 stimuli. These results assert the selectivity of Syt7 on release site during prolonged stimuli.

However, long-term synaptic plasticity at a single synaptic contact relies on the sustained release to build-up the postsynaptic $Ca^{2+}$ level [44]. In this case, the average release profiles across the 100 APs of LFS and HFS are more important. We calculate the average normalized response and a significant reduction in the average normalized response is revealed in Syt7 KO only for the large $P_0$ (>0.5) during LFS (Fig. 5.7 (a)). In HFS, there is a significant reduction in Syt7 KO for the whole range of $P_0$ (Fig. 5.7 (b)). However, the average normalized response in WT is similar across the whole range of $P_0$ while that in Syt7 KO exhibits a great drop approaching the low side of $P_0$ (Fig. 5.7 (b)). Given that the exact threshold and mechanism of postsynaptic response are unclear, if the postsynaptic response to multiple-AP stimulations is strongly linked to the presynaptic release profile, then we can expect Syt7 KO to cause a reduced postsynaptic response and synaptic strength.

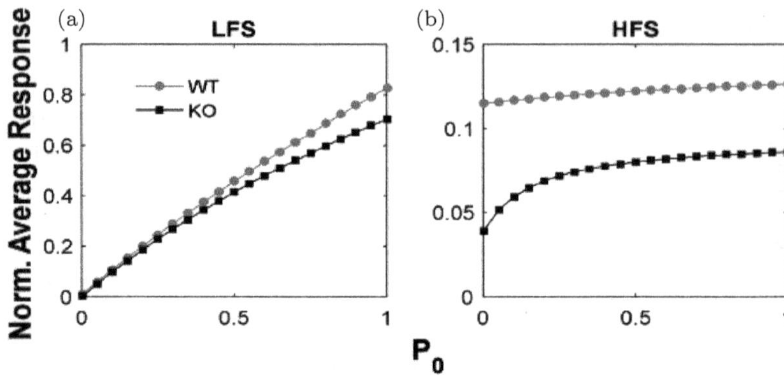


Fig. 5.7. Normalized average response across the 100 APs of LFS and HFS. The normalized average response of the release sites with different $P_0$ in both WT and Syt7 KO during (a) LFS and (b) HFS.

## 5.4. Discussion and Conclusion

We use a mathematical model to study the possible role of Syt7 and its driven short-term synaptic plasticity in the synapse, as well as the possible connection between presynaptic release and postsynaptic response. We observe using a mathematical model that release sites have non-uniform release probability in SC and different release sites contribute differently to transmission and plasticity. Having these properties, synaptic plasticity may function as a high-pass filter. But, to be able to adapt these release sites, a selection process is required to select the appropriate release sites and train them with respect to a target frequency for the filter and Syt7 may be involved.

### 5.4.1. *Short-term plasticity as a high-pass filter*

Short-term plasticity can function as a high-pass filter [45] to allow signals with a frequency higher than a threshold to trigger a postsynaptic AP. As shown in Fig. 5.4, the balance of the facilitation and the depression requires the proper coordination between the inter-AP interval and the release probability. Hence, for carefully trained release probability, the target inter-AP interval is right on the balance between the facilitation and the depression. The dominance in the facilitation or the depression indicates a smaller or a larger

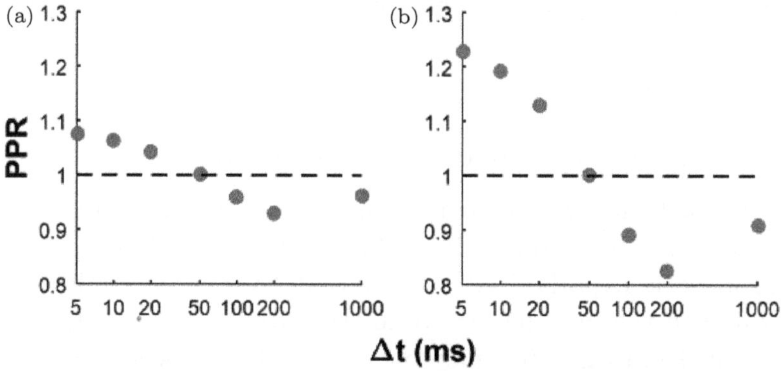


Fig. 5.8. High-pass filtering. An example is shown to filter an inter-AP interval of 50 ms. (a) The filter is based on a uniform release probability model ($P_H = 1$, high $P_0 = 0.313$). (b) The filter is based on a non-uniform release probability model ($P_H = 0.2225$, high $P_0 = 0.8$, and low $P_0 = 0.03$).

interval to the target, respectively. However, the basal transmission is not affected since it is fully recovered from the facilitation and the depression. Setting a threshold of postsynaptic potential at the basal transmission to generate an AP would only allow the higher frequency burst (smaller-AP interval) and filter out the low-frequency burst.

Figure 5.8 shows an example of a high-pass filter with a target inter-AP interval of 50 ms.

Both the uniform and non-uniform models distinguish smaller and larger inter-AP intervals. But, the PPR difference in the uniform model is rather small which may be difficult for a postsynaptic neuron to detect. The non-uniform model shows a significant PPR differences in both smaller and larger intervals. The reason is that the uniform model requires a high $P_0$ to induce sufficient depression to balance the strong facilitation. But, a high $P_0$ diminishes the potential for the facilitation as well as the availability of release sites for the next responses. The non-uniform model does not have this problem as the depression is contributed mainly from the minority of high sites that will leave sufficient potential for the facilitation by the low sites. Hence, by a proper coordination of $P_H$ and high $P_0$, the balance can be reached without having to deplete release sites too much.

Furthermore, the amplitude of mEPSC may be non-uniform [31] that may add another dimension of adaptability to have an even higher degree of differentiation for the high-pass filter.

### 5.4.2. *Activity-dependent selection of release sites*

However, a training process is required to selectively train release sites to adapt to the target inter-AP interval. Based on our results, we propose a hypothesized system that Syt7 selects release sites according to the presynaptic stimulation frequency and the expression of Syt7 in target sites may be a tag to direct the site to enhance postsynaptic response and modulate synaptic strength (Fig. 5.9).

Syt7 KO causes reductions in PPR, but not basal transmission (Fig. 5.5). However, the reduction is on high sites at the large inter-AP intervals because depletion in these sites is difficult to recover from by the reduced fast recovery rate of Syt7 KO within these intervals. On the other hand, the majority of the reduction is in low $P_0$ release sites at small inter-AP intervals because the lack of facilitation significantly reduces the response of these sites.

Over the first 10 APs of LFS, Syt7 KO reduces the responses of the high sites while the low sites remain unchanged (Fig. 5.6(a)). As for HFS, the responses of the high sites are similar between WT and Syt7 KO before the 5th AP and are slightly different afterwards (Fig. 5.6(b)). The low sites display a significant difference before the 6th AP and the difference becomes smaller afterwards (Fig. 5.6(b)). Responses reach a steady level after the 7th AP in HFS. The steady levels are similar between two sets of release sites in WT, while the levels are slightly different in Syt7 KO.

Syt7 KO causes reductions in average responses across the 100 APs during HFS and LFS (Fig. 5.7). Similarly, the reduction is on the high sites during LFS and the low sites reveal a greater reduction than the high sites during HFS.

All these results suggest that Syt7, together with the non-uniform release structure, is a candidate to control the long-term changes (For example, long-term potentiation (LTP) and long-term depression LTD) which are induced by multi-AP stimulations. The Syt7 KO causes consistent alterations across the PPR, the first 10 responses

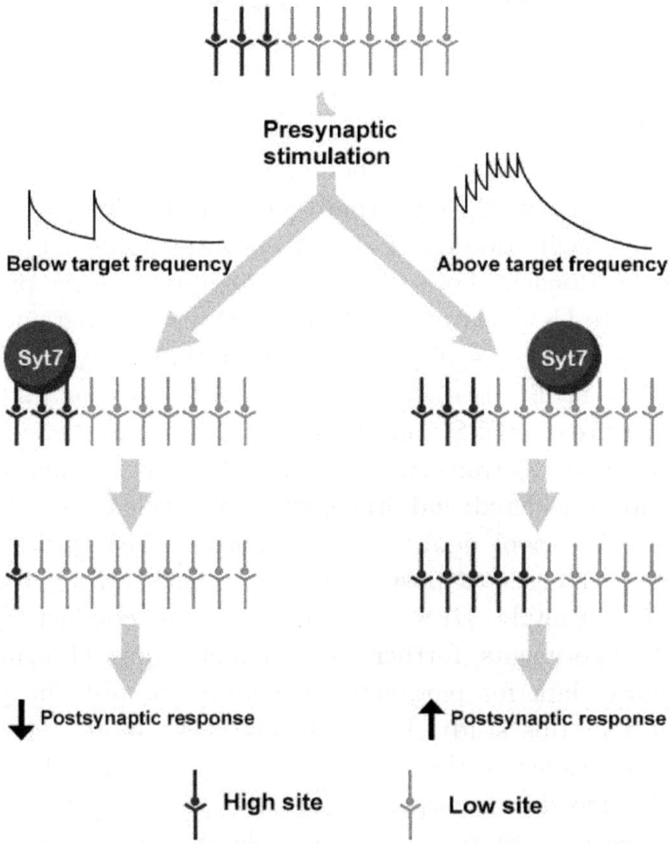


Fig. 5.9. Schematic representation of the hypotheses proposed for activity-dependent selection of release sites by Syt7. When presynaptic stimulation frequency bellows the target frequency, Syt7 controls the high sites to reduce postsynaptic response. Otherwise, Syt7 controls the low sites to increase postsynaptic response.

and the average response. In LFS, Syt7 controls the high sites to reduce postsynaptic response and synaptic strength. In HFS, Syt7 controls the low sites to increase postsynaptic response and synaptic strength. If the reduction and increase cause changes in $P_H$ by changing the ratio of high sites to low sites, i.e., through activation of silent synapses and inactivation of synapses to be silent [46], or changes the $P_0$ of high sites by presynaptic mechanisms [3], then we have a theoretical basis for the high-pass filter as well as the regulation of the target of the filter (Fig. 5.9).

## 5.5. Limitation

The current work has a number of limitations: (1) the previous experimental study [22] states that there is a supralinear relationship between the facilitation and $Ca^{2+}$ pre-elevation, however we predict a linear relationship between the presynaptic release probability and moderate $Ca^{2+}$ elevation. Hence, postsynaptic parameters and mechanisms, which we didn't consider in this study, are likely to contribute to facilitation: for example, the EPSC dynamics, the non-uniform nature of mEPSC and postsynaptic receptor desensitization and saturation; (2) transient activation and inactivation of release sites are not considered; and (3) depletion of vesicle pool and different rates of vesicle pool depletion are ignored. These processes may cause large changes to postsynaptic responses during continuous stimuli, for example, HFS. It is important to conduct Syt7 WT and KO experiments further on different types of synapses to obtain more data for parameter estimation to test the proposed hypotheses in this study. In the future, with more knowledge of the mechanisms across the synapse, a new mechanistic driven model can be developed to incorporate pre- and post-processes, so that a comprehensive understanding of the relationship between synaptic plasticity and synaptic functions can be obtained. Understanding of the relationships is essential to deduce the causal relationships between brain diseases and synaptic malfunctions.

## Appendix A

### A.1. Processes included in the model

It is certain that Syt7 KO reduces facilitation significantly [9]. Let us consider two hypotheses: (1) facilitation lasts for more than 0.5s, and (2) facilitation lasts for less than 0.5s. For the first hypothesis, there must be a depression event in WT cells which is perfectly balanced by facilitation after 0.5s (Fig. A.1). For the second hypothesis, the greater depression observed in Syt7 KO cells after 0.5s cannot be

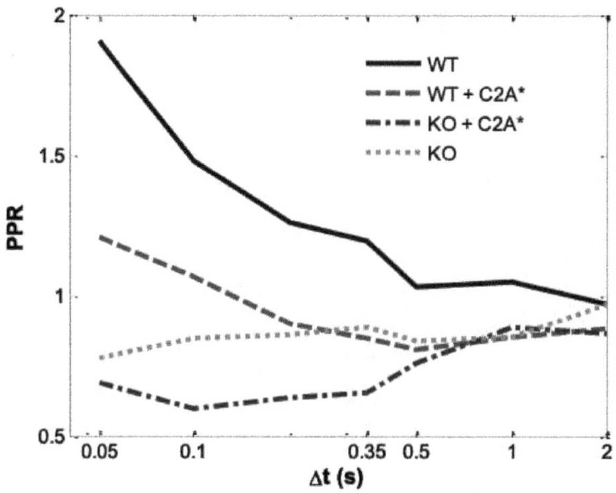


Fig. A.1. Pair–Pulse ratios (PPRs) across WT and Syt7 KO neurons. Experimental data is retrieved from [9]. C2A* is a mutated $Ca^{2+}$-insensitive C2A domain. Only WT expresses the strong facilitation, while the other three express the reduced or impaired facilitation.

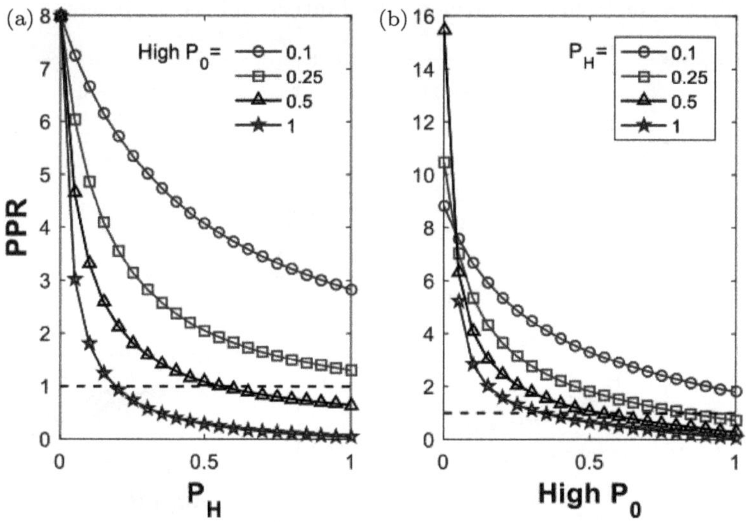


Fig. A.2. PPR at 0.01s inter-AP interval as a function of $P_H$ and high $P_0$. (a) PPR as a function of $P_H$ (0 – 1) and high $P_0$ (four levels, 0.1, 0.25, 0.5 and 1). (b) PPR as a function of high $P_0$ (0 – 1) and $P_H$ (four levels, 0.1, 0.25, 0.5 and 1). Other parameters are held at their estimated levels. The black dashed line indicates the balance between the facilitation and the depression.

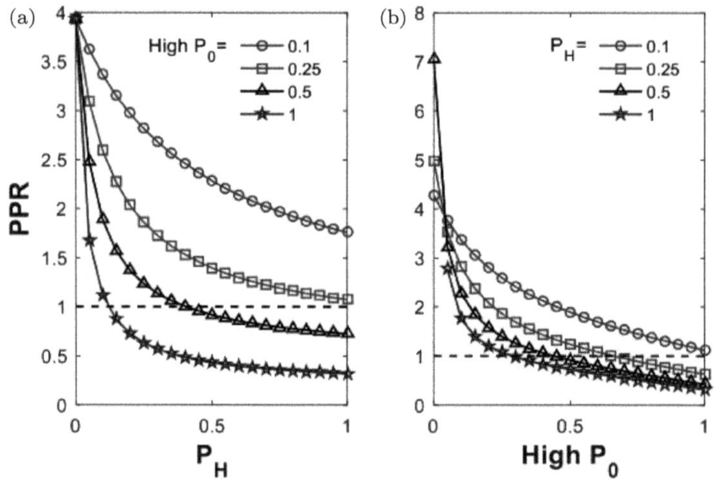


Fig. A.3. PPR at 0.1s inter-AP interval as a function of $P_H$ and high $P_0$. (a) PPR as a function of $P_H$ $(0-1)$ and high $P_0$ (four levels, 0.1, 0.25, 0.5 and 1). (b) PPR as a function of high $P_0$ $(0-1)$ and $P_H$ (four levels, 0.1, 0.25, 0.5 and 1). Other parameters are held at their estimated levels. The black dashed line indicates the balance between the facilitation and the depression.

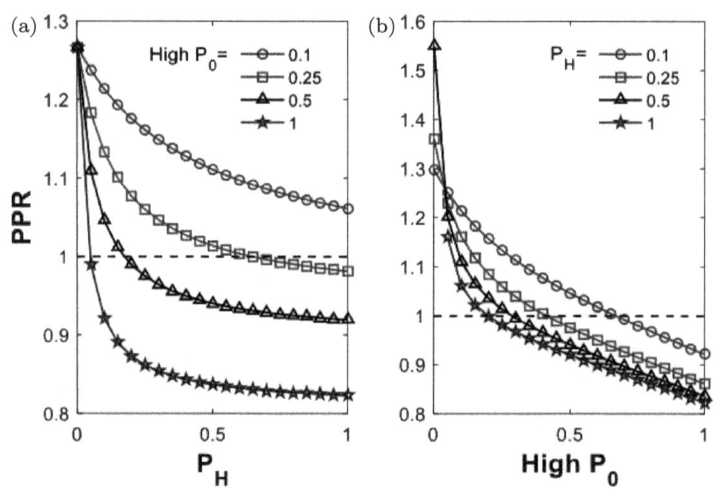


Fig. A.4. PPR at 1s inter-AP interval as a function of $P_H$ and high $P_0$. (a) PPR as a function of $P_H$ $(0-1)$ and high $P_0$ (four levels, 0.1, 0.25, 0.5 and 1). (b) PPR as a function of high $P_0$ $(0-1)$ and $P_H$ (four levels, 0.1, 0.25, 0.5 and 1). Other parameters are held at their estimated levels. The black dashed line indicates the balance between the facilitation and the depression.

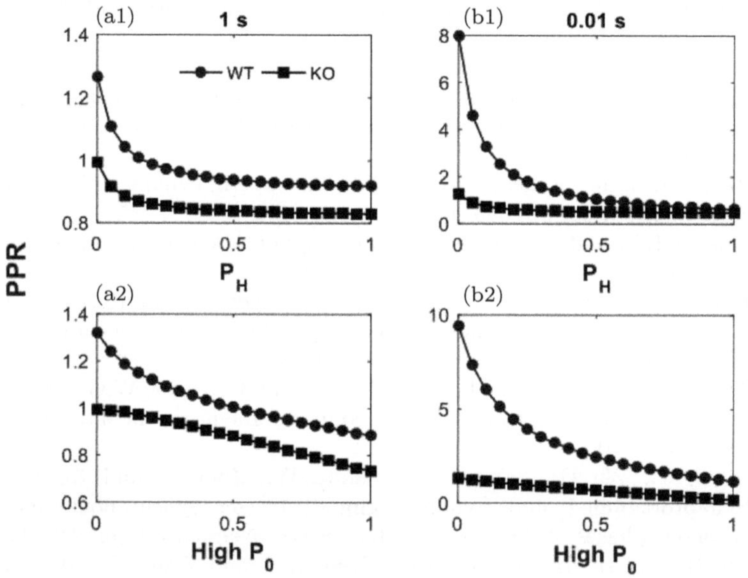


Fig. A.5. PPRs between WT and Syt7 KO. The difference in PPR initially decreases and then becomes constant for (a1) PPR at 1s inter-AP interval as a function of $P_H$ (0 – 1) and (a2) PPR at 1s inter-AP interval as a function of high $P_0$ (0 – 1). The difference in PPR always decreases for (b1) PPR at 0.01s inter-AP interval as a function of $P_H$ (0 – 1) and (b2) PPR at 0.01s inter-AP interval as a function of high $P_0$ (0 – 1).

explained by the dysfunction of facilitation which lasts only for 0.5s (Fig. A.1). A depression process, whose recovery may be controlled by Syt7, is required to explain the data in both hypotheses. Hence, at least a facilitation process and a depression process need to be included in the model. In addition, the process of Syt7 activation by residue $Ca^{2+}$ is also required.

## References


[1] Citri A. and Malenka R.C. (2008). Synaptic plasticity: multiple forms, functions, and mechanisms. *Neuropsychopharmacology*, *33*, pp. 18–41.

[2] He Y., Kulasiri D. and Liang J. (2019). A mathematical model of synaptotagmin 7 revealing functional importance of short-term synaptic plasticity. *Neural Regen. Res.*, *14*(4), pp. 621–631.

[3] Regehr W.G. (2012). Short-term presynaptic plasticity. *Cold Spring Harb. Perspect., Biol.* 4, a005702.

[4] Klyachko V.A. and Stevens C.F. (2006). Excitatory and feed-forward inhibitory hippocampal synapses work synergistically as an adaptive filter of natural spike trains. *PLoS Biol.*, *4*, p. e207.
[5] Martin S.J., Grimwood P.D. and Morris R.G.M. (2000). Synaptic plasticity and memory: an evaluation of the hypothesis. *Ann. Rev. Neurosci.*, *23*, pp. 649–711.
[6] Zucker R.S. and Regehr W.G. (2002). Short-term synaptic plasticity. *Ann. Rev. Physiol.*, *64*, pp. 355–405.
[7] Abbott L.F. and Regehr W.G. (2004). Synaptic computation. *Nature 431*, pp. 796–803.
[8] Blatow M., Caputi A., Burnashev N., Monyer H. and Rozov A. (2003). $Ca^{2+}$ Buffer saturation underlies paired pulse facilitation in calbindin-D28k-containing terminals. *Neuron 38*, pp. 79–88.
[9] Jackman S.L., Turecek J., Belinsky J.E. and Regehr W.G. (2016). The calcium sensor synaptotagmin 7 is required for synaptic facilitation. *Nature*, *529*, pp. 88–91.
[10] Bacaj T., Wu D., Yang X., Morishita W., Zhou P. and Xu W.(2013). Synaptotagmin-1 and Synaptotagmin-7 trigger synchronous and asynchronous phases of neurotransmitter release. *Neuron*, *80*, pp. 947–959.
[11] Liu H., Bai H., Hui E., Yang L., Evans C.S. and Wang Z. (2014). Synaptotagmin 7 functions as a $Ca^{2+}$-sensor for synaptic vesicle replenishment. *Elife*, *3*, p. e01524.
[12] Gustavsson N., Lao Y., Maximov A., Chuang J.C., Kostromina E. and Repa J.J. (2008). Impaired insulin secretion and glucose intolerance in synaptotagmin-7 null mutant mice. *Proc. Natl. Acad. Sci. USA.*, *105*, pp. 3992–3997.
[13] Martinez I., Chakrabarti S., Hellevik T., Morehead J., Fowler K. and Andrews N.W. (2000). Synaptotagmin VII regulates Ca(2+)-dependent exocytosis of lysosomes in fibroblasts. *J Cell Biol.*, *148*, pp. 1141–49.
[14] Murthy V.N., Sejnowski T.J., Stevens C.F., Allen C., Stevens C.F., Bekkers J.M. (1997). Heterogeneous release properties of visualized individual hippocampal synapses. *Neuron*, *18*, pp. 599–612.
[15] Dobrunz L.E. and Stevens C.F. (1997). Heterogeneity of release probability, facilitation, and depletion at central synapses. *Neuron*, *18*, pp. 995–1008.
[16] Bliss T.V., Collingridge G.L., Kerchner G., Nicoll R., Kessels H. and Malinow R. (2013). Expression of NMDA receptor-dependent LTP in the hippocampus: bridging the divide. *Mol. Brain*, *6*, pp. 5.
[17] Neves G. and Lagnado L. (1999). The kinetics of exocytosis and endocytosis in the synaptic terminal of goldfish retinal bipolar cells. *J. Physiol.*, *515*, pp. 181–202.
[18] Denker A. and Rizzoli S.O. (2010). Synaptic vesicle pools: an update. *Front Synaptic Neurosci.*, *2*, p. 135.
[19] Rizzoli S.O., Betz W.J. (2005). Synaptic vesicle pools. *Nat. Rev. Neurosci.*, *6*, pp. 57–69.
[20] del Castillo J. and Katz B. (1954). Quantal components of the end-plate potential. *J. Physiol.*, *124*, pp. 560–573.

[21] Dittman J.S. and Regehr W.G. (1998). Calcium dependence and recovery kinetics of presynaptic depression at the climbing fiber to Purkinje cell synapse. *J. Neurosci., 18*, pp. 6147–6162.
[22] Felmy F., Neher E. and Schneggenburger R. (2003). The timing of phasic transmitter release is $Ca^{2+}$-dependent and lacks a direct influence of presynaptic membrane potential. *Natl. Acad. Sci. USA*, *100*(25), pp. 15200–15205.
[23] Stevens C.F. and Wang Y. (1995). Facilitation and depression at single central synapses. *Neuron, 14*, pp. 795–802.
[24] Biró A.A., Holderith N.B. and Nusser Z. (2005). Quantal size is independent of the release probability at hippocampal excitatory synapses. *J. Neurosci.*, *25*, pp. 223–232.
[25] Schneggenburger R., Sakaba T. and Neher E. (2002). Vesicle pools and short-term synaptic depression: lessons from a large synapse. *Trends Neurosci.*, *25*, pp. 206–212.
[26] Conti R. and Lisman J. (2003). The high variance of AMPA receptor- and NMDA receptor-mediated responses at single hippocampal synapses: evidence for multiquantal release. *Proc. Natl. Acad. Sci., USA, 100*, pp. 4885–4890.
[27] Christie J.M. and Jahr C.E. (2006). Multivesicular release at Schaffer collateral-CA1 hippocampal synapses. *J. Neurosci.*, *26*, pp. 210–216.
[28] Dittman J.S., Kreitzer A.C. and Regehr W.G. (2000). Interplay between facilitation, depression, and residual calcium at three presynaptic terminals. *J. Neurosci.*, *20*, pp. 1374–85.
[29] Sun H.Y., Lyons S.A. and Dobrunz L.E. (2005). Mechanisms of target-cell specific short-term plasticity at Schaffer collateral synapses onto interneurones versus pyramidal cells in juvenile rats. *J. Physiol.*, *568*, pp. 815–840.
[30] Rosenmund C., Clements J.D. and Westbrook G.L. (1993). Nonuniform probability of glutamate release at a hippocampal synapse. *Science*, *262*, pp. 754–757.
[31] Silver R.A. (2003). Estimation of nonuniform quantal parameters with multiple-probability fluctuation analysis: theory, application and limitations. *J. Neurosci. Methods*, *130*, pp. 127–141.
[32] Walmsley B., Edwards F.R. and Tracey D.J. (1988). Nonuniform release probabilities underlie quantal synaptic transmission at a mammalian excitatory central synapse. *J. Neurophysiol.*, *60*, pp. 889–908.
[33] Trommershäuser J., Schneggenburger R., Zippelius A. and Neher A. (2003). Heterogeneous presynaptic release probabilities: Functional relevance for short-term plasticity. *Biophysical J., 84*(3), pp. 1563–1579.
[34] Collin T., Chat M., Lucas M.G., Moreno H., Racay P. and Schwaller B. (2005). Developmental changes in parvalbumin regulate presynaptic $Ca^{2+}$ signaling. *J. Neurosci., 25*, pp. 96–107.
[35] Chen C., Blitz D.M. and Regehr W.G. (2002). Contributions of receptor desensitization and saturation to plasticity at the retinogeniculate synapse. *Neuron*, *33*, pp. 779–788.

[36] Sun Y.G. and Beierlein M. (2011). Receptor saturation controls short-term synaptic plasticity at corticothalamic synapses. *J. Neurophysiol., 105*, pp. 2319–2329.
[37] Dobrunz L.E., Huang E.P. and Stevens C.F. (1997). Very short-term plasticity in hippocampal synapses. *Proc. Natl. Acad. Sci. USA*, *94*, pp. 14843–14847.
[38] Hjelmstad G.O., Nicoll R.A. and Malenka R.C. (1997). Synaptic refractory period provides a measure of probability of release in the hippocampus. *Neuron, 19*, pp. 1309–1318.
[39] Haario H., Laine M., Mira A., and Saksman E. (2006). DRAM: efficient adaptive MCMC. *Stat. Comput.*, *16*, pp. 339–354.
[40] McDougal R.A., Morse T.M. and Carnevale T. (2017). Twenty years of ModelDB and beyond: building essential modeling tools for the future of neuroscience. *J. Comput. Neurosci.*, *42*(1), pp. 1–10.
[41] Branco T. and Staras K. (2009). The probability of neurotransmitter release: variability and feedback control at single synapses. *Nat. Rev. Neurosci.*, *10*, pp. 373–383.
[42] Catterall W.A. and Few A.P. (2008). Calcium channel regulation and presynaptic plasticity. *Neuron*, *59*, pp. 882–901.
[43] Mochida S., Few A.P., Scheuer T., Catterall W.A., Abbott L.F. and Regehr W.G. (2008). Regulation of presynaptic Ca(V)2.1 channels by $Ca^{2+}$ sensor proteins mediates short-term synaptic plasticity. *Neuron*, *57*, pp. 210–216.
[44] Bliss T.V.P. and Collingridge G.L. (1993). A synaptic model of memory: long-term potentiation in the hippocampus. *Nature, 361*, pp. 31–39.
[45] Rothman J.S., Cathala L., Steuber V. and Silver R.A. (2009). Synaptic depression enables neuronal gain control. *Nature*, *457*, pp. 1015–1018.
[46] Isaac J.T.R., Nicoll R.A. and Malenka R.C. (1995). Evidence for silent synapses: implications for the expression of LTP. *Neuron, 15*, pp. 427–434.

# Chapter 6

# Role of Calcium in AD, Related Pathways and Hypotheses

## 6.1. Introduction

$Ca^{2+}$ is the most universal second messenger in a variety of cells. The ubiquitous nature of $Ca^{2+}$ accounts for its diverse functions in cell regulation [1]. In the nervous system, $Ca^{2+}$ signaling plays essential roles in brain physiology, including neuronal development, synaptic plasticity and neuron survival/death. Therefore, maintaining $Ca^{2+}$ homeostasis is crucial for normal brain activity in the healthy state [2]. Dysregulation of intracellular $Ca^{2+}$ signaling has been observed as an early event prior to the presence of clinical symptoms of AD and is believed to be a crucial factor contributing to its pathogenesis [3, 4]. In Section 6.2 of this chapter, we first introduce the mechanisms of NMDAR-mediated $Ca^{2+}$ signalling in the dendritic spine under healthy conditions. Then, in Section 6.6, we review the experimental findings and conflicting interpretations about $Ca^{2+}$ dysregulation in AD, including the latest publications, with an emphasis on its effect on the factors involved in NMDAR-mediated $Ca^{2+}$ signaling. We also discuss computational modeling approaches in $Ca^{2+}$ signaling studies in Section 6.4, and the existing models and their potential to understand the underlying causes of AD in Section 6.5.

## 6.2. NMDAR-Mediated $Ca^{2+}$ Transients in the Dendritic Spine

In the nervous system, a chemical synapse is a junction that comprises a presynaptic terminal, a postsynaptic component and a synaptic cleft between them (Fig. 6.1). Through a synapse, the electrical signal from a presynaptic cell is converted into a chemical signal (neurotransmitter release), which can be transferred to the postsynaptic cell. At the distal dendritic spine, a region called the postsynaptic density (PSD) attached to the postsynaptic site of the spine head is enriched with large numbers of proteins, including

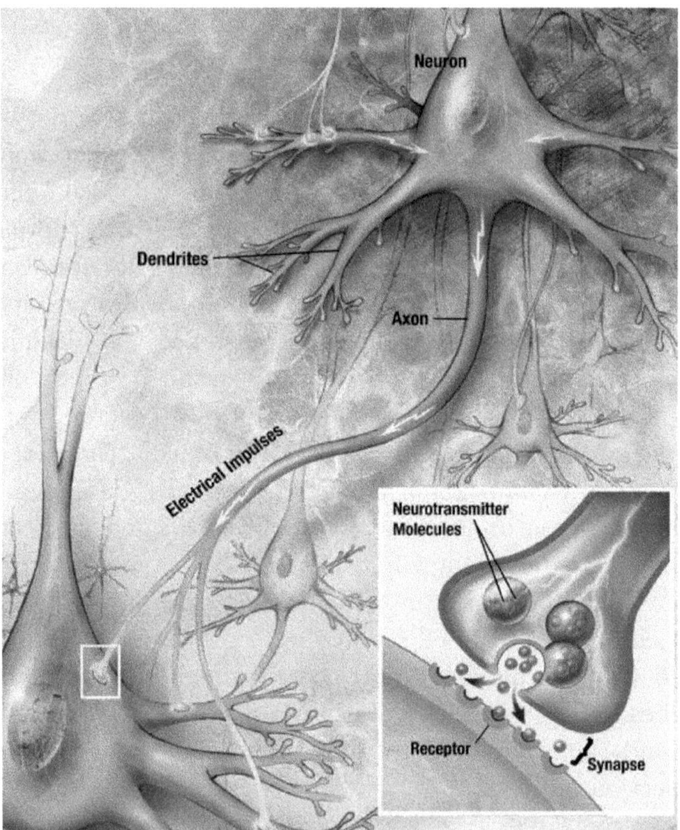


Fig. 6.1. Structure of a typical chemical synapse (public domain in US).

neurotransmitter receptors, cytoskeleton proteins, ion channels and other signalling proteins. These proteins function collaboratively to translate the chemical signal from the presynaptic terminal into a postsynaptic response [5].

The NMDAR-mediated $Ca^{2+}$ transient in the dendritic spine is initiated by the presynaptic neurotransmitter release induced by the action potential (AP) at the presynaptic terminal. The neurotransmitter released activates postsynaptic receptors, such as NMDAR and AMPAR. The $Ca^{2+}$ influx through NMDAR into the cytosol determines various neuronal activities, such as synaptic remodeling, synaptic plasticity and cognitive functions [6].

### 6.2.1. *Glutamate*

When the action potential reaches the presynaptic terminal it will promote the release of neurotransmitter vesicles into the synaptic cleft [7]. Glutamate is the major excitatory neurotransmitter in the central nervous system and each vesicle contains thousands of glutamate molecules [8]. The glutamate molecules released diffuse across the synaptic cleft and into the extrasynaptic space. In addition, glutamate can also be released from the glial cells, induced by their intracellular $Ca^{2+}$ oscillations, which initiates the neuronal synchrony by activating extrasynaptic NMDARs [9].

The clearance of glutamate is achieved by high affinity glutamate transporters (i.e., sodium and potassium coupled glutamate transporters or excitatory amino acid transporters (EAATs)) on surrounding glial cells. To date, five subtypes of these have been identified at different locations (for reviews see [10, 11]). The glutamate transporters play a role in maintaining the extrasynaptic glutamate levels to avoid the pathologic accumulation of extrasynaptic glutamate, which may cause excessive activation of the glutamate receptors and, ultimately, neuronal death [12].

### 6.2.2. *Glutamate receptors*

Once released, glutamate can activate both metabotropic glutamate receptors (mGluRs) and ionotropic glutamate receptors (iGluRs)

[13, 14]. The mGluRs themselves do not function as ion channels, instead, they trigger other signaling pathways, indirectly leading to cellular events, such as the opening of ion channels. The opening of intracellular $Ca^{2+}$ channels initiated by the activation of mGluRs is discussed in Section 6.2.3. The iGluRs are located at chemical synapses, both pre and postsynaptically, and mediate vesicle fusion and postsynaptic response, respectively [15]. iGluRs have also been found in the extrasynaptic space, where they can be activated by glutamate spillover from the synaptic cleft or by the ectopic release of glutamate, such as release from the adjacent glia [9, 16]. In the mammalian central nervous system, there are three major types of iGluRs: AMPAR, NMDAR and the kainate receptor. The kainate receptor is less well studied than the other two and is not the focus of this research. AMPAR and NMDAR are mobile and their trafficking from their synthesis sites and between the synaptic and extrasynaptic zones is crucial to their synaptic transmission and the formation of different forms of synaptic plasticity [17].

#### 6.2.2.1. *NMDAR*

NMDAR is the major type of $Ca^{2+}$ channel contributing to the postsynaptic $Ca^{2+}$ response in pyramidal neurons of the hippocampus [18]. Moreover, it has also been found at extrasynaptic locations, such as the dendritic spine neck, the dendritic shaft, and in neuron bodies. It is a heterotetramer, mostly comprising two NR1 and two NR2 subunits [19]. There are eight and four splice variants of NR1 and NR2 subunits, respectively. Those subunits share similar membrane topologies in that they contain an extracellular N-terminal domain with three transmembrane regions (M1, M3 and M4), a pore lining region (M2), an extracellular N-terminus and an intracellular C-terminus [20]. The agonists bind to the their extracellular binding domains; in NR1 subunits it is the glycine-binding site, whereas in NR2, it is the glutamate-binding sites. To activate NMDAR, all binding sites at the four subunits need to be occupied. Therefore, it requires the binding of two molecules of glutamate to the NR2 subunits and two molecules of agonist to the NR1 subunits.

Recent research has found that NMDARs at different locations are gated by different co-agonists: D-serine for the synaptic NMDARs and glycine for the extrasynaptic NMDARs [21]. The affinity for glutamate by NMDAR depends on their NR2 subunit composition. The NMDAR subunit compositions at different locations change during postnatal development [22]. The ratio of NR2A to NR2B increases at the synaptic site and decreases at the extrasynaptic site during postnatal development. In mature synapses, NR2A–NMDARs are predominant at the synaptic sites, which take about 60% of the total synaptic NMDARs [23]. In contrast, NMDARs located outside the synaptic region are mainly NR1/NR2B–NMDARs. They are proposed to play opposite physiological roles in mediating intracellular signaling and death pathways: activation of synaptic NMDARs shows neuroprotective effects, whereas stimulation of extrasynaptic NMDARs contributes to cell death [24].

In addition to the binding of glutamates and their co-agonists, $Ca^{2+}$ entry through NMDAR requires the relief of $Mg^{2+}$ blocks [25]. The $Ca^{2+}$ permeation of NMDAR is mainly mediated by the M2 and M4 regions. The receptor is voltage-dependent and blocked by physiological concentrations of $Mg^{2+}$ at resting membrane potential. During stimulation, the $Mg^{2+}$ block is relieved by membrane depolarization, which can be produced by the opening of AMPARs. NMDARs are often found to be co-localized with AMPARs at the central synapses [14].

#### 6.2.2.2. *AMPAR*

AMPAR is not permeable or less permeable to $Ca^{2+}$ in comparison to NMDAR [13]. The activation of AMPAR by glutamate is fast and leads to a brief depolarization that only lasts for a few milliseconds. When the so-called excitatory postsynaptic potentials (EPSPs) generated by AMPAR are large enough, they will generate an action potential at the postsynaptic neuron. Therefore, the number of AMPAR in the PSD is crucial for synaptic transmission and different types of synaptic plasticity formation. Their numbers in the PSD are not fixed but are precisely mediated by the trafficking mechanisms [17].

### 6.2.3. *ER*

Intraneuronally, NMDAR-dependent $Ca^{2+}$ influx is augmented via $Ca^{2+}$-induced $Ca^{2+}$ release (CICR) from intracellular $Ca^{2+}$ stores. In the ER, $Ca^{2+}$ homeostasis and signaling are precisely mediated by release and sequestration mechanisms, via $Ca^{2+}$ channels and pumps, to accomplish the regulation of numerous downstream activities [26, 27]. The mechanism of intracellular $Ca^{2+}$ signaling is briefly illustrated in Fig. 6.2. ER $Ca^{2+}$ handling involves $Ca^{2+}$ release from the ER through two types of $Ca^{2+}$ channels, inositol 1,4,5-trisphosphate receptors ($IP_3Rs_3Rs$) and ryanodine receptors (RyRs), and $Ca^{2+}$ uptake via the $Ca^{2+}$ pumps, the sarcoendoplasmic reticulum $Ca^{2+}$ ATPase (SERCA) pumps. Two separate mechanisms involved in the process of $Ca^{2+}$ release from the ER via $IP_3Rs$ and RyRs are inositol 1,4,5-trisphosphate ($IP_3$)-induced $Ca^{2+}$ release and $Ca^{2+}$-induced $Ca^{2+}$ release (CICR), respectively.

$IP_3R$ is a ligand-gated $Ca^{2+}$ channel embedded in the ER membrane and is activated by the intracellular second messenger, $IP_3$ [29]. $IP_3$-mediated $Ca^{2+}$ release is initiated by the agonist stimulation of G protein-coupled plasma membrane receptors, such as mGluRs. Subsequently, mGluRs catalyze the activation of G-proteins, which, in turn, activate phospholipase C (PLC). PLC catalyzes a membrane phospholipid, phosphatidyl inositol-bisphosphate ($PIP_2$), to release $IP_3$ and diacyl glycerol (DAG) [30]. In addition, $IP_3Rs$ are also regulated by cytosolic $Ca^{2+}$ after activation, in a rapidly stimulated and slowly inhibited, fashion [31, 32]. They can interact locally with nearby channels to coordinate $Ca^{2+}$ release in clusters through the CICR process [33]. This will generate the hierarchical recruitment of elementary $Ca^{2+}$ release events and lead to the propagation of regenerative $Ca^{2+}$ waves through cells in response to strong extracellular stimulation [31, 34].

The other $Ca^{2+}$ channel, RyR, is the largest ion channel protein to date and exists in three isoforms, RyR1, RyR2 and RyR3, all of which can be found in the brain [35]. These RyR subtypes seem to have different involvement in memory processing, synaptic plasticity and other events in the brain [35, 36]. $Ca^{2+}$ release through RyRs is stimulated by cytosolic $Ca^{2+}$ ions via the process of CICR.

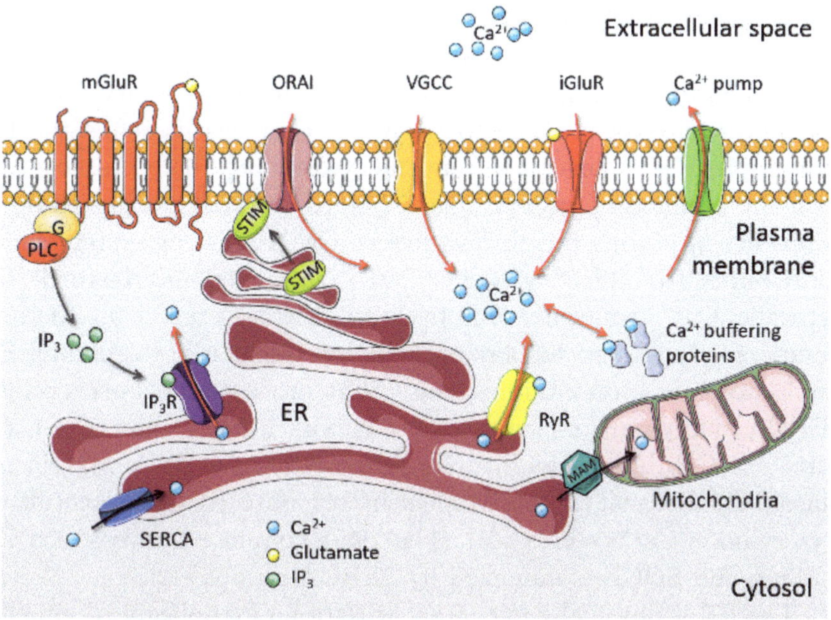


Fig. 6.2. Neuronal ER $Ca^{2+}$ signaling. Intracellular $Ca^{2+}$ concentration is regulated by ion channels (e.g., voltage-gated $Ca^{2+}$ channels (VGCCs) and ionotropic glutamate receptors (iGluRs)) and transporters in the plasma membrane. Intracellularly, the regulation of $Ca^{2+}$ homeostasis involves $Ca^{2+}$ ions bind to $Ca^{2+}$ buffer proteins and release from organelles, especially ER. $Ca^{2+}$ ions release from ER via $IP_3Rs$ and RyRs and uptake by ER via SERCA. Activation of $IP_3Rs$ requires $IP_3$, which is generated by stimulation of metabotropic glutamate receptors (mGluRs) in the plasma membrane. Depletion in ER $Ca^{2+}$ level promotes $Ca^{2+}$ influx via membrane store-operated $Ca^{2+}$ channels, such as ORAI, by translocation of the stromal interaction molecules (STIMs) into the ER–plasma membrane junctions. Besides, $Ca^{2+}$ ions in ER can be transported into mitochondria via mitochondria-associated ER membrane (MAM) to stimulate mitochondrial metabolism. This figure is produced using Servier Medical Art (http://www.servier.com/Powerpoint-image-bank). Also acknowledgment to [28]: Liang, J., Kulasiri D., and Samarasinghe S. (2015) $Ca^{2+}$ dysregulation in the endoplasmic reticulum related to Alzheimer's disease: A review on experimental progress and computational modeling. *Biosystems 134*, pp. 1–15.

Although $IP_3Rs$ can also be activated by $Ca^{2+}$ ions, as mentioned above, CICR is more usually related to RyR-mediated $Ca^{2+}$ release. As an inherent, positive feedback mechanism, CICR can greatly amplify the initial $Ca^{2+}$ signals and contribute to subsequent neuronal events, such as neurotransmitter release and $Ca^{2+}$-dependent gene expression [37].

The SERCA pump of the ER has the highest affinity for $Ca^{2+}$ and plays an opposite role to the $Ca^{2+}$ channels: it refills the ER by pumping $Ca^{2+}$ ions back from the cytosol [38]. SERCAs, together with $Ca^{2+}$-buffering proteins, $Ca^{2+}$ pumps and transporters on other organelles or plasma membrane, form a $Ca^{2+}$ buffering system which contributes to buffer cytosolic $Ca^{2+}$ transients and restores the cytosolic $Ca^{2+}$ concentration back to resting levels [39]. SERCA pumps are suggested to play an interdependent role to refilling ER $Ca^{2+}$ stores and maintaining the CICR process [40]. Furthermore, SERCA pumps are one of the key factors in store-operated $Ca^{2+}$ entry (SOCE), a process involving extracellular $Ca^{2+}$ influx via the plasma membrane $Ca^{2+}$ channels in response to the depletion of intracellular $Ca^{2+}$ stores [41]. The tight coupling between the ER refilling and SOCE is managed by SERCA pumps [42].

The ER is physically in contact with other cytoplasmic organelles and the plasma membrane. Through these contact sites, ER $Ca^{2+}$ release is closely coupled with other cellular events via communication with other cytoplasmic organelles, especially mitochondria, and with $Ca^{2+}$ channels in the plasma membrane [43]. Depletion in the ER $Ca^{2+}$ level will cause SOCE and $Ca^{2+}$ refilling via store-operated $Ca^{2+}$ channels (SOCs) located in the plasma membrane. In the ER, a family of $Ca^{2+}$ sensor proteins, the stromal interaction molecules (STIMs), will be activated by the decreased ER $Ca^{2+}$ level and translocate into the ER–plasma membrane junctions. STIMs can then activate SOCs, such as the ORAI channels, and induce $Ca^{2+}$ influx via them (reviewed in [44]). The ER is physically connected to the mitochondria by a lipid, raft-like structure, called mitochondria-associated ER membrane (MAM) [45]. The MAM plays important roles in multiple cellular functions, including maintaining intracellular $Ca^{2+}$ homeostasis. It allows selective transmission of $Ca^{2+}$ ions from the ER to the mitochondria and, thereby, regulates the functions of the mitochondria and cell survival. This process is regulated by a set of proteins localized in the MAMs (reviewed in [46]).

## 6.3. AD and $Ca^{2+}$ Dysregulation

As mentioned in Chapter 1, AD was first described by Alois Alzheimer in 1906 [47]. It is the most common form of all neurodegenerative diseases, and it is the leading cause of dementia, resulting in progressive loss of memory, cognitive function and, ultimately, death. Aging is the most important risk factor for developing AD. Based on the age of onset, AD is often divided into two categories: early-onset and late-onset. Less than 5 percent of all AD cases are early-onset AD, which affects people younger than 65. About 13% of early-onset AD cases is familial AD (FAD), because of inherited genetic mutations. In contrast, the rest of AD cases, including the late-onset ones which have an age of onset of over 65, are called sporadic AD (SAD) and they are not inherited and caused by unestablished factors [48]. Noticeably, although FAD and SAD have different onset times and development speeds, they present similar neurodegenerative processes, such as a loss of memory and neuronal cell death [41].

AD is characterized by the abnormal accumulation of harmful proteins, A$\beta$ protein plaques and neurofibrillary tangles in the nervous system [49]. It is widely believed that these hallmarks are in connection with a massive loss of neurons and synapses in the patients' brains. AD has been identified for more than 100 years, however, to date, the cause of AD is still not fully understood. Scientists believe that AD is a result of multiple factors and to explain the underlying mechanisms of AD pathophysiology, various hypotheses have been proposed, such as the amyloid hypothesis, the tau hyperphosphorylation hypothesis, the oxidative stress hypothesis, the cholesterol hypothesis and the vascular hypothesis [50–53]. The amyloid hypothesis is the dominant hypothesis (see Chapter 2) and many other hypotheses are variants of it [54, 55]. In this chapter, we focus on the $Ca^{2+}$ hypothesis.

### 6.3.1. *$Ca^{2+}$ hypothesis*

A$\beta$ is proposed to cause neuronal toxicity via a variety of pathways, and several hypotheses extending from the amyloid hypothesis attempt to further elucidate the alterations resulting from amyloid pathology. The most important, and feasible, pathway is through the $Ca^{2+}$ signaling systems by triggering deregulation of $Ca^{2+}$ homeostasis. The $Ca^{2+}$ hypothesis that links altered $Ca^{2+}$ signaling to A$\beta$-induced neuronal dysfunction is becoming more popular, and is supported by accumulating experimental evidence [56]. The deregulation of $Ca^{2+}$ homeostasis is suggested to account for the decline in cognitive function in the early stages of AD, and neuronal loss in the late stage of AD. Therefore, the focus of this research is mainly on the amyloid and $Ca^{2+}$ hypotheses.

A$\beta$ oligomers may interact with neuronal membranes and result in a significant elevation of intracellular $Ca^{2+}$ levels in AD. High levels of $Ca^{2+}$ are toxic and will induce neural apoptosis, the most common form of programmed cell death, leading to neuronal dysfunction and neurodegeneration in both SAD and FAD [57]. Meanwhile, alteration of $Ca^{2+}$ signaling may accelerate A$\beta$ formation, thus, a degenerative feed-forward cycle of A$\beta$ formation and dysregulation of $Ca^{2+}$ signaling is formed.

Since 1987, when Khachaturian [58] first proposed the $Ca^{2+}$ hypothesis, a growing body of evidence reveals that dysregulation of $Ca^{2+}$ signaling plays a crucial role in the initiation and development of AD [57, 59, 60]. A rise in the resting level of cytosolic $Ca^{2+}$ in neurons has been reported in both transgenic AD animal models and autopsies of brains from patients who have died from AD, which suggests it is a causal factor in neuronal excitotoxicity, synaptic loss and cell death during the development of AD [61].

### 6.3.2. *Dysregulation of glutamatergic transmission by A$\beta$ in AD*

The extracellular A$\beta$ oligomers have been shown to impair intracellular $Ca^{2+}$ homeostasis by promoting an influx of extracellular $Ca^{2+}$ through the plasma membrane [62, 63]. This can disrupt the plasma membrane $Ca^{2+}$ permeability, particularly by forming $Ca^{2+}$-permeable pores that affect certain $Ca^{2+}$-permeable channels and/or interact with membrane lipids and affect their integrity. In this section, we review dysregulation related to the NMDAR-mediated

$Ca^{2+}$ response in the dendritic spine as observed under AD conditions.

#### 6.3.2.1. *Glutamate excitotoxicity in AD*

An excessive extracellular glutamate concentration will cause a mild, chronic activation of glutamate receptors, especially NMDARs, and lead to neurodegeneration, which is also called excitotoxicity [64]. Several studies have shown that a dysfunctional glutamatergic system in AD could be a critical upstream event, which leads to the overactivation of NMDARs and pathological NMDAR-mediated $Ca^{2+}$ influx into the neuron that, subsequently, triggers apoptotic pathways [65, 66]. An elevated resting level of extracellular glutamate has been found in AD patients [67]. Researchers have linked A$\beta$ to upregulation in glutamate availability and excitotoxicity in AD, suggesting that A$\beta$ is the culprit behind enhanced excitotoxicity, either by increasing glutamate release or by reducing glutamate uptake by transporters [64, 68, 69].

A$\beta$ is suggested to promote glutamate release from presynaptic terminal [70–72]. A$\beta$ may participate and act as a positive regulator in the glutamate release under healthy conditions [73]. In aged neurons, $A\beta_{23-25}$ has been found to augment glutamate release [74] Puzzo, Privitera, Leznik, Fa, Staniszewski, Palmeri and Arancio [75] reported that a picomolar level of $A\beta_{42}$ could enhance LTP by increasing neurotransmitter release, whereas low nanomolar levels of $A\beta_{42}$ lead to synaptic depression. Palop and Mucke [76] proposed a bell-shaped relationship between extracellular A$\beta$ and the synaptic transmission that low and high levels of A$\beta$ depress the postsynaptic transmission while intermediate levels facilitate the presynaptic vesicle release.

In addition, A$\beta$ is suggested to potentiate the release of glutamate from glia cells [77, 78]. A$\beta$ induces excitotoxic levels of glutamate release from astrocytes, which can lead to the activation of extrasynaptic NMDARs and synaptic loss [79, 80]. This may be a result of the abnormal $Ca^{2+}$ signaling in the glial cells observed in AD [81].

In contrast, Aβ oligomers may disturb glutamate clearance mechanisms by affecting the glutamate transporters [82–86]. A number of studies have revealed reduced levels of vesicular glutamate transporters (VGLUT) and excitatory amino acid transporters (EAAT) in AD [87–89]. Moroever, the decreased glutamate transporter activity found in AD is suggested to be associated with this excitotoxicity and neurodegeneration [89]. The lack of transporters and/or reduced activity of these transporters contributes to the increase in glutamate availability in the synaptic cleft and extrasynaptic space [64]. Persistent activation of postsynaptic NMDAR may lead to receptor desensitization and affect synaptic functions, such as synaptic plasticity [65], whereas prolonged extrasynaptic NMDAR promotes neuronal Aβ production [90, 91].

#### 6.3.2.2. *Dysregulation of NMDAR in AD*

Several research studies have shown the colocalization of Aβ oligomers with NMDARs, suggesting that they may directly interact with NMDARs [92–94]. Aβ oligomers are reported to directly interact with NMDARs [95] and activate NR2B-NMDAR, leading to an increase in cytosolic $Ca^{2+}$ levels [96]. The precise underlying mechanisms are yet to be uncovered. Besides, Aβ may also cause a slowed desensitization by disturbing the cellular prion protein-mediated NMDAR activity [97]. Furthermore, the increase of NMDAR activity may inhibit the α-cleavage of APP and promotes the amyloidogenic processing of APP [98–100].

Aβ oligomers may also disturb the distribution of NMDARs [101]. Interestingly, the full-length APP is found to increase surface NR2B-containing NMDAR (but not NR2A-containing one) and decrease its internalization in primary hippocampal neurons [100]. However, following exposure of $A\beta_{25-35}$ and full-length $A\beta_{1-40}$ deposits, a decrease in the number of NMDARs positive cells is observed in the immediate surrounding of $A\beta_{25-35}$ and full-length $A\beta_{1-40}$ deposits in mature hippocampal slice cultures [102]. Pretreatment of cultured hippocampal neurons with $A\beta_{1-42}$ significantly reduces surface expression of NR1, without affecting the total NR1 expression [103].

Treatment of mature hippocampal cells with A$\beta$ oligomers rapidly decreases the expression of surface NMDARs [104]. In cultured cortical neurons, A$\beta_{1-42}$ has been shown to reduce the surface expression of NR1 and NR2B subunits of the whole cell as well as at the synaptic site, without any changes in the total level of synaptic NMDARs (including both internal and surface receptors) [101]. Similarly, a decrease in the surface expression of synaptic NR2B containing NMDAR is observed in primary hippocampal neurons pre-incubated with A$\beta$ oligomers and in APP transgenic mice [93]. In contrast, extrasynaptic NMDARs are not affected by the presence of A$\beta$ oligomers [101].

In addition, A$\beta$-induced surface expression of NMDAR can be restored by the $\gamma$-secretases inhibitor [101]. Taken all together, it is suggested that A$\beta$ may play a role in mediating the trafficking of NMDARs, especially by promoting synaptic NR2B-NMDAR endocytosis. A$\beta$ can bind to 7$\alpha$-nicotinic acetylcholine receptors (nAchRs) with high affinity [105]. Snyder, Nong, Almeida, Paul, Moran, Choi, Nairn, Salter, Lombroso, Gouras and Greengard [101] hypothesized that the A$\beta$-dependent endocytosis of NMDARs is initiated by A$\beta$ binding to the nAchRs, which leads to activation of the protein phosphatase 2B and dephosphorylation and activation of tyrosine phosphatase STEP. The activation of STEP may promote dephosphorylation of the phosphotyrosine residue Tyr1472 of NMDAR, which is located in a region which regulates NMDAR endocytosis and interaction with synaptic scaffolding proteins. Kessels, Nabavi and Malinow [106] showed that A$\beta$ oligomers promote the switch in subunit composition from NR2B- to NR2A-NMDAR, which normally occurs during development.

The loss of synaptic NMDARs may also contribute to the depression of glutamatergic transmission and reductions in memory formation. A$\beta$ is reported to decrease NMDAR-mediated synaptic responses and inhibit NMDAR-dependent LTP [107, 108]. However, to some extent, the internalization of synaptic NMDARs may be a neuroprotective mechanism against the glutamate-induced neurotoxicity and excessive $Ca^{2+}$ influx [103, 109].

### 6.3.3. *Dysregulation of ER* $Ca^{2+}$ *handling in AD*

The ER, as the major $Ca^{2+}$ storage compartment, has been studied with the focus on its effects in $Ca^{2+}$ signaling in AD [28, 110]. Research carried on stabilizing ER $Ca^{2+}$ signaling in AD cells or animal models have offered new strategies for the prevention of disease progression [111–114]. The mechanism causing these alterations in ER $Ca^{2+}$ signaling is not fully understood, while current research is mainly on the disturbing effects from mutations in presenillin or/and deposition of A$\beta$ oligomers.

#### 6.3.3.1. *Disruption of ER* $Ca^{2+}$ *homeostasis by the* A$\beta$ *peptide*

A$\beta$ can affect the ER $Ca^{2+}$ signaling by promoting an influx of extracellular $Ca^{2+}$ through the plasma membrane [62, 63]. For example, there is growing evidence that links overactivation of NMDARs to ER $Ca^{2+}$ dysregulation [115, 116]. A$\beta$ oligomers can induce ER stress through interaction with NMDARs, resulting in a massive $Ca^{2+}$ influx [116]. This will subsequently activate downstream NADPH oxidase (NOX)-mediated superoxide production and, consequently, impair ER $Ca^{2+}$ homeostasis and even trigger ER stress-mediated apoptotic pathway [117]. Costa *et al.* have demonstrated that A$\beta$-induced ER stress can be effectively inhibited by ifenprodil, an antagonist of NR2B subunits, which could be a potential therapeutic strategy for AD [116].

Intracellular A$\beta$ oligomers are also shown to directly disrupt $Ca^{2+}$ handling of intracellular $Ca^{2+}$ stores, especially ERs (Fig 6.3) [118]. They are believed to be produced from APP residing on the ER and in other intracellular compartments, as well as by uptake from the extracellular space [119]. The accumulation of intracellular A$\beta$ oligomers has been observed prior to the presence of extracellular A$\beta$ deposition in the cerebral cortex and hippocampus in AD brains, suggesting that intracellular $A\beta_{42}$-induced $Ca^{2+}$ dysregulation may play a crucial role in early AD pathogenesis [3, 4].

$A\beta_{42}$ oligomer pores can also be formed in the membranes of intracellular organelles and result in $Ca^{2+}$ leakage into the cytosol; the disruption is comparatively mild [62, 118]. Most research has

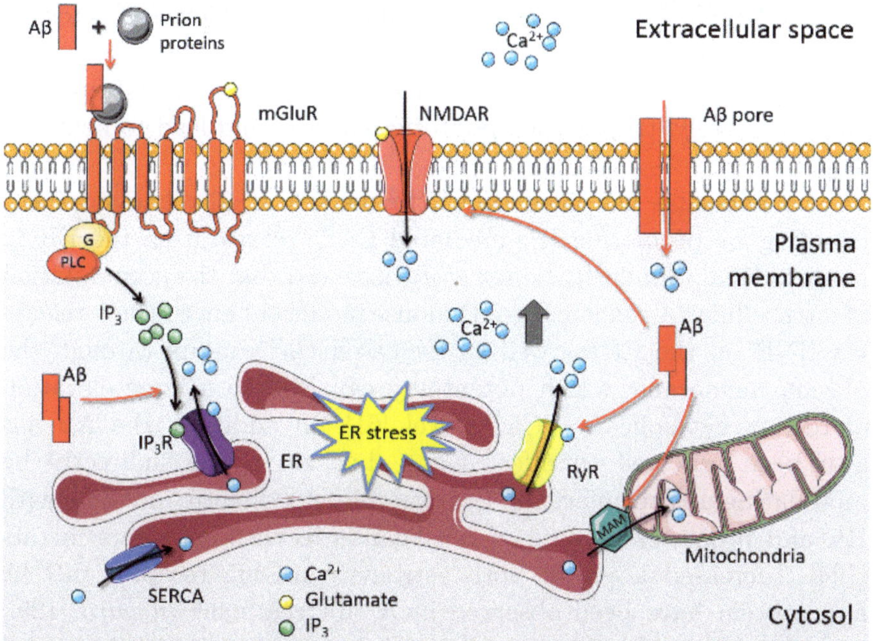


Fig. 6.3. Mechanisms that might contribute to intracellular $Ca^{2+}$ dysregulation in AD. Evidences show that A$\beta$ oligomers may disturb intracellular $Ca^{2+}$ homeostasis and result in ER stress, by disturbing membrane conductance, forming $Ca^{2+}$-permeable pores as well as perturbing ER $Ca^{2+}$ handling. A$\beta$ oligomers may directly modulate activity of $IP_3$Rs and RyRs. A$\beta$ oligomers are also suggested to accelerate the production of $IP_3$, and therefore, induce $IP_3$Rs opening and $Ca^{2+}$ release from ER. Besides, A$\beta$ oligomers may promote ER $Ca^{2+}$ transmission to mitochondria via MAM and lead to mitochondrial dysfunction. Furthermore, alteration of intracellular $Ca^{2+}$ homeostasis may in turn promote the production of A$\beta$ peptides. This figure is produced using Servier Medical Art (http://www.servier.com/Powerpoint-image-bank). Also acknowledgment to [28]: Liang, J., Kulasiri, D., and Samarasinghe, S. (2015) $Ca^{2+}$ dysregulation in the endoplasmic reticulum related to Alzheimer's disease: A review on experimental progress and computational modeling. *Biosystems*, 134, pp. 1–15.

demonstrated that A$\beta$ oligomers enhance $Ca^{2+}$ liberation from the ER, mainly by altering $Ca^{2+}$ release via $IP_3$Rs and RyRs [120–123]. Pereira's group suggested an ER-specific apoptotic pathway trigged by A$\beta$-induced perturbation of ER $Ca^{2+}$ homeostasis [117]. Their research showed that continuous production could induce ER stress, perturb ER $Ca^{2+}$ homeostasis and lead to increased levels of cytosolic $Ca^{2+}$. Together with other A$\beta$-induced alterations, such as increased activity of caspase-3, it would further affect mitochondrial function and, ultimately, trigger the apoptotic pathway and lead to neuronal death.

A$\beta$ exposure has been shown to disturb $IP_3$R-regulated $Ca^{2+}$ handling by promoting $IP_3$-mediated $Ca^{2+}$ release from the ER in neurons [123]. Similarly, Lopez *et al.* observed that the accumulation of intracellular A$\beta$ in a 3xTg-AD mouse model enhanced $Ca^{2+}$ release via $IP_3$R on the ER as well as increased $Ca^{2+}$ influx through the plasma membrane, which potentially can lead to a large elevation of resting cytosolic free $Ca^{2+}$ level [61]. In addition, the A$\beta$ has also been reported to affect $Ca^{2+}$ flux via $IP_3$Rs indirectly by modulating upstream events. For example, A$\beta$ directly interacts with $IP_3$ and promotes binding of $IP_3$ ligands to their receptors in rats [124]. Increased levels of both astrocytic mGlu5 receptor mRNA and protein have been observed in A$\beta$40 treatment *in vitro* [125]. Likewise, Renner *et al.* reported that A$\beta$ oligomers altered the lateral diffusion of mGluR5 within the plasma membrane and stimulated their signaling activity which, therefore, affected downstream $IP_3$ production [126]. A$\beta$ oligomers can bind to cellular prion proteins and active mGluR5, which may also lead to increased production of $IP_3$ [127]. In addition, recent research has reported that intracellular $A\beta_{42}$ oligomers induce $Ca^{2+}$ liberation from the ER via $IP_3$Rs by stimulating PLC-mediated IP3 production in *Xenopus* oocytes [118]. Another study demonstrated that $A\beta_{42}$-induced $Ca^{2+}$ release from ER may mostly depend on activating PLC and, subsequent, IP3 generation rather than by direct interaction with $IP_3$Rs [128].

Similarly, intracellular $A\beta_{42}$ oligomers also disrupt RyR-regulated $Ca^{2+}$ signals. $A\beta_{42}$ oligomers appear to increase the channel open probability of RyRs and change their gating kinetics, leading to enhanced $Ca^{2+}$ liberation from the ER. Shtifman *et al.* demonstrated that A$\beta$ can directly modify and activate RyR type1 by reconstituting RyR in a planar lipid bilayer to study inclusion body myositis, a skeletal muscle disorder, which may share a similar pathogenesis with AD [129]. This is consistent with their earlier study on the inclusion body myositis showing that A$\beta$ overexpression increases RyRs' sensitivity to the agonist caffeine and susceptibility to CICR [130]. In addition, the elevation of RyR expression has also been reported to result in the accumulation of intracellular A$\beta$ oligomers. For example, Supnet *et al.* reported increased levels of RyR type3 in non-TgCRND8 neurons after treatment with A$\beta_{42}$ oligomers, suggesting that A$\beta_{42}$ may potentiate $Ca^{2+}$ release from the ER and disturb intracellular $Ca^{2+}$ homeostasis by directly increasing RyR expression [131, 132].

Moreover, A$\beta$ oligomers also upregulate the MAM function in post-mortem AD brains and APPSwe/Lon mice [133]. Increased expression of MAM-associated proteins and a number of contact sites have been observed in primary hippocampal neurons. This may, consequently, promote $Ca^{2+}$ transmission from the ER to mitochondria and may lead to mitochondrial dysfunction and AD pathology.

Interestingly, the APP intracellular domain (AICD), which is generated together with A$\beta$ by $\gamma$-secretases cleavage, may also be associated with ER signaling. AICD has been suggested to be involved in regulating $IP_3$-mediated $Ca^{2+}$ signaling, although the precise mechanism is unknown [134]. AICD may promote the expression of $IP_3R$ based on its transcriptional activity [135]. Alternatively, in another study, AICD was shown to activate glycogen synthase kinase-3$\beta$ (GSK3$\beta$), a modulator of the $IP_3R$ expression, which may be a potential mechanism by which AICD mediated ER $Ca^{2+}$ signaling [136]. In addition, Oule's *et al.* observed disrupted ER $Ca^{2+}$ homeostasis contributed to the increased expression of RyR levels in APP-overexpressing transgenic mice, suggesting AICD may also mediate RyR expression [137].

#### 6.3.3.2. *Presenilin mutation and its roles in ER* $Ca^{2+}$ *homeostasis*

Presenilins (PSs), a family of highly conserved membrane proteins with multi-transmembrane domains, are primarily located in the ER membrane and can be found in the cell body and in the dendrites of neurons [138]. Missense mutations in the genes encoding presenilin-1

(*PSEN1*) and presenilin-2 (*PSEN2*) account for the majority of the inherited forms of AD, FAD [139, 140]. *PSEN1* mutations are the most common genetic cause and, to date, 185 *PSEN1* mutations have been identified in patients with FAD, listed on the Alzheimer Disease & Frontotemporal Dementia Mutation Database [141]. In contrast, the *PSEN2* mutation is a relatively rare cause of FAD, with only 13 mutations have been identified [141]. Mutations in PS genes are suggested to play a causal role in the cognitive impairment and pathogenesis of FAD [142]. Much research has reported that mutations in both *PSEN1* and *PSEN2* are associated with the dysregulation of $Ca^{2+}$ signaling, disturbance of A$\beta$ production and neuronal death, although the underlying mechanisms are still under investigation. Proposed effects of PS mutation are summarized in Fig 6.4.

However, unlike FAD, causes of SAD are much more complex, involving both genetic and environmental risk factors. Therefore, it remains a great challenge to study the key factors in SAD [143]. Research on microRNA (miRNA) may be helpful to explain the pathogenesis of SAD as well as FAD. miRNAs are small noncoding RNAs, which negatively regulate gene expression at the posttranscriptional level [144]. Dysregulation of several miRNAs have been observed in AD brains and are suggested to relate to AD pathology [145]. Increase in hallmark genes such as *APP* and *PSEN1* observed in SAD brains may be a result of dysregulation in miRNAs which target these genes (reviewed in [146]). Thus, an abnormal level of miRNAs may indirectly affect $Ca^{2+}$ signaling and A$\beta$ production in AD. Therefore, identifying specific miRNAs for AD, especially SAD, will be useful for modeling SAD in the laboratory, which remains underdeveloped. Besides, miRNAs may serve as potential biomarkers in AD, which may contribute to the early diagnosis and therapeutic research [143].

#### 6.3.3.3. *Role of PS as a subunit of $\gamma$-secretases*

The PS protein is synthesized in the ER and can be transported to the plasma membrane or endosomal structure compartments to form

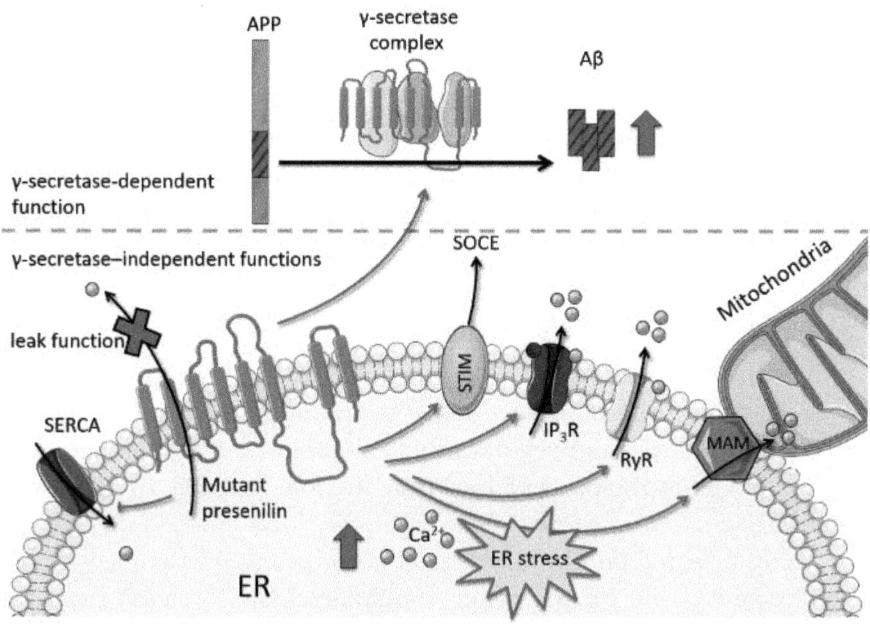


Fig. 6.4. Proposed effects of PS mutation. PS mutation affects both its $\gamma$-secretase-dependent function and $\gamma$-secretase-independent function. Mutant PS disturbs the proteolytic cleavage of the APP and promotes A$\beta$ production. Inside of the neuron, mutant PS may physically associate with ER membrane receptors and SERCA pumps to disturb their functions. Furthermore, PS is proposed to function as an ER $Ca^{2+}$ leak channel. Mutations on it may diminish this function and lead to ER $Ca^{2+}$ dysregulation. This figure is produced using Servier Medical Art (http://www.servier.com/Powerpoint-image-bank). Also acknowledgment to [28]: Liang, J., Kulasiri, D., and Samarasinghe, S. (2015) $Ca^{2+}$ dysregulation in the endoplasmic reticulum related to Alzheimer's disease: A review on experimental progress and computational modeling. *Biosystems 134*, pp. 1–15.

a catalytic subunit of $\gamma$-secretases, together with nicastrin, APH-1 and PEN-2 [147]. $\gamma$-secretases mediate the proteolytic cleavage of the APP via the amyloidogenic pathway to release A$\beta$ peptide fragments. PS mutations are suggested to promote the amyloidogenic processing of APP and can result in the increasing formation of A$\beta_{42}$ relative to A$\beta_{40}$, which has been found both in the fibroblasts of AD patients and animal models [148]. Compared with A$\beta_{40}$, A$\beta_{42}$ is less soluble and is considered to be the most toxic, pathogenic form of A$\beta$, which contributes to synaptic dysfunction, neuronal toxicity and the formation of senile plaques [149]. Likewise, Shen and Kelleher proposed in the presenilin hypothesis that the loss of $\gamma$-secretase cleavage function in PS mutations in AD may lead to increased production of A$\beta_{42}$ relative to A$\beta_{40}$ [139]. The effects of A$\beta$ on ER $Ca^{2+}$ regulation are discussed in later sections.

#### 6.3.3.4. *Roles of PS mutations in ER $Ca^{2+}$ signaling*

As it is independent of the $\gamma$-secretase activity, PS is also suggested to play other physiological roles, including mediation of neurotransmitter release and regulation of intracellular $Ca^{2+}$ homeostasis [147, 148]. This review will focus on its role relating to ER $Ca^{2+}$ signaling.

Early research on fibroblasts from asymptomatic members of AD families linked the PS mutation to enhanced $Ca^{2+}$ release from the ER [111]. Alterations in ER $Ca^{2+}$ signaling have been consistently observed in later research on cells expressing mutant PS [150] or tissues from animal models expressing mutant PS genes [151, 152]. Mutations in PS are also reported to directly attenuate SOCE. Normal presenillins may negatively regulate SOCE, which could explain the attenuated SOCE found in cells with FAD PS mutations [151, 153, 154]. Furthermore, Bojarski *et al.* have demonstrated that FAD mutations in PS affect the cellular levels of STIM1 and STIM2 differently and result in the attenuation of SOCE [155]. They suggest that SOCE in the neurons of FAD patients may be affected by the disturbed expression of STIM genes or proteins. A recent study in mutant presenilin mice showed that downregulation of STIM2 protein impaired the SOCE in mature spines [156].

Inhibited SOCE, together with the reduced ER $Ca^{2+}$ content by PS mutations, would ultimately result in disruption of intracellular $Ca^{2+}$ homeostasis [157]. Increased vulnerability of neurons to excitotoxicity, to mitochondrial impairment and to apoptosis have been observed in PS1 mutant knock-in mice, which are suggested to be associated with enhanced $Ca^{2+}$ release from the ER [158]. Presenilins have also been observed to be highly enriched in MAM [159].

This suggests a potential role for PS in ER–mitochondria $Ca^{2+}$ cross-talk and possible links between FAD mutations in PS to mitochondrial dysfunctions, which could be another key deficit in AD [160, 161]. FAD mutations in PS have shown to upregulate MAM function and ER–mitochondria communication, which may play central role in AD pathogenesis (reviewed in [162]).

The mechanisms by which PS mutations disturb intracellular $Ca^{2+}$ signalling remain controversial. Current research focused on the influence of PS mutations on $Ca^{2+}$ uptake or leak functions by the ER, suggests that the dysregulation is caused by an enhanced expression of ER $Ca^{2+}$ release channels or alterations in the properties of $Ca^{2+}$ channels and pumps by FAD mutants in PS1 or PS2 [163–165]. Enhanced $IP_3$-evoked $Ca^{2+}$ responses related to PS mutation have been observed in several cell lines and brain slices from AD animal models [150, 165–167]. Increased proportions of neurons responding to the $IP_3$ have also been observed in cortical neurons of mutant PS1 knock-in mice [152]. Rather than changing the level of $IP_3Rs$, both FAD mutants, PS1 and PS2, have been observed to strongly enhance the gating activity for $Ca^{2+}$ release of $IP_3Rs$, which might be a potential mechanism for how PS mutations alters ER $Ca^{2+}$ signaling. PSs have been shown to influence the activity of $IP_3Rs$ by physically interacting with them directly to enhance $IP_3R$-evoked $Ca^{2+}$ signaling [164]. Foskett and collaborators found the sensitivity of $IP_3R$ to a low level of cytosolic $IP_3$ increased in cells from FAD patients and PS1-AD mice [164, 165]. They suggested that the mutations in PS activate $IP_3R$ gating and shift it toward a high open-probability burst mode by affecting the gating kinetics through $IP_3R$-PS interaction. Mattson's group observed that PS1 mutations in rat neural cells perturbed ER $Ca^{2+}$ regulation by activating $IP_3Rs$ and this may further sensitize neurons to apoptosis and increase neuronal vulnerability to excitotoxic and ischemic injury [166, 168]. A recent study in PS FAD mouse models demonstrated that reduction of $IP_3R1$ expression can rescue neurons from exaggerated $Ca^{2+}$ release and attenuate $A\beta$ and tau accumulation [169].

Similar to $IP_3R$, PS1 has been reported to physically interact with RyRs and may regulate intracellular $Ca^{2+}$ signaling through modulation of the RyR ion channel gating by protein-protein interactions [163]. Goussakov *et al.* suggested that the reduced CICR threshold for RyR-mediated $Ca^{2+}$ signaling in mutant PS neurons could be responsible for the enhanced RyR-evoked $Ca^{2+}$ response and increased basal $Ca^{2+}$ level [170]. In addition to the alteration of RyRs activity, many researchers have demonstrated the upregulation of RyR expression levels in cells expressing PS mutations [163, 167, 171, 172]. For example, enhanced expression of RyR type 3 with an increased amplitude of caffeine-induced $Ca^{2+}$ release has been observed in rat neural cells with the PS1 mutation and primary hippocampal neurons from PS1 mutant knock-in mice [167]. Likewise, increased $Ca^{2+}$ release through RyRs has been observed with increased RyR expression across all ages in FAD-PS1 knock-in mice, which may be responsible for the dysregulation of cytosolic $Ca^{2+}$ homeostasis [122, 171]. Lee *et al.* reported increased RyR expression in mutant PS2 transgenic mice and demonstrated that PS2 mutations sensitized neurons by enhancing $Ca^{2+}$ release through RyRs and, ultimately, activating the caspase-3-dependent apoptotic pathway [112]. Surprisingly, recent research has found reduced levels of RyR proteins with an unaffected mRNA expression level in hippocampal slices from PS conditional double (PS1 and PS2) knock-out mice [173]. They concluded that PS mediates intracellular $Ca^{2+}$ homeostasis via regulation of RyR expression and function, while the underlying mechanism is undetermined. They suggested that PS regulates RyR expression but not at the mRNA transcription level.

PS also has been reported to physically associate with the SERCA pump and may act as a positive modulator for the normal functioning of this $Ca^{2+}$ pump [38]. La Ferla's group reported that overexpression of wild-type PS (PS1 and PS2) and FAD PS1 mutations could stimulate SERCA activity and lead to $Ca^{2+}$ overfilling in the ER, and the absence of both types of presenilins would affect SERCA function [38]. The opposite results by another group showed that both wild-type and mutant PS2 could reduce SERCA activity [174]. They suggest PS2 is stronger in inhibiting SERCA activity compared with PS1. These different conclusions may be because of the different cell types they used or other unknown mechanisms involved.

6.3.3.5. *PS as a $Ca^{2+}$ leaking channel of the ER*

Besides interaction with ER $Ca^{2+}$ channels and pumps, it has been proposed by Bezprozvanny's group that PSs themselves may act as $Ca^{2+}$-leak channels on the ER membrane [175]. FAD mutations in PSs may diminish this function and lead to ER $Ca^{2+}$ dysregulation. They observed $Ca^{2+}$ permeable channel formation in planar lipid bilayers by wild-type PS1 and PS2 (but not FAD mutants). They suggested PSs might act as low-conductance, passive $Ca^{2+}$ leak channels at the ER membrane in hippocampal neurons, whereas FAD PS mutations might disrupt this function and lead to overfilling of ER $Ca^{2+}$ stores [172, 175]. They also observed that the cysteine point mutations in selected transmembranes (TMs) of PS1 in mouse models led to a loss of its ER $Ca^{2+}$ leak function, suggesting that the hydrophilic water-filled catalytic cavity (between TM6 and TM7) of PS may play a role in forming a low conductance $Ca^{2+}$-permeable pore [147]. In Bezprozvanny's latest review [176], he points out that this hypothesis can be explained and supported by two other recent independent research papers on the crystal structure of an archaeal presenilin homolog PSH1 [177] and a quantitative model of cellular $Ca^{2+}$ signal combined with siRNA perturbations to identify key proteins in $Ca^{2+}$ homeostasis regulation [178].

Increased expression in RyRs has been observed to partially compensate for the loss of the ER $Ca^{2+}$ leak function in PS, suggesting a protective role for RyRs in AD [172, 175]. This is consistent with a previous study carried by the Supnet group, which suggests that the observed upregulation of RyR type3 in neurons of TgCRND8 AD mice may protect neurons against the increased excitotoxicity in AD [131]. The results from their subsequent study have revealed that upregulation of RyR type3 does not affect neuronal health or global $Ca^{2+}$ homeostasis, while knockdown of RyR type3 promotes neuronal death of TgCRND8-cultured cortical neurons, but not in non-transgenic neurons, further supporting the hypothesis of the compensation and neuroprotective roles of RyRs in

AD [132]. This hypothesis could be an explanation for how neurons maintain $Ca^{2+}$ homeostasis and neuronal function against the effects of mutations in the early stages of AD [179].

However, the hypothesis of PS as an ER $Ca^{2+}$ leaking channel has been challenged by several recent studies. Cheung *et al.* argue that PS may not participate in the $Ca^{2+}$ permeability of ER membranes and, instead of a $Ca^{2+}$ overload, FAD mutant PS reduced or had no effect on the $Ca^{2+}$ levels in the ER [164]. Likewise, Shilling *et al.* investigated ER $Ca^{2+}$ filling rates, basal ER $Ca^{2+}$ fill levels and ER $Ca^{2+}$ leak rates from three different cell systems using different indicators, and concluded that the FAD mutant PS2 may not induce ER $Ca^{2+}$ overfilling and PS may not function as an ER $Ca^{2+}$ leak channel [180]. A reduced ER $Ca^{2+}$ load has also been reported for both PS1 and PS2 mutations in other research [174]. Furthermore, a recent study carried by Wu *et al.* showed that in the hippocampus of PS conditional double knock-out mice, the ER $Ca^{2+}$ level stayed at a normal level while the levels of RyR proteins were reduced with impaired RyR function for regulating $Ca^{2+}$ release [173].

### 6.3.4. *ER alteration may influence A$\beta$ production*

Alteration of intracellular $Ca^{2+}$ homeostasis shows conflicting effects on the production of A$\beta$ peptides. Sustained elevation in the cytosolic $Ca^{2+}$ level has been reported to accelerate A$\beta$ formation, which suggests that reduction of $Ca^{2+}$ released from the ER could be a potential neuroprotective strategy against A$\beta$ peptide neurotoxicity [121, 181, 182]. Increased expression and activity of RyRs have been found in the brains of patients in the early stages of AD and also in mice models, and this may contribute to the progression of AD pathogenesis [120, 183, 184]. Early research showed that the elevation of intracellular $Ca^{2+}$ levels after treatment with $Ca^{2+}$ ionophore led to increased A$\beta$ generation in human cell lines [181]. Later, the same group reported that RyR activation by caffeine resulted in an elevation of intracellular $Ca^{2+}$ level, which further enhanced the release of A$\beta$ from APP in HEK293 cells [185]. Likewise, a study of tissues from post-mortem AD brains has shown the loss of RyRs is correlated with amyloid deposition [120]. Recent research on

Tg2576 mouse models has revealed that reduced RyR-mediated $Ca^{2+}$ released after treatment with dantrolene, a RyR inhibitor, could reduce both intracellular and extracellular A$\beta$ loads and slow down the loss of memory and cognitive function. This possibly resulted from the decreasing Thr-668-dependent APP phosphorylation and $\beta$- and $\gamma$-secretases activities by interaction with $Ca^{2+}$[137]. However, there are conflicting results from other groups who reported that knock-out of RyR type3 neurons or ones after long-term exposure to the RyR inhibitor, dantrolene, presented an increased amyloid pathologic condition, suggesting a potential neuroprotective role for RyR [132, 172].

Likewise, $IP_3R$ and its regulation of cytosolic $Ca^{2+}$ signaling may also affect the generation of A$\beta$. Cheung *et al.* reported reduced A$\beta$ production in $IP_3R$-deficient cells, which suggested that amyloidogenic processing of APP might be closely related to $IP_3R$-mediated $Ca^{2+}$ release [164]. Pierrot *et al.* suggested increased $Ca^{2+}$ liberation from the ER alone may not be enough to trigger intraneuronal A$\beta$ production, which requires influx of extracellular $Ca^{2+}$ and a sustained increase in cytosolic $Ca^{2+}$ concentration [186]. This is consistent with results from a recent study, which investigated the interactions between genes of different $Ca^{2+}$ channels in the AD pathway from a genetic perspective using an online database [187]. In their study, Gene *RYR3* and *CACNA1C* have shown to interact significantly with each other, suggesting this interaction may result in $Ca^{2+}$ dysregulation, leading to enhanced A$\beta$ production and deposition.

However, there are studies which show that an increase in the cytosolic $Ca^{2+}$ level reduces A$\beta$ production. The role of $IP_3R$ in A$\beta$ generation is controversial. Severely reduced levels of both $IP_3$ and $IP_3Rs$ have been observed in AD brains, and the loss of $IP_3Rs$ is suggested to be important for amyloid production and deposition by altering $Ca^{2+}$ homeostasis [188]. Other studies have demonstrated that the elevation of intracellular $Ca^{2+}$ can decrease amyloid production and promote the anti-amyloidogenic processing of APP [189, 190]. For example, an increased level of cytosolic $Ca^{2+}$ followed treatment with high levels of thapsigargin, an inhibitor of SERCA, leading to decreased A$\beta$ formation [189]. This is consistent with results from a recent study, which suggests increased $Ca^{2+}$ influx via SOCE significantly by overexpression of STIM1 can actually inhibit A$\beta$ secretion [191]. These contradicting observations may result from different experimental conditions, such as cell lines chosen, or different detection methods [188].

In addition, Dreses-Werringloer *et al.* identified a gene called $Ca^{2+}$ homeostasis modulator 1 (*CALHM1*), which was mostly expressed in the hippocampus and was located in AD linkage regions *190*. It encodes a multipass transmembrane glycoprotein, CALHM1, the majority of which is localized in the ER and plasma membrane. CALHM1 may function as a component of a cerebral $Ca^{2+}$ channel, which controls cytosolic $Ca^{2+}$ concentrations and APP processing [190, 192]. A recent study showed *CALHM1* P86L polymorphism modulates A$\beta$ levels in the cerebrospinal fluid of cognitively healthy individuals at risk of AD, which further confirmed that CALHM1 is involved in AD pathogenesis [193]. Dreses-Werringloer *et al.* also proposed that a P86L polymorphism in the *CALHM1* gene may interfere with CALHM1-mediated $Ca^{2+}$ permeability to increase A$\beta$ production, and may be associated with SAD [190]. This notion is supported by subsequent research carried out in populations from countries such as China [194], Spain [195] and Iran [196]. However, it has been challenged by studies carried out in Caucasian [197] and Asian populations [198], which were unable to show an association between *CALHM1* polymorphism and the risk of AD. Lambert *et al.* suggested that *CALHM1* P86L polymorphism may modulate the age of onset of AD by interacting with the $\varepsilon$4 allele of the apolipoprotein E gene, rather than be an independent risk genetic determinant of AD [199].

## 6.4. Modeling $Ca^{2+}$ Dynamics in Dendritic Spines

A number of models for the $Ca^{2+}$ dynamics in the dendritic spines have been proposed for different types of neurons. Some of them have been modified and extended to study $Ca^{2+}$-mediated downstream events, such as the induction of different types of synaptic plasticity [200]. These $Ca^{2+}$ models generally contain similar components, such as $Ca^{2+}$ influxes from extracellular spaces and/or release from an internal $Ca^{2+}$ source, $Ca^{2+}$ extrusion from the cytosol and interactions between $Ca^{2+}$ and various intracellular proteins. In multi-compartment models, which take the geometry of the spine head and dendrites into consideration, and simulate the whole model as a series of compartments, among which $Ca^{2+}$ ions diffuse. In this section, we review the existing models related to the NMDAR-mediated $Ca^{2+}$ response in dendritic spines regarding these components.

### 6.4.1. *Simulation of membrane $Ca^{2+}$ influx from extracellular spaces*

#### 6.4.1.1. *Simulation of glutamate release, uptake and diffusion*

There are a number of models for presynaptic neurons that contain components related to glutamate vesicle release with different levels of detail [201, 202] that are beyond the scope of this thesis. In contrast, most models of the postsynaptic $Ca^{2+}$ response to the presynaptic stimulation assume the release of glutamate from a vesicle is instantaneous and, thus, uses a constant glutamate concentration per release [203]. Other models use simple equations to represent glutamate release from a vesicle as a non-instantaneous process [204].

The spatiotemporal profiles of the glutamate concentration in the synaptic cleft and the extrasynaptic space following release are due to their diffusion and uptake. The transient of a single pulse of glutamate at certain locations can be simply modeled as a process with a fast rise time and a slow decay time or a square-wave pulse [205]. Complex models make an approximation of the synaptic cleft and the extracellular space. They then simulate the glutamate concentration using diffusion equations, which are either partial differential equations or analytical solutions for them. The synaptic cleft can be approximated as a two-dimensional disc or a three-dimensional flat cylinder whereas the extrasynaptic space is a spherical, isotropic porous medium containing obstacles [203, 204]. The diffusion space is divided into different small diffusion compartments, such as concentric cylindrical or spherical shells [206] and, in the diffusion compartments of the extrasynaptic space, the gluatmate concentration with time is simulated as a diffusion–reaction process, which accounts for the diffusion of glutamate as well as the uptake of glutamate by glutamate transporters in the adjacent glia. The kinetics of binding glutamate to different glutamate transporters can be modeled as a chain reaction [204], or follow the proposed schemes that contain more reaction details [207].

#### 6.4.1.2. *Simulation of kinetics of glutamate receptors*

The $Ca^{2+}$ influx depends on the number of opened $Ca^{2+}$ channels as well as the difference between the membrane potential and the reversal potential. Markov's kinetic models are mostly used to describe the state-transition of AMPAR and NMDAR by glutamates. The receptor between the different states includes a closed state, a ligand-binding state (single-bound and double-bound, respectively), a desensitization state and an open state, which are determined by various rate constants. The values of these rate constants are estimated by fitting the specific model schemes to the single-channel experimental data. For example, there are six-state [208], seven-state [209] and twelve-state [210] models for AMPAR, and a five-state model [211] for NMDAR. In these models, the fully bound receptors by two glutamates can directly transition to the open state. In contrast, there are a number of NMDAR models that take into consideration the conformational changes of the fully bound receptors by including two desensitised states before moving to the open state [212–216]. Most NMDAR models only consider the glutamate binding steps and exclude the co-agonist binding steps by fitting the data obtained from experiments under the condition of a saturated concentration of glycine [217].

The relief of the $Mg^{2+}$ blockage of NMDAR is generally treated as an instantaneous event and the model proposed by Jahr and Stevens [218] is widely used for simulating it as a membrane voltage-dependent process. Moreover, there are few models that simulate the relief of $Mg^{2+}$ blockage as a slow component and include it into their kinetic models [219, 220].

### 6.4.2. *Simulation of $Ca^{2+}$ release from ER*

The ER as an important compartment appears in numerous synaptic and neuronal models, because of its crucial role of internal $Ca^{2+}$ source and its ubiquitous distribution across the neuron. At the synaptic level, models mostly include $Ca^{2+}$ ions flux via proteins (such as NMDARs and sodium $Ca^{2+}$ exchangers) on the post synaptic plasma membrane and $Ca^{2+}$-mediated membrane electrical activity as the aspect of contribution from $Ca^{2+}$ dynamics [221–223]. In spite of this, several synaptic models include the ER as an internal $Ca^{2+}$ source to study its response to synaptic $Ca^{2+}$ fluctuation during the signal transduction process in synapses [224–226]. In contrast, the ER as an internal compartment is generally included in the most models to study its contribution to cytosolic events such as intracellular propagation of $Ca^{2+}$ waves [227, 228].

In $Ca^{2+}$ signaling models of the ER, the uptake mechanism is basically governed by modeling the gating behavior of its two $Ca^{2+}$ channels, $IP_3R$ and RyR. At the molecular level, different kinetic schemas have been developed for these channels.

Models of the $IP_3R$-regulated $Ca^{2+}$ release have been the focus of much research on account of their significance to $Ca^{2+}$ oscillations and synaptic plasticity. Modeling of the $IP_3R$ shows more complexity in its dynamics as a result of its activation and inhibition, and the involvement of a G-protein-coupled receptor-induced $IP_3$-production process. The first model for $IP_3$-induced $Ca^{2+}$ release was proposed by De Young and Keizer, which assumed that there were three equivalent and independent subunits of $IP_3R$ involved and all of them have to be in particular conducting states to open the channel [229]. Similar to De Young and Keizer's four-state model, there were other models that used Markov kinetic schemes to describe transitions between states. They are based on the assumption that each subunit of the $IP_3R$ receptor has different multiple transitional states; for example, three [230] and seven transitional states [225]. A well-known simplified two-state model based on De Young and Keizer's model was developed by Li and Rinzel, who used Hodgkin–Huxley style equations based on the assumption of instantaneous activation following slow inactivation of $IP_3R$ by $Ca^{2+}$ [231]. Many follow-up models have been developed based on, or influenced by, the De Young–Keizer model or the Li–Rinzel simplified model. In addition, the models have been also developed based on other different assumptions such as sequential binding behavior [232], receptor conformation changes [233], adaptation of $IP_3R$ and saturable fashion of $IP_3$ [233], or built to investigate different aspects, such as open probability of $IP_3R$ on steady states [234, 235] and $IP_3$ metabolism [236]. The development of $IP_3R$ models has been reviewed in detail by Sneyd and Falcke [237].

RyR-regulated $Ca^{2+}$ signaling has been ignored in some models of intracellular $Ca^{2+}$ dynamics, since $IP_3R$ itself is enough to be responsible for CICR as it is sensitive to both $IP_3$ and $Ca^{2+}$ ions. However, it is important to include RyRs to simulate the $Ca^{2+}$ signaling more realistically. For example, by using Chemesis, a neural simulation software, Blackwell and Kotaleski demonstrated that the involvement of RyR channels affects the model simulation behavior of $Ca^{2+}$ wave propagation [238]. The models, which included RyR-regulated $Ca^{2+}$ signaling, have considered that both RyR and $IP_3R$ contribute to $Ca^{2+}$ release from the ER and the CICR is governed by RyR [200, 239]. Similar strategies to $IP_3R$ modeling have been applied in RyR models. For example, there are models of RyRs using Markov kinetic schemes which describe the transitions between activated and inactivated states or multi-states [240–242].

### 6.4.3. *Simulation of $Ca^{2+}$ pumps and membrane leakage*

$Ca^{2+}$ pumps include plasma membrane $Ca^{2+}$ pumps, which remove $Ca^{2+}$ from the cytosol, and SERCA pumps, which are the only contributors to the elevation of ER $Ca^{2+}$ levels. These pumps have been present in cytosolic $Ca^{2+}$ signaling models and have been shown to modify $Ca^{2+}$ transient and oscillation. In most models, they have been modeled as unidirectional pumps that are activated by the increase in cytosolic $Ca^{2+}$ and defined by a simple Hill equation or Michaelis–Menten kinetics with a Hill coefficient [243–246]. A few complex models have taken into consideration the multiple transition states of SERCA activation [247, 248] or the effect of $Ca^{2+}$-buffering on SERCA [249]. A leak influx of $Ca^{2+}$ through the membrane into the cytosol is always used to balance the $Ca^{2+}$ eflux through the pumps in the resting state. The membrane leakage is modeled as either a constant leak influx [246] or as a process with a constant rate and in proportion to the concentration gradient across the membrane [250].

### 6.4.4. *Simulation of $Ca^{2+}$ buffering and diffusion*

$Ca^{2+}$ ions also interact with a variety of buffer proteins, such as calmodulin and calbindin, and fluorescent dyes in both the cytosolic space and the ER lumina [26]. In order to study the buffering effects on $Ca^{2+}$ dynamics and make the simulation more reliable, various assumptions relating to buffering behavior have been made and added to the $Ca^{2+}$ signaling model. It would obviously make the model more complex and larger to include details of binding events, such as the effects from different buffers. Therefore, simplified theories for $Ca^{2+}$ buffering have been applied. For example, the excess buffer approximation (EBA) and the rapid buffer approximation (RBA) are two important simplifications, which assume the mobile buffers are exceeded and cannot be saturated, and binding of $Ca^{2+}$ ions to buffers is much faster compared to the changes in cytosolic $Ca^{2+}$ concentration, respectively [205]. Besides, diffusion of $Ca^{2+}$ ions has been included in some models which consider the effect of concentration gradients from geometry facts [226]. The intracellular compartment is divided into multiple pools and movement of $Ca^{2+}$ ions between two adjacent subcompartments is driven by Brownian motion [205]. Radial and/or longitudinal diffusion will be considered based on model assumptions. Either deterministic or stochastic strategies can be applied to simulation of diffusion processing by using partial differential equations [251].

### 6.4.5. *Simulation software*

There are a great deal of simulation tools available to date, and they can be generally classified into two categories: general purpose software and biological simulation software (reviewed in [251–253])

Table 6.1. Simulation tools. [28].

(A) *General purpose simulation tools*

| Tool | Freeware | Website |
|---|---|---|
| MATLAB | | http://www.mathworks.com |
| Mathematica | | http://www.wolfram.com/mathematica |
| Python | √ | https://www.python.org |
| XPPAUT | √ | http://www.math.pitt.edu/~bard/xpp/xpp.html |

(B) *Biological specific simulation tools*

| Tool | Simulation algorithm | | Interface for model development | Freeware | Website |
|---|---|---|---|---|---|
| | Deterministic | Stochastic | | | |
| CellDesigner | √ | | GUI | √ | a |
| COPASI | √ | | GUI | √ | b |
| VCell | √ | √ | GUI | √ | c |
| MCell | | √ | CellBlender | √ | d |
| Smoldyn | | √ | CLI | √ | e |
| Chemesis | √ | | GENESIS | √ | f |
| MOOSE | √ | √ | Python | √ | g |
| NEURON | √ | √* | GUI;CLI | √ | h |
| NeuroRD | | √ | xml files | √ | i |

a. http://www.celldesigner.org.
b. http://www.copasi.org.
c. https://vcell.org/.
d. http://www.mcell.org.
e. http://www.smoldyn.org.
f. http://krasnow1.gmu.edu/CENlab/software.html.
g. http://moose.ncbs.res.in.
h. http://www.neuron.yale.edu.
i. http://krasnow1.gmu.edu/CENlab/software.html.
*Channel states and total conductance can be deterministic or stochastic [253].

(Table 6.1). Tools that belong to the former category, such as MATLAB, XPPAUT, Mathematica and Python, require the user have some programming capability. The power of these tools provides the user great opportunities to analyze the model freely. Tools that fall into the second category include common biological simulation tools such as CellDesigner, VCell, MCell and COPASI, as well as specific software packages for neuron simulation, such as GENESIS and NeuroRD. These tools have built-in capabilities for simulation of biological processes and are widely used by biologists. Most of these software have friendly graphical user interfaces (GUIs) for model development, while others have command-line interfaces (CLIs) or require scripting languages (Fig. 6.5).

## 6.5. Modeling Intracellular Signaling Related to AD

Modeling intracellular signaling in computational neuroscience has been used to study complex temporal and spatial characteristics of nerve systems for decades. Well-established modeling methods of $Ca^{2+}$ signaling provide good foundations for modeling studies of AD and $Ca^{2+}$-related disease. However, only a few models have been developed to simulate or partly simulated the dysregulation of $Ca^{2+}$ signaling in neurons related to AD. There has been some progress in the mathematical modeling of A$\beta$-induced alterations as a function of ion channels in neuronal plasma membranes. For example, based on the observation that A$\beta$ induced a voltage-dependent decrease in membrane conductance in rat hippocampal neurons, Good and Murphy developed a mathematical model of A$\beta$-mediated blockages of fast-inactivating $K^+$ channels [254]. The simulation results show good agreement with experiments in the aspect of stimulation responses and $Ca^{2+}$ buffering capacity. Through simulation they hypothesized that blockages of a fast-inactivating $K^+$ current by A$\beta$ as an early event plays a crucial role in the neurotoxicity of A$\beta$ in AD and could lead to increased intracellular $Ca^{2+}$ levels and membrane excitability resulting in neuronal neurotoxicity and, eventually, death. A recently published model based on Good and Murphy's model has been developed to further explore A$\beta$-induced blockages of fast-inactivating $K^+$ channels as well as to study the A$\beta$-induced increase of membrane conductance during A$\beta$-neuron interaction in a short timescale [255]. This model included the voltage-clamp and the membrane conductance mechanisms, which enabled it to simulate A$\beta$-neuron interactions under various experimental conditions, make comparisons with available data and

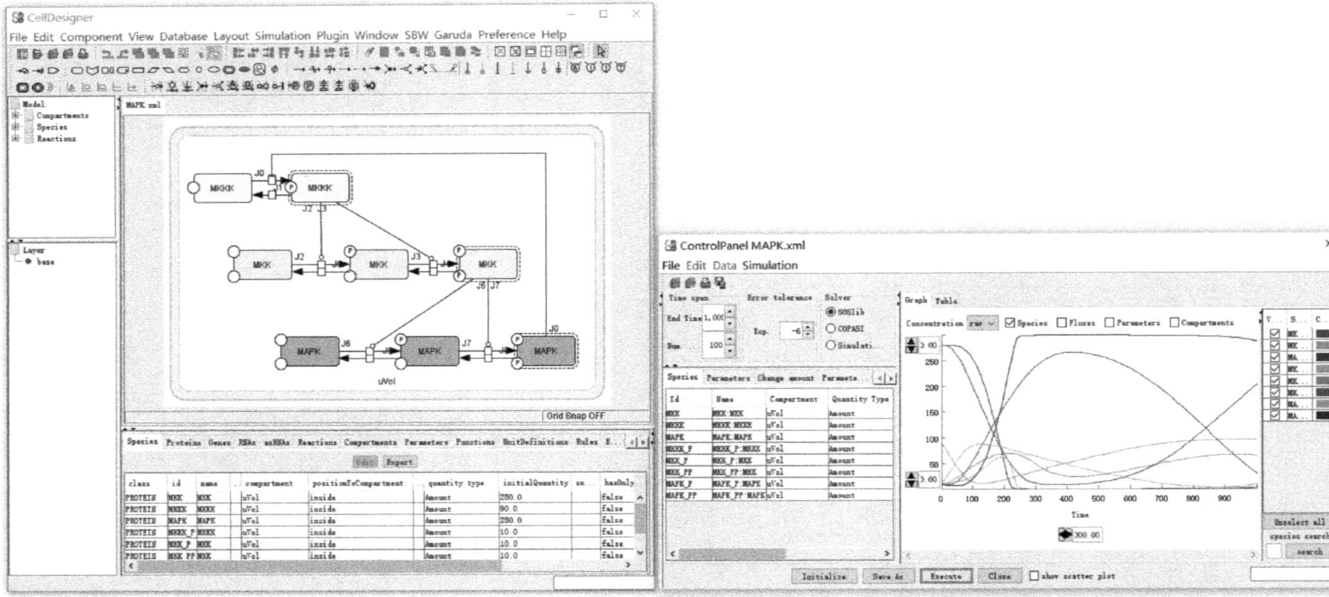


(a)

Fig. 6.5. Interface captures of simulation tools. Biological simulation tools (e.g., CellDesigner) with user-friendly GUI allow users to draw biochemical networks of their model (a, left) and run models with the simulation control panel (a, right). Neuronal simulation tools (e.g., NEURON (b)) provide several built-in builders, with which users can set the precise parameters of neuronal specific features for their model. (a) CellDesigner interface capture for a sample model (MAPK.xml). (b) NEURON interface capture of a sample demonstration of a hippocampal pyramidal neuron. All sample models in this figure are pre-assembled in these software. Also acknowledgment to [28]: Liang, J., Kulasiri, D. and Samarasinghe S. (2015) $Ca^{2+}$ dysregulation in the endoplasmic reticulum related to Alzheimer's disease: A review on experimental progress and computational modeling. *Biosystems* 134, pp. 1–15.

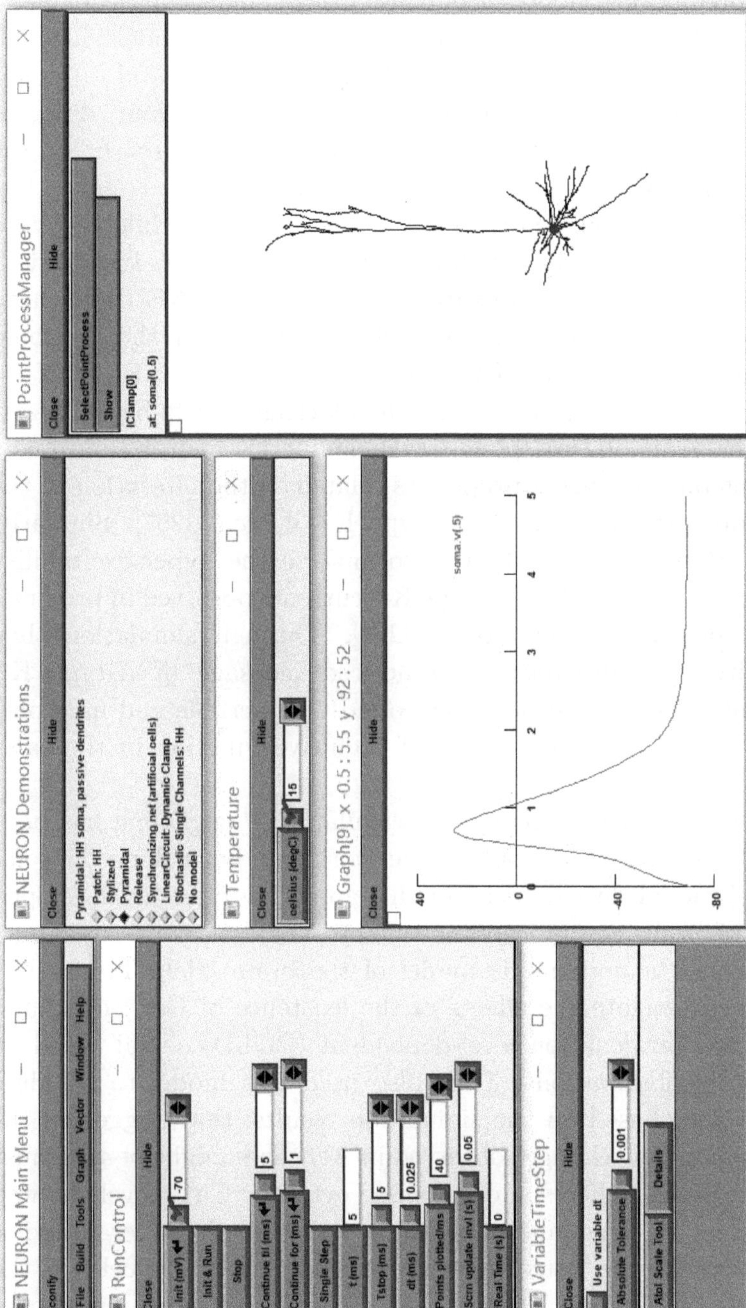


(b)

Fig. 6.5. *(Continued)*

generate predictions. Good's group also utilized mathematical kinetic analysis methods to study the potential mechanism underlying A$\beta$-induced G-protein activation [256]. By using experimental rate data to examine the proposed underlying mechanisms, four different possible mechanisms have been tested to find the best fit of the rate expression with experimental data. Based on the results, they suggested that A$\beta$ oligomers activate the G-protein while excessive A$\beta$ aggregation might inhibit further GTPase activity. In addition, models using simulation software packages such as NEURON have been constructed to simulate the effects of A$\beta$ on the neuronal membrane. For example, Kidd and Sattelle used NEURON to construct a model neuron to study the blockage effects by A$\beta$ on A-type $K^+$ currents of *Drosophila* larval cholinergic neurons and made predictions on their firing properties related to the alteration of the steady-state properties of the A-type $K^+$ current [257]. Similarly, Morse *et al.* also used NEURON to analyze the hyper-excitability induced by A$\beta$ via blocking A-type $K^+$ currents observed in proximal dendrites in AD animal models [258]. Through simulation they hypothesized the disruption A$\beta$-induced blockage of A-type $K^+$ currents at oblique branches may be most vulnerable and may play an important role in the decline of cognitive function in the early stage of AD.

Although dysregulation of intracellular $Ca^{2+}$ signaling has been widely studied experimentally, little work has been done on computational model development in this area. Tiveci *et al.* developed a model of brain energy metabolism by including $Ca^{2+}$ dynamics to an existing hemodynamic model of the brain [259]. This model is able to investigate the effects of the existence of $Ca^{2+}$ dynamics on the blood oxygenation level-dependent (BOLD) signal based on experimental observations. They also used this model to simulate AD cases and, based on the simulation results, they suggested the cerebral blood flow changes observed in AD cases might be the cause of negative BOLD effects and increased cytosolic $Ca^{2+}$ level. As well as modeling $Ca^{2+}$ dynamics in neurons, there are also several models of $Ca^{2+}$ signaling in astrocytes, the predominant glial cells in the central nervous system. Toivari *et al.* developed a computational stochastic model of $Ca^{2+}$ signaling in rat cortical astrocytes [260]. This model was used to study the effects of $A\beta_{25-35}$ and transmitters on intracellular $Ca^{2+}$ signalling in astrocytes. The simulation results were consistent with their experimental findings on $A\beta_{25-35}$ and transmitters induced $Ca^{2+}$ transient. Another example is a model constructed by Riera *et al.* This model was based on spontaneous $Ca^{2+}$ oscillations in astrocytes and used it to simulate $Ca^{2+}$ dynamics in wild-type (WT) and Tg2576 mice [261]. Through simulations, they suggested an increased $Ca^{2+}$ influx from the extracellular space in APP transgenic mice might activate astrocytes and play a crucial role in intracellular $Ca^{2+}$ dysregulation. Furthermore, De Caluwé and Dupont recently developed a simple theoretical model to study the positive feedback loop between $A\beta$ and cytosolic $Ca^{2+}$[262]. Despite this being a simplified model, which has excluded detailed molecular mechanisms, it was still able to reveal a bistable switch between 'healthy' and the 'pathological' states, depending on the concentration of $A\beta$ and cytosolic $Ca^{2+}$. A model which concentrated more on the effects of $A\beta$ on $Ca^{2+}$ signalling in a single neuron is yet to be developed.

## References

[1] Toescu E.C. and Verkhratsky A. (2007). The importance of being subtle: small changes in calcium homeostasis control cognitive decline in normal aging. *Aging Cell*, *6*(3), pp. 267–273.

[2] Zundorf G. and Reiser G. (2011). Calcium dysregulation and homeostasis of neural calcium in the molecular mechanisms of neurodegenerative diseases provide multiple targets for neuroprotection. *Antioxid Redox Signal*, *14*(7), pp. 1275–1288.

[3] Gouras G.K. *et al.* (2000). Intraneuronal Abeta42 accumulation in human brain. *Amer. J. Pathology*, *156*(1), pp. 15–20.

[4] Oddo S., Caccamo A., Smith I.F., Green K.N. and LaFerla F.M. (2006). A dynamic relationship between intracellular and extracellular pools of Abeta. *The Amer. J. Pathology*, *168*(1), pp. 184–194.

[5] Ghosh A. and Greenberg M.E. (1995). Calcium signaling in neurons: molecular mechanisms and cellular consequences. *Science (N.Y.)*, *268*(5208), p. 239.

[6] Sabatini B.L., Maravall M. and Svoboda K. (2001). $Ca^{2+}$ signaling in dendritic spines. *Curr. Opinion neurobiol.*, *11*(3), pp. 349–356.

[7] Pinheiro P.S. and Mulle C. (2008). Presynaptic glutamate receptors: physiological functions and mechanisms of action. *Nature Rev. Neurosci.*, *9*(6), pp. 423–436.

[8] Nakanishi S. (1992). Molecular diversity of glutamate receptors and implications for brain function. *Science*, *258*(5082), pp. 597–603.
[9] Fellin T. *et al.* (2004). Neuronal synchrony mediated by astrocytic glutamate through activation of extrasynaptic NMDA receptors. *Neuron*, *43*(5), pp. 729–743.
[10] Danbolt N.C. (2001). Glutamate uptake. *Prog. Neurobiol.*, *65*(1), pp. 1–105.
[11] Vandenberg R.J. and Ryan R.M. (2013). Mechanisms of glutamate transport. *Physiological Reviews*, *93*(4), pp. 1621–1657.
[12] Sheldon A.L. and Robinson M.B. (2007). The role of glutamate transporters in neurodegenerative diseases and potential opportunities for intervention. *Neurochem. Int.*, *51*(6–7), pp. 333–355.
[13] Michaelis E.K. (1998). Molecular biology of glutamate receptors in the central nervous system and their role in excitotoxicity, oxidative stress and aging. *Prog. Neurobiol.*, *54*(4), pp. 369–415.
[14] Dingledine R., Borges K., Bowie D. and Traynelis S.F. (1999). The glutamate receptor ion channels. *Pharmacological Rev.*, *51*(1), pp. 7–62.
[15] Engelman H.S. and MacDermott A.B. (2004). Presynaptic ionotropic receptors and control of transmitter release. *Nature Rev. Neurosci.*, *5*(2), pp. 135–145.
[16] Petralia R.S. *et al.* (2010). Organization of NMDA receptors at extrasynaptic locations. *Neuroscience*, *167*(1), pp. 68–87.
[17] Collingridge G.L., Isaac J.T. and Wang Y.T. (2004). Receptor trafficking and synaptic plasticity. *Nature Rev. Neurosci.*, *5*(12), pp. 952–962.
[18] Bloodgood B.L. and Sabatini B.L. (2007). Nonlinear regulation of unitary synaptic signals by CaV(2.3) voltage-sensitive calcium channels located in dendritic spines. *Neuron*, *53*(2), pp. 249–260.
[19] Furukawa H., Singh S.K., Mancusso R. and Gouaux E. (2005). Subunit arrangement and function in NMDA receptors. *Nature*, *438*(7065), pp. 185–192.
[20] Cull-Candy S.G. (2001). NMDA receptors. In *eLS*. John Wiley & Sons, Ltd.
[21] Papouin T. *et al.* (2012). Synaptic and extrasynaptic NMDA receptors are gated by different endogenous coagonists. *Cell*, *150*(3), pp. 633–646.
[22] Petralia R.S. (2012). Distribution of Extrasynaptic NMDA Receptors on Neurons. *Sci. World J.*, *2012*, Article 267120, p. 11.
[23] Cheng D. *et al.* (2006). Relative and absolute quantification of postsynaptic density proteome isolated from rat forebrain and cerebellum. *Mol. Cellular Proteomics*, *5*(6), pp. 1158–1170.
[24] Hardingham G.E. and Bading H. (2010). Synaptic versus extrasynaptic NMDA receptor signalling: implications for neurodegenerative disorders. *Nature rev. Neurosci.*, *11*(10), pp. 682–696.
[25] Mayer M.L., Westbrook G.L. and Guthrie P.B. (1984). Voltage-dependent block by $Mg^{2+}$ of NMDA responses in spinal cord neurones. *Nature*, *309*(5965), pp. 261–263.

[26] Berridge M.J., Bootman M.D., and Roderick H.L. (2003). Calcium signalling: dynamics, homeostasis and remodelling. *Nature Rev. Mole. Cell Biol.*, *4*(7), pp. 517–529.
[27] Verkhratsky A. (2005). Physiology and pathophysiology of the calcium store in the endoplasmic reticulum of neurons. *Physiological Rev.*, *85*(1), pp. 201–279.
[28] Liang J., Kulasiri D. and Samarasinghe S. (2015). $Ca^{2+}$ dysregulation in the endoplasmic reticulum related to Alzheimer's disease: A review on experimental progress and computational modeling. *Biosystems*, *134*, pp. 1–15.
[29] Bezprozvanny I. (2005). The inositol 1,4,5-trisphosphate receptors. *Cell Calcium*, *38*(3–4), pp. 261–272.
[30] Mattson M.P. *et al.* (2000). Calcium signaling in the ER: its role in neuronal plasticity and neurodegenerative disorders. *Trends Neurosci.*, *23*(5), pp. 222–229.
[31] Foskett J.K., White C., Cheung K.H. and Mak D.O. (2007). Inositol trisphosphate receptor $Ca^{2+}$ release channels. *Physiological Rev.*, *87*(2), pp. 593–658.
[32] Taufiq Ur R., Skupin A., Falcke M. and Taylor C.W. (2009). Clustering of InsP3 receptors by InsP3 retunes their regulation by InsP3 and $Ca^{2+}$. *Nature*, *458*(7238), pp. 655–659.
[33] Berridge M.J. (1997). Elementary and global aspects of calcium signalling. *J. Physiol.*, *499*(Pt 2), pp. 291–306.
[34] Rahman T. and Taylor C.W. (2009). Dynamic regulation of $IP_3$ receptor clustering and activity by $IP_3$. *Channels*, *3*(4), pp. 226–232.
[35] Galeotti N., Vivoli E., Bartolini A., and Ghelardini C. (2008). A gene-specific cerebral types 1, 2, and 3 RyR protein knockdown induces an antidepressant-like effect in mice. *J. Neurochem.*, *106*(6), pp. 2385–2394.
[36] Adasme T. *et al.* (2011). Involvement of ryanodine receptors in neurotrophin-induced hippocampal synaptic plasticity and spatial memory formation. *Proc. Natl. Acad. Sci. USA*, *108*(7), pp. 3029–3034.
[37] Hidalgo C., Bull R., Behrens M.I., and Donoso P. (2004). Redox regulation of RyR-mediated $Ca^{2+}$ release in muscle and neurons. *Biological Res.*, *37*(4), pp. 539–552.
[38] Green K.N. *et al.* (2008). SERCA pump activity is physiologically regulated by presenilin and regulates amyloid beta production. *J. Cell. Biol.*, *181*(7), pp. 1107–1116.
[39] Wuytack F., Raeymaekers L. and Missiaen L. (2002). Molecular physiology of the SERCA and SPCA pumps. *Cell Calcium*, *32*(5–6), pp. 279–305.
[40] Buchholz J.N., Pottorf J.N., Vanterpool C.K., Behringer E.J. and Duckles S.P. (2012). Calcium regulation in neuronal function with advancing age: limits of homeostasis. In T. Nagata (Ed.), *Senescence* (pp. 531–558). InTech.
[41] Stutzmann G.E. and Mattson M.P. (2011). Endoplasmic reticulum Ca(2+) handling in excitable cells in health and disease. *Pharmacol Rev.*, *63*(3), pp. 700–727.

[42] Manjarres I.M., Rodriguez-Garcia A., Alonso M.T. and Garcia-Sancho J. (2010). The sarco/endoplasmic reticulum Ca(2+) ATPase (SERCA) is the third element in capacitative calcium entry. *Cell Calcium*, *47*(5), pp. 412–418.
[43] Elbaz Y. and Schuldiner M. (2011). Staying in touch: the molecular era of organelle contact sites. *Trends Biochem. Sci.*, *36*(11), pp. 616–623.
[44] Soboloff J., Rothberg B.S., Madesh M., and Gill D.L. (2012). STIM proteins: dynamic calcium signal transducers. *Nature Rev. Mol. Cell Biol.*, *13*(9), pp. 549–565.
[45] Rowland A.A. and Voeltz G.K. (2012). Endoplasmic reticulum-mitochondria contacts: function of the junction. *Nature Rev. Mol. Cell biol.*, *13*(10), pp. 607–625.
[46] Hayashi T., Rizzuto R., Hajnoczky G. and Su T.P. (2009). MAM: more than just a housekeeper. *Trends Cell Biol.*, *19*(2), pp. 81–88.
[47] Goedert M. and Spillantini M.G. (2006). A century of Alzheimer's disease. *Science (N.Y.)*, *314*(5800), pp. 777–781.
[48] Thies W. and Bleiler L. (2013). World Alzheimer report 2016. *Alzheimer's & Dement: J. Alzheimer's Asso.*, *9*(2), pp. 208–245.
[49] Mattson M.P. (2004). Pathways towards and away from Alzheimer's disease. *Nature*, *430*(7000), pp. 631–639.
[50] Francis P.T., Palmer A.M., Snape M. and Wilcock G.K. (1999). The cholinergic hypothesis of Alzheimer's disease: a review of progress. *J. Neurology, Neurosurgery Psychiatry*, *66*(2), pp. 137–147.
[51] Spires-Jones T.L., Stoothoff W.H., de Calignon A., Jones P.B. and Hyman B.T. (2009). Tau pathophysiology in neurodegeneration: a tangled issue. *Trends Neurosci.*, *32*(3), pp. 150–159.
[52] Selkoe D.J. (1991). The molecular pathology of Alzheimer's disease. *Neuron*, *6*(4), pp. 487–498.
[53] Hardy J. and Higgins G. (1992). Alzheimer's disease: the amyloid cascade hypothesis. *Science (N.Y.)*, *256*(5054), pp. 184–185.
[54] Hardy J. and Selkoe D.J. (2002). The amyloid hypothesis of Alzheimer's disease: progress and problems on the road to therapeutics. *Science (N.Y.)*, *297*(5580), pp. 353–356.
[55] Selkoe D.J. and Hardy J. (2016). The amyloid hypothesis of Alzheimer's disease at 25 years. *EMBO Mol. Medi.*, *8*(6), pp. 595–608.
[56] Van Dam D. and De Deyn P.P. (2006). Drug discovery in dementia: the role of rodent models. *Nat. Rev. Drug Discov.*, *5*(11), pp. 956–970.
[57] Berridge M.J. (2010). Calcium hypothesis of Alzheimer's disease. *Pflugers Archiv: European J. Physiol.*, *459*(3), pp. 441–449.
[58] Khachaturian Z.S. (1987). Hypothesis on the regulation of cytosol calcium concentration and the aging brain. *Neurobiol. Aging*, *8*(4), pp. 345–346.
[59] LaFerla F.M. (2002). Calcium dyshomeostasis and intracellular signalling in alzheimer's disease. *Nature Rev. Neurosci.*, *3*(11), pp. 862–872.
[60] Berridge M.J. (2013). Dysregulation of neural calcium signaling in Alzheimer disease, bipolar disorder and schizophrenia. *Prion*, *7*(1), pp. 2–13.

[61] Lopez J.R. *et al.* (2008). Increased intraneuronal resting [$Ca^{2+}$] in adult Alzheimer's disease mice. *J Neurochem*, *105*(1), pp. 262–271.
[62] Demuro A. *et al.* (2005). Calcium dysregulation and membrane disruption as a ubiquitous neurotoxic mechanism of soluble amyloid oligomers. *J. Biol. Chem.*, *280*(17), pp. 17294–17300.
[63] Kawahara M., Ohtsuka I., Yokoyama S., Kato-Negishi M. and Sadakane Y. (2011). Membrane incorporation, channel formation, and disruption of calcium homeostasis by Alzheimers Amyloid protein. *Int. J. Alzheimer's Dis.*, *2011*, 304583.
[64] Danysz W. and Parsons C.G. (2012). Alzheimer's disease, beta-amyloid, glutamate, NMDA receptors and memantine–searching for the connections. *British J. Pharmacology*, *167*(2), pp. 324–352.
[65] Li S. *et al.* (2009). Soluble oligomers of amyloid $\beta$ protein facilitate hippocampal long-term depression by disrupting neuronal glutamate uptake. *Neuron*, *62*(6), pp. 788–801.
[66] Hardingham G.E. and Bading H. (2010). Synaptic versus extrasynaptic NMDA receptor signalling: implications for neurodegenerative disorders. *Nat. Rev. Neurosci.*, *11*(10), pp. 682–696.
[67] Talantova M. *et al.* (2013). A$\beta$ induces astrocytic glutamate release, extrasynaptic NMDA receptor activation, and synaptic loss. *Proc. Acad. Sci. USA*, *110*(27), pp. E2518–E2527.
[68] Ondrejcak T. *et al.* (2010). Alzheimer's disease amyloid beta-protein and synaptic function. *Neuromolecular Med.*, *12*(1), pp. 13–26.
[69] Butterfield D.A. and Pocernich C.B. (2003). The glutamatergic system and Alzheimer's disease. *CNS Drugs*, *17*(9), pp. 641–652.
[70] Bobich J.A., Zheng Q. and Campbell A. (2004). Incubation of nerve endings with a physiological concentration of $A\beta_{1-42}$ activates CaV2.2(N-Type)-voltage operated calcium channels and acutely increases glutamate and noradrenaline release. *J. Alzheimer's Dis.*, *6*(3), pp. 243–255.
[71] Kabogo D., Rauw G., Amritraj A., Baker G., and Kar S. (2010). $\beta$-amyloid -related peptides potentiate K+-evoked glutamate release from adult rat hippocampal slices. *Neurobiolo. Aging*, *31*(7), pp. 1164–1172.
[72] Chin J.H., Ma L., MacTavish D. and Jhamandas J.H. (2007). Amyloid $\beta$ protein modulates glutamate-mediated neurotransmission in the rat basal forebrain: involvement of presynaptic neuronal nicotinic acetylcholine and metabotropic glutamate receptors. *J. Neurosci.*, *27*(35), pp. 9262–9269.
[73] Abramov E. *et al.* (2009). Amyloid-beta as a positive endogenous regulator of release probability at hippocampal synapses. *Nat. Neurosci.*, *12*(12), pp. 1567–1576.
[74] Arias C., Arrieta I. and Tapia R. (1995). $\beta$-Amyloid peptide fragment 25–35 potentiates the calcium-dependent release of excitatory amino acids from depolarized hippocampal slices. *J. Neurosci. Res.*, *41*(4), pp. 561–566.
[75] Puzzo D. *et al.* (2008). Picomolar amyloid-$\beta$ positively modulates synaptic plasticity and memory in hippocampus. *J. Neurosci.*, *28*(53), pp. 14537–14545.

[76] Palop J.J. and Mucke L. (2010). Amyloid-$\beta$ induced neuronal dysfunction in Alzheimer's Disease: from synapses toward neural networks. *Nature Neurosci.*, *13*(7), pp. 812–818.
[77] Noda M., Nakanishi H. and Akaike N. (1999). Glutamate release from microglia via glutamate transporter is enhanced by amyloid-beta peptide. *Neurosci.*, *92*(4), pp. 1465–1474.
[78] Orellana J.A. *et al.* (2011). Amyloid $\beta$-induced death in neurons involves glial and neuronal hemichannels. *J. Neurosci.*, *31*(13), pp. 4962–4977.
[79] Talantova M. *et al.* (2013). A$\beta$ induces astrocytic glutamate release, extrasynaptic NMDA receptor activation, and synaptic loss. *Proc. Natl. Acad. Sci. USA*, *110*(27), pp. E2518–E2527.
[80] Barger S.W. and Basile A.S. (2001). Activation of microglia by secreted amyloid precursor protein evokes release of glutamate by cystine exchange and attenuates synaptic function. *J. neurochem.*, *76*(3), pp. 846–854.
[81] Kuchibhotla K.V., Lattarulo C.R., Hyman B.T. and Bacskai B.J. (2009). Synchronous hyperactivity and intercellular calcium waves in astrocytes in Alzheimer mice. *Science (N.Y.)*, *323*(5918), pp. 1211–1215.
[82] Harris M.E. *et al.* (1995). Beta-Amyloid peptide-derived, oxygen-dependent free radicals inhibit glutamate uptake in cultured astrocytes: implications for Alzheimer's disease. *Neuroreport*, *6*(14), pp. 1875–1879.
[83] Harris M.E. *et al.* (1996). Amyloid $\beta$ Peptide (25–35) Inhibits $Na^+$-Dependent Glutamate Uptake in Rat Hippocampal Astrocyte Cultures. *J. Neurochem.*, *67*(1), pp. 277–286.
[84] Parpura-Gill A., Beitz D. and Uemura E. (1997). The inhibitory effects of $\beta$-amyloid on glutamate and glucose uptakes by cultured astrocytes. *Brain Res.*, *754*(1–2), pp. 65–71.
[85] Matos M., Augusto E., Oliveira C. and Agostinho P. (2008). Amyloid-beta peptide decreases glutamate uptake in cultured astrocytes: involvement of oxidative stress and mitogen-activated protein kinase cascades. *Neurosci.*, *156*(4), pp. 898–910.
[86] Fernández-Tomé P., Brera B., Arévalo M.A.-A. and de Ceballos M.A.L. (2004). $\beta$-Amyloid 25-35 inhibits glutamate uptake in cultured neurons and astrocytes: modulation of uptake as a survival mechanism. *Neurobiolo. Dis.*, *15*(3), pp. 580–589.
[87] Kirvell S.L., Esiri M. and Francis P.T. (2006). Down-regulation of vesicular glutamate transporters precedes cell loss and pathology in Alzheimer's disease. *J. Neurochem.*, *98*(3), pp. 939–950.
[88] Masliah E. *et al.* (2000). Abnormal glutamate transport function in mutant amyloid precursor protein transgenic mice. *Exp. Neurology*, *163*(2), pp. 381–387.
[89] Masliah E., Hansen L., Alford M., Deteresa R. and Mallory M. (1996). Deficient glutamate Transport is associated with neurodegeneration in Alzheimer's disease. *Ann. Neurology*, *40*(5), pp. 759–766.
[90] Bordji K., Becerril-Ortega J. and Buisson A. (2011). Synapses, NMDA receptor activity and neuronal Abeta production in Alzheimer's disease. *Rev. Neurosci.*, *22*(3), pp. 285–294.

[91] Bordji K., Becerril-Ortega J., Nicole O. and Buisson A. (2010). Activation of extrasynaptic, but not synaptic, NMDA receptors modifies amyloid precursor protein expression pattern and increases amyloid-$\beta$ production. *J. Neurosci.*, *30*(47), pp. 15927-15942.
[92] Texidó L., Martín-Satué M., Alberdi E., Solsona C. and Matute C. (2011). Amyloid $\beta$ peptide oligomers directly activate NMDA receptors. *Cell Calcium*, *49*(3), pp. 184–190.
[93] Dewachter I. *et al.* (2009). Deregulation of NMDA-receptor function and down-stream signaling in APP[V717I] transgenic mice. *Neurobiol. Aging*, *30*(2), pp. 241–256.
[94] Alberdi E. *et al.* (2010). Amyloid beta oligomers induce Ca2+ dysregulation and neuronal death through activation of ionotropic glutamate receptors. *Cell Calcium*, *47*(3), pp. 264–272.
[95] De Felice *et al.* (2007). A$\beta$ oligomers induce neuronal oxidative stress through an N-methyl-D-aspartate receptor-dependent mechanism that is blocked by the Alzheimer drug memantine. *J. Biol. Chem.*, *282*(15), pp. 11590–11601.
[96] Ferreira I.L. *et al.* (2012). Amyloid beta peptide 1–42 disturbs intracellular calcium homeostasis through activation of GluN2B-containing N-methyl-D-aspartate receptors in cortical cultures. *Cell Calcium*, *51*(2), pp. 95–106.
[97] You H. *et al.* (2012). A$\beta$ neurotoxicity depends on interactions between copper ions, prion protein, and N-methyl-D-aspartate receptors. *Proc. Natl. Acad. Sci. USA*, *109*(5), pp. 1737–1742.
[98] Kamenetz F. *et al.* (2003). APP processing and synaptic function. *Neuron*, *37*(6), pp. 925–937.
[99] Lesné S. *et al.* (2005). NMDA receptor activation inhibits $\alpha$-secretase and promotes neuronal amyloid-$\beta$ production. *J. Neurosci.*, *25*(41), pp. 9367–9377.
[100] Hoe H.-S. *et al.* (2009). The effects of amyloid precursor protein on postsynaptic composition and activity. *J. Biol. Chem.*, *284*(13), pp. 8495–8506.
[101] Snyder E.M. *et al.* (2005). Regulation of NMDA receptor trafficking by amyloid-beta. *Nat. Neurosci.*, *8*(8), pp. 1051–1058.
[102] Johansson S., Radesäter A.-C., Cowburn R.F., Thyberg J. and Luthman J. (2006). Modelling of amyloid $\beta$-peptide induced lesions using roller-drum incubation of hippocampal slice cultures from neonatal rats. *Exp. Brain Res.*, *168*(1–2), pp. 11–24.
[103] Goto Y., Niidome T. Akaike A., Kihara T., and Sugimoto H. (2006). Amyloid $\beta$-peptide preconditioning reduces glutamate-induced neurotoxicity by promoting endocytosis of NMDA receptor. *Biochem. Biophy. Res. Commun.*, *351*(1), pp. 259–265.
[104] Lacor P.N. *et al.* (2007). A$\beta$ oligomer-induced aberrations in synapse composition, shape, and density provide a molecular basis for loss of connectivity in Alzheimer's disease. *J. Neurosci.*, *27*(4), pp. 796–807.
[105] Wang H.-Y. *et al.* (2000). *$\beta$-Amyloid1–42 binds to $\alpha$7 nicotinic acetylcholine receptor with* affinity implications for Alzheimer's Disease Pathology. *J. Biol. Chem.*, *275*(8), pp. 5626–5632.

[106] Kessels H.W., Nabavi S. and Malinow R. (2013). Metabotropic NMDA receptor function is required for beta-amyloid-induced synaptic depression. *Proc. Natl. Acad. Sci. USA*, *110*(10), pp. 4033–4038.

[107] Chen Q.-S., Wei W.-Z., Shimahara T. and Xie C.-W. (2002). Alzheimer amyloid $\beta$-peptide inhibits the late phase of long-term potentiation through calcineurin-dependent mechanisms in the hippocampal dentate gyrus. *Neurobiol. Learning Memory*, *77*(3), pp. 354–371.

[108] Li S. *et al.* (2011). Soluble A$\beta$ oligomers inhibit long-term potentiation through a mechanism involving excessive activation of extrasynaptic NR2B-containing NMDA receptors. *J. Neurosci.*, *31*(18), pp. 6627–6638.

[109] Chuang D., Gao X.-M. and Paul S.M. (1992). N-methyl-D-aspartate exposure blocks glutamate toxicity in cultured cerebellar granule cells. *Mol. Pharmacology*, *42*(2), pp. 210–216.

[110] Popugaeva E. and Bezprozvanny I. (2013). Role of endoplasmic reticulum $Ca^{2+}$ signaling in the pathogenesis of Alzheimer disease. *Frontiers in Mole. Neurosci.*, *6*, p. 29.

[111] Ito E. *et al.* (1994). Internal $Ca^{2+}$ mobilization is altered in fibroblasts from patients with Alzheimer disease. *Proc. Natl. Acad. Sci. USA*, *91*(2), pp. 534–538.

[112] Lee S. *et al.* (2006). PS2 mutation increases neuronal cell vulnerability to neurotoxicants through activation of caspase-3 by enhancing of ryanodine receptor-mediated calcium release. *FASEB J.*, *20*(1), pp. 151–153.

[113] Thibault O., Gant J.C. and Landfield P.W. (2007). Expansion of the calcium hypothesis of brain aging and Alzheimer's disease: minding the store. *Aging Cell*, *6*(3), pp. 307–317.

[114] Chakroborty S. *et al.* (2012). Stabilizing ER $Ca^{2+}$ channel function as an early preventative strategy for Alzheimer's disease. *PLoS One*, *7*(12), p. e52056.

[115] Kelly B.L. and Ferreira A. (2006). beta-Amyloid-induced dynamin 1 degradation is mediated by N-methyl-D-aspartate receptors in hippocampal neurons. *J. Biol. Chem.*, *281*(38), pp. 28079–28089.

[116] Costa R.O. *et al.* (2012). Endoplasmic reticulum stress occurs downstream of GluN2B subunit of N-methyl-D-aspartate receptor in mature hippocampal cultures treated with amyloid-$\beta$ oligomers. *Aging Cell*, *11*(5), pp. 823–833.

[117] Ferreiro E., Resende R., Costa R., Oliveira C.R. and Pereira C.M.F. (2006). An endoplasmic-reticulum-specific apoptotic pathway is involved in prion and amyloid-beta peptides neurotoxicity. *Neurobiol. of Dis.*, *23*(3), pp. 669–678.

[118] Demuro A. and Parker I. (2013). Cytotoxicity of intracellular abeta42 amyloid oligomers involves $Ca^{2+}$ release from the endoplasmic reticulum by stimulated production of inositol trisphosphate. *J. Neurosci.*, *33*(9), pp. 3824–3833.

[119] Mohamed A. and Posse de Chaves E. (2011). A$\beta$ Internalization by neurons and Glia. *Int. J. Alzheimer's Dis.*, *2011*, 127984.

[120] Kelliher M. *et al.* (1999). Alterations in the ryanodine receptor calcium release channel correlate with Alzheimer's disease neurofibrillary and beta-amyloid pathologies. *Neurosci.*, *92*(2), pp. 499–513.
[121] Ferreiro E., Oliveira C.R., and Pereira C. (2004). Involvement of endoplasmic reticulum $Ca^{2+}$ release through ryanodine and inositol 1,4,5-triphosphate receptors in the neurotoxic effects induced by the amyloid-beta peptide. *J. Neurosci. Res.*, *76*(6), pp. 872–880.
[122] Stutzmann G.E. *et al.* (2006). Enhanced ryanodine receptor recruitment contributes to $Ca^{2+}$ disruptions in young, adult, and aged Alzheimer's disease mice. *J. Neurosci.*, *26*(19), pp. 5180–5189.
[123] Schapansky J., Olson K., Van Der Ploeg R. and Glazner G. (2007). NF-$\kappa$B activated by ER calcium release inhibits A$\beta$-mediated expression of CHOP protein: enhancement by AD-linked mutant presenilin 1. *Exp. Neurology*, *208*(2), pp. 169–176.
[124] Cowburn R.F., Wiehager B. and Sundström E. (1995). $\beta$-amyloid peptides enhance binding of the calcium mobilising second messengers, inositol(1,4,5)trisphosphate and inositol-(1,3,4,5)tetrakisphosphate to their receptor sites in rat cortical membranes. *Neurosci. Lett.*, *191*(1–2), pp. 31–34.
[125] Casley C.S. *et al.* (2009). Up-regulation of astrocyte metabotropic glutamate receptor 5 by amyloid-beta peptide. *Brain Res.*, *1260*, pp. 65–75.
[126] Renner M. *et al.* (2010). Deleterious effects of amyloid beta oligomers acting as an extracellular scaffold for mGluR5. *Neuron*, *66*(5), pp. 739–754.
[127] Um J.W. *et al.* (2013). Metabotropic glutamate receptor 5 is a coreceptor for Alzheimer abeta oligomer bound to cellular prion protein. *Neuron*, *79*(5), pp. 887–902.
[128] Jensen L.E. *et al.* (2013). Alzheimer's Disease-associated peptide Aß42 mobilises ER $Ca^{2+}$ via $InsP_3R$-dependent and -independent mechanisms. *Frontiers Mol. Neurosci.*, *6*, 36.
[129] Shtifman A. *et al.* (2010). Amyloid-beta protein impairs $Ca^{2+}$ release and contractility in skeletal muscle. *Neurobiol. Aging*, *31*(12), pp. 2080–2090.
[130] Christensen R.A., Shtifman A., Allen P.D., Lopez J.R. and Querfurth H.W. (2004). Calcium dyshomeostasis in beta-amyloid and tau-bearing skeletal myotubes. *J. Biol. Chem.*, *279*(51), pp. 53524–53532.
[131] Supnet C., Grant J., Kong H., Westaway D. and Mayne M. (2006). Amyloid-beta-(1–42) increases ryanodine receptor-3 expression and function in neurons of TgCRND8 mice. *J. Biol. Chem.*, *281*(50), pp. 38440–38447.
[132] Supnet C., Noonan C., Richard K., Bradley J. and Mayne M. (2010). Up-regulation of the type 3 ryanodine receptor is neuroprotective in the TgCRND8 mouse model of Alzheimer's disease. *J. Neurochem*,*112*(2), pp. 356–365.
[133] Hedskog L. *et al.* (2013). Modulation of the endoplasmic reticulum-mitochondria interface in Alzheimer's disease and related models. *Proc. Natl. Acad. Sci. USA*, *110*(19), pp. 7916–7921.

[134] Smith I.F., Green K.N. and LaFerla F.M. (2005). Calcium dysregulation in Alzheimer's disease: Recent advances gained from genetically modified animals. *Cell Calcium*, *38*(3–4), pp. 427–437.

[135] Kasri N.N. *et al.* (2006). Up-regulation of inositol 1,4,5-trisphosphate receptor type 1 is responsible for a decreased endoplasmic-reticulum $Ca^{2+}$ content in presenilin double knock-out cells. *Cell Calcium*, *40*(1), pp. 41–51.

[136] Ryan K.A. and Pimplikar S.W. (2005). Activation of GSK-3 and phosphorylation of CRMP2 in transgenic mice expressing APP intracellular domain. *J. Cell Biol.*, *171*(2), pp. 327–335.

[137] Oule's B. *et al.* (2012). Ryanodine receptor blockade reduces amyloid-beta load and memory impairments in Tg2576 mouse model of Alzheimer disease. *J. Neurosci.*, *32*(34), pp. 11820–11834.

[138] Brunkan A.L. and Goate A.M. (2005). Presenilin function and $\gamma$- secretase activity. *J. Neurochem.*, *93*(4), pp. 769–792.

[139] Shen J. and Kelleher R.J., 3rd. (2007). The presenilin hypothesis of Alzheimer's disease: evidence for a loss-of-function pathogenic mechanism. *Proc. Natl. Acad. Sci. USA*, *104*(2), pp. 403–409.

[140] Supnet C. and Bezprozvanny I. (2011). Presenilins function in ER calcium leak and Alzheimer's disease pathogenesis. *Cell Calcium*, *50*(3), pp. 303–309.

[141] Cruts M., Theuns J. and Van Broeckhoven C. (2012). Locus-specific mutation databases for neurodegenerative brain diseases. *Human Mutation*, *33*(9), pp. 1340–1344.

[142] Saura C.A. *et al.* (2004). Loss of presenilin function causes impairments of memory and synaptic plasticity followed by age-dependent neurodegeneration. *Neuron*, *42*(1), pp. 23–36.

[143] Young J.E. and Goldstein L.S.B. (2012). Alzheimer's disease in a dish: promises and challenges of human stem cell models. *Human Mol. Genetics*, *21*(R1), pp. R82–R89.

[144] He L. and Hannon G.J. (2004). MicroRNAs: small RNAs with a big role in gene regulation. *Nat. Rev. Genet.*, *5*(7), pp. 522–531.

[145] Van den Hove D.L. *et al.* (2014). Epigenetically regulated microRNAs in Alzheimer's disease. *Neurobiol. Aging*, *35*(4), pp. 731–745.

[146] Maes O.C., Chertkow H.M., Wang E. and Schipper H.M. (2009). MicroRNA: implications for Alzheimer disease and other human CNS disorders. *Curr. Genomics*, *10*(3), pp. 154–168.

[147] Nelson O. *et al.* (2011). Mutagenesis mapping of the presenilin 1 calcium leak conductance pore. *J. Biol. Chem.*, *286*(25), pp. 22339–22347.

[148] De Strooper B., Iwatsubo T. and Wolfe M.S. (2012). Presenilins and gamma-secretase: structure, function, and role in Alzheimer disease. *Cold Spring Harb. Perspect. Med.*, *2*(1), p. a006304.

[149] Hardy J. and Selkoe D.J. (2002). The amyloid hypothesis of Alzheimer's disease: progress and problems on the road to therapeutics. *Science*, *297*(5580), pp. 353–356.

[150] Leissring M.A., Paul B.A., Parker I., Cotman C.W. and LaFerla F.M. (1999). Alzheimer's presenilin-1 mutation potentiates inositol 1,4,5-trisphosphate-mediated calcium signaling in Xenopus oocytes. *J. Neurochem.*, *72*(3), pp. 1061–1068.

[151] Leissring M.A. *et al.* (2000). Capacitative calcium entry deficits and elevated luminal calcium content in mutant presenilin-1 knockin mice. *J. Cell Biol*, *149*(4), pp. 793–798.

[152] Stutzmann G.E., Caccamo A., LaFerla F.M. and Parker I. (2004). Dysregulated $IP_3$ signaling in cortical neurons of knock-in mice expressing an Alzheimer's-linked mutation in presenilin1 results in exaggerated $Ca^{2+}$ signals and altered membrane excitability. *J. Neurosci.*, *24*(2), pp. 508–513.

[153] Herms J. *et al.* (2003). Capacitive calcium entry is directly attenuated by mutant presenilin-1, independent of the expression of the amyloid precursor protein. *J. Biol. Chem.*, *278*(4), pp. 2484–2489.

[154] Bojarski L., Herms J. and Kuznicki J. (2008). Calcium dysregulation in Alzheimer's disease. *Neurochem. Int.*, *52*(4–5), pp. 621–633.

[155] Bojarski L. *et al.* (2009). Presenilin-dependent expression of STIM proteins and dysregulation of capacitative $Ca^{2+}$ entry in familial Alzheimer's disease. *Biochim. Biophys. Acta (BBA) Molecular Cell Res.*, *1793*(6), pp. 1050–1057.

[156] Sun S. *et al.* (2014). Reduced synaptic STIM2 expression and impaired store-operated calcium entry cause destabilization of mature spines in mutant presenilin mice. *Neuron*, *82*(1), pp. 79–93.

[157] Giacomello M. *et al.* (2005). Reduction of $Ca^{2+}$ stores and capacitative $Ca^{2+}$ entry is associated with the familial Alzheimer's disease presenilin-2 T122R mutation and anticipates the onset of dementia. *Neurobiol Dis*, *18*(3), pp. 638–648.

[158] Guo Q. *et al.* (1999). Increased vulnerability of hippocampal neurons to excitotoxic necrosis in presenilin-1 mutant knock-in mice. *Nature Med.*, *5*(1), pp. 101–106.

[159] Area-Gomez E. *et al.* (2009). Presenilins are enriched in endoplasmic reticulum membranes associated with mitochondria. *Amer. J. Pathology*, *175*(5), pp. 1810–1816.

[160] Zampese E. *et al.* (2011). Presenilin 2 modulates endoplasmic reticulum (ER)–mitochondria interactions and $Ca^{2+}$ cross-talk. *Proc. Natl. Acad. Sci. USA*, *108*(7), pp. 2777–2782.

[161] Area-Gomez E. *et al.* (2012). Upregulated function of mitochondria-associated ER membranes in Alzheimer disease. *EMBO J.*, *31*(21), pp. 4106–4123.

[162] Schon E.A. and Area-Gomez E. (2013). Mitochondria-associated ER membranes in Alzheimer disease. *Mol. Cell. Neurosci.*, *55*(0), pp. 26–36.

[163] Rybalchenko V., Hwang S.Y., Rybalchenko N. and Koulen P. (2008). The cytosolic N-terminus of presenilin-1 potentiates mouse ryanodine receptor single channel activity. *Int. J. Biochem. Cell Biol.*, *40*(1), pp. 84–97.

[164] Cheung K.H. *et al.* (2008). Mechanism of $Ca^{2+}$ disruption in Alzheimer's disease by presenilin regulation of InsP3 receptor channel gating. *Neuron, 58*(6), pp. 871–883.
[165] Cheung K.H. *et al.* (2010). Gain-of-function enhancement of $IP_3$ receptor modal gating by familial Alzheimer's disease-linked presenilin mutants in human cells and mouse neurons. *Science Signaling, 3*(114), p. ra22.
[166] Guo Q. *et al.* (1997). Alzheimer's presenilin mutation sensitizes neural cells to apoptosis induced by trophic factor withdrawal and amyloid beta-peptide: involvement of calcium and oxyradicals. *J. Neurosci., 17*(11), pp. 4212–4222.
[167] Chan S.L., Mayne M., Holden C.P., Geiger J.D. and Mattson M.P. (2000). Presenilin-1 mutations increase levels of ryanodine receptors and calcium release in PC12 cells and cortical neurons. *J. Biol. Chem., 275*(24), pp. 18195–18200.
[168] Mattson M.P., Zhu H., Yu J. and Kindy M.S. (2000). Presenilin-1 mutation increases neuronal vulnerability to focal ischemia *in vivo* and to hypoxia and glucose deprivation in cell culture: involvement of perturbed calcium homeostasis. *The J. Neurosci, 20*(4), pp. 1358–1364.
[169] Shilling D. *et al.* (2014). Suppression of InsP3 receptor-mediated $Ca^{2+}$ signaling alleviates mutant presenilin-linked familial Alzheimer's disease pathogenesis. *J. Neurosci., 34*(20), pp. 6910–6923.
[170] Goussakov I., Miller M.B. and Stutzmann G.E. (2010). NMDA-mediated Ca(2+) influx drives aberrant ryanodine receptor activation in dendrites of young Alzheimer's disease mice. *J. Neurosci., 30*(36), pp. 12128–12137.
[171] Stutzmann G.E., *et al.* (2007). Enhanced ryanodine-mediated calcium release in mutant PS1-expressing Alzheimer's mouse models. *Ann. New York Acad. Sci., 1097*, pp. 265–277.
[172] Zhang H., Sun S., Herreman A., De Strooper B. and Bezprozvanny I. (2010). Role of presenilins in neuronal calcium homeostasis. *J. Neurosci., 30*(25), pp. 8566–8580.
[173] Wu B., Yamaguchi H., Lai F.A. and Shen J. (2013). Presenilins regulate calcium homeostasis and presynaptic function via ryanodine receptors in hippocampal neurons. *Proc. Natl. Acad. Sci. USA, 110*(37), pp. 15091–15096.
[174] Brunello L. *et al.* (2009). Presenilin-2 dampens intracellular Ca2+ stores by increasing $Ca^{2+}$ leakage and reducing $Ca^{2+}$ uptake. *J. Cell. Mol. Med., 13*(9b), pp. 3358–3369.
[175] Tu H.P. *et al.* (2006). Presenilins form ER $Ca^{2+}$ leak channels, a function disrupted by familial Alzheimer's disease-linked mutations. *Cell, 126*(5), pp. 981–993.
[176] Bezprozvanny I. (2013). Presenilins and calcium signaling–systems biology to the rescue. *Science Signaling, 6*(283), p. pe24.
[177] Li X. *et al.* (2013). Structure of a presenilin family intramembrane aspartate protease. *Nature, 493*(7430), pp. 56–61.

[178] Bandara S. Malmersjo S. and Meyer T. (2013). Regulators of calcium homeostasis identified by inference of kinetic model parameters from live single cells perturbed by siRNA. *Science Signaling*, *6*(283), p. ra56.
[179] Supnet C. and Bezprozvanny I. (2010). The dysregulation of intracellular calcium in Alzheimer disease. *Cell Calcium*, *47*(2), pp. 183–189.
[180] Shilling D., Mak D.O., Kang D.E. and Foskett J.K. (2012). Lack of evidence for presenilins as endoplasmic reticulum Ca2+ leak channels. *J. Biol. Chem.*, *287*(14), pp. 10933–10944.
[181] Querfurth H.W. and Selkoe D.J. (1994). Calcium ionophore increases amyloid beta peptide production by cultured cells. *Biochemistry*, *33*(15), pp. 4550–4561.
[182] Isaacs A.M., Senn D.B., Yuan M., Shine J.P. and Yankner B.A. (2006). Acceleration of amyloid beta-peptide aggregation by physiological concentrations of calcium. *J. Biol. Chem.*, *281*(38), pp. 27916–27923.
[183] Bruno A.M. *et al.* (2012). Altered ryanodine receptor expression in mild cognitive impairment and Alzheimer's disease. *Neurobiol. Aging*, *33*(5), pp. 1001.e1001–1001.e1006.
[184] Chakroborty S. *et al.* (2012). Early Presynaptic and postsynaptic calcium signaling abnormalities mask underlying synaptic depression in presymptomatic Alzheimer's disease mice. *J. Neurosci.*, *32*(24), pp. 8341–8353.
[185] Querfurth H.W., Jiang J., Geiger J.D. and Selkoe D.J. (1997). Caffeine stimulates amyloid beta-peptide release from beta-amyloid precursor protein-transfected HEK293 cells. *J. Neurochem.*, *69*(4), pp. 1580–1591.
[186] Pierrot N., Ghisdal P., Caumont A.-S. and Octave J.-N. (2004). Intraneuronal amyloid-$\beta$1-42 production triggered by sustained increase of cytosolic calcium concentration induces neuronal death. *J. Neurochem.*, *88*(5), pp. 1140–1150.
[187] Koran M.E., Hohman T.J. and Thornton-Wells T.A. (2014). Genetic interactions found between calcium channel genes modulate amyloid load measured by positron emission tomography. *Human Genetics*, *133*(1), pp. 85–93.
[188] Kurumatani T. *et al.* (1998). Loss of inositol 1,4,5-trisphosphate receptor sites and decreased PKC levels correlate with staging of Alzheimer's disease neurofibrillary pathology. *Brain Res.*, *796*(1–2), pp. 209–221.
[189] Buxbaum J.D., Ruefli A.A., Parker C.A., Cypess A.M. and Greengard P. (1994). Calcium regulates processing of the Alzheimer amyloid protein precursor in a protein kinase C-independent manner. *Proc. Natl. Acad. Sci. USA*, *91*(10), pp. 4489–4493.
[190] Dreses-Werringloer U. *et al.* (2008). A polymorphism in CALHM1 influences $Ca^{2+}$ homeostasis, Abeta levels, and Alzheimer's disease risk. *Cell*, *133* (7), pp. 1149–1161.
[191] Zeiger W. *et al.* (2013). $Ca^{2+}$ influx through store-operated $Ca^{2+}$ channels reduces Alzheimer disease beta-amyloid peptide secretion. *J. Biol. Chem.*, *288*(37), pp. 26955–26966.
[192] Gallego-Sandin S., Alonso M.T. and Garcia-Sancho J. (2011). Calcium homoeostasis modulator 1 (CALHM1) reduces the calcium content of the

endoplasmic reticulum (ER) and triggers ER stress. *Biochemical J.*, *437*(3), pp. 469–475.
[193] Koppel J. *et al.* (2011). CALHM1 P86L polymorphism modulates CSF Abeta levels in cognitively healthy individuals at risk for Alzheimer's disease. *Mol. Med. (Cambridge, Mass.)*, *17*(9–10), pp. 974–979.
[194] Cui P.J. *et al.* (2010). CALHM1 P86L polymorphism is a risk factor for Alzheimer's disease in the Chinese population. *J. Alzheimer's Dis. JAD*, *19*(1), pp. 31–35.
[195] Boada M. *et al.* (2010). CALHM1 P86L polymorphism is associated with late-onset Alzheimer's disease in a recessive model. *J. Alzheimer's Dis.*, *20*(1), pp. 247–251.
[196] Aqdam M.J. *et al.* (2010). Association of CALHM1 gene polymorphism with late onset Alzheimer's disease in Iranian Population. *J. Med. Biotechnol.*, *2*(3), pp. 153–157.
[197] Minster R.L., Demirci F.Y., DeKosky S.T., and Kamboh M.I. (2009). No association between CALHM1 variation and risk of Alzheimer disease. *Human Mutation*, *30*(4), pp. E566–E569.
[198] Tan E.K. *et al.* (2011). CALHM1 variant is not associated with Alzheimer's disease among Asians. *Neurobiol Aging*, *32*(3), pp. 546.e511–546.e512.
[199] Lambert J.C. *et al.* (2010). The CALHM1 P86L polymorphism is a genetic modifier of age at onset in Alzheimer's disease: a meta-analysis study. *J. Alzheimer's Dis.*, *22*(1), pp. 247–255.
[200] De Schutter E. and Smolen P. (1998). Calcium dynamics in large neuronal models. In Koch C. and Segev, I. (Eds.), *Methods Neuronal Model*, 2nd Ed. MIT press.
[201] Senn W., Markram H. and Tsodyks M. (2001). An algorithm for modifying neurotransmitter release probability based on pre-and postsynaptic spike timing. *Neural Comput.*, *13*(1), pp. 35–67.
[202] Nadkarni S., Bartol T.M., Sejnowski T.J. and Levine H. (2010). Modelling vesicular release at hippocampal synapses. *PLoS Comput. Biol.*, *6*(11), p. e1000983.
[203] Barbour B. and Häusser M. (1997). Intersynaptic diffusion of neurotransmitter. *Trends Neurosci*, *20*(9), pp. 377–384.
[204] Rusakov D.A. and Kullmann D.M. (1998). Extrasynaptic glutamate diffusion in the hippocampus: ultrastructural constraints, uptake, and receptor activation. *J. Neurosci.*, *18*(9), pp. 3158–3170.
[205] Sterratt D., Graham B., Gillies A. and Willshaw D. (2011). Intracellular mechanisms. In *Principles of Computational Modelling in Neuroscience*. Cambridge University Press.
[206] Rusakov D.A. (2001). The role of perisynaptic glial sheaths in glutamate spillover and extracellular $Ca^{2+}$ depletion. *Biophys. J.*, *81*(4), pp. 1947–1959.
[207] Grewer C. *et al.* (2008). Glutamate forward and reverse transport: From molecular mechanism to transporter-mediated release after ischemia. *IUBMB Life*, *60*(9), pp. 609–619.

[208] Patneau D.K. and Mayer M.L. (1991). Kinetic–analysis of interactions between kainate and AMPA: evidence for activation of a single receptor in mouse hippocampal neurons. *Neuron*, *6*(5), pp. 785–798.
[209] Jonas P., Major G. and Sakmann B. (1993). Quantal components of unitary EPSCs at the mossy fibre synapse on CA3 pyramidal cells of rat hippocampus. *J. Physiol.*, *472*, pp. 615–663.
[210] Nielsen T.A., DiGregorio D.A. and Silver R.A. (2004). Modulation of glutamate mobility reveals the mechanism underlying slow-rising AMPAR EPSCs and the diffusion coefficient in the synaptic cleft. *Neuron*, *42*(5), pp. 757–771.
[211] Lester R. and Jahr C.E. (1992). NMDA channel behavior depends on agonist affinity. *J. Neurosci.*, *12*(2), pp. 635–643.
[212] Erreger K., Dravid S.M., Banke T.G., Wyllie D.J. and Traynelis S.F. (2005). Subunit-specific gating controls rat NR1/NR2A and NR1/NR2B NMDA channel kinetics and synaptic signalling profiles. *J. Physiol.*, *563*(Pt 2), pp. 345–358.
[213] Kussius C.L. and Popescu G.K. (2009). Kinetic basis of partial agonism at NMDA receptors. *Nature Neurosci.*, *12*(9), pp. 1114–1120.
[214] Schorge S., Elenes S. and Colquhoun D. (2005). Maximum likelihood fitting of single channel NMDA activity with a mechanism composed of independent dimers of subunits. *J. Physiology*, *569*(2), pp. 395–418.
[215] Banke T.G. and Traynelis S.F. (2003). Activation of NR1/NR2b NMDA receptors. *Nature Neuroscience*, *6*(2), pp. 144–152.
[216] Popescu G., Robert A., Howe J.R. and Auerbach A. (2004). Reaction mechanism determines NMDA receptor response to repetitive stimulation. *Nature*, *430*(7001), pp. 790–793.
[217] Ambert N. *et al.* (2010). Computational studies of NMDA receptors: differential effects of neuronal activity on efficacy of competitive and non-competitive antagonists. *Open Access Bioinformatics*, *2*, pp. 113–125.
[218] Jahr C.E. and Stevens C.F. (1990). Voltage dependence of NMDA-activated macroscopic conductances predicted by single-channel kinetics. *J. Neuroscience* *10*(9), pp. 3178–3182.
[219] Kampa B.M., Clements J., Jonas P. and Stuart G.J. (2004). Kinetics of Mg(2+) unblock of NMDA receptors: implications for spike-timing dependent synaptic plasticity. *J. Physiology*, *556*(Pt 2), pp. 337–345.
[220] Vargas-Caballero M. and Robinson H.P. (2004). Fast and slow voltage-dependent dynamics of magnesium block in the NMDA receptor: the asymmetric trapping block model. *J. Neurosci.*, *24*(27), pp. 6171–6180.
[221] Holmes W.R. and Levy W.B. (1990). Insights into associative long-term potentiation from computational models of NMDA receptor-mediated calcium influx and intracellular calcium concentration changes. *J. neurophysiology*, *63*(5), pp. 1148–1168.
[222] Holcman D., Schuss Z. and Korkotian E. (2004). Calcium dynamics in dendritic spines and spine motility. *Biophy. J.*, *87*(1), pp. 81–91.
[223] Rubin J.E., Gerkin R.C., Bi G.Q. and Chow C.C. (2005). Calcium time course as a signal for spike-timing-dependent plasticity. *J. Neurophysiology*, *93*(5), pp. 2600–2613.

[224] Schiegg A., Gerstner W., Ritz R. and van Hemmen J.L. (1995). Intracellular $Ca^{2+}$ stores can account for the time course of LTP induction: a model of $Ca^{2+}$ dynamics in dendritic spines. *J. Neurophysiology, 74*(3), pp. 1046–1055.
[225] Doi T., Kuroda S., Michikawa T. and Kawato M. (2005). Inositol 1,4,5-trisphosphate-dependent $Ca^{2+}$ threshold dynamics detect spike timing in cerebellar Purkinje cells. *J. Neurosci., 25*(4), pp. 950–961.
[226] Volfovsky N., Parnas H., Segal M. and Korkotian E. (1999). Geometry of dendritic spines affects calcium dynamics in hippocampal neurons: theory and experiments. *J. Neurophysiology, 82*(1), pp. 450–462.
[227] Fink C.C. *et al.* (2000). An image-based model of calcium waves in differentiated neuroblastoma cells. *Biophys. J., 79*(1), pp. 163–183.
[228] Sneyd J., Girard S. and Clapham D. (1993). Calcium wave propagation by calcium-induced calcium release: an unusual excitable system. *Bull. Math. Biol., 55*(2), pp. 315–344.
[229] De Young G.W. and Keizer J. (1992). A single-pool inositol 1,4,5-trisphosphate-receptor-based model for agonist-stimulated oscillations in $Ca^{2+}$ concentration. *Proc. Natl. Acad. Sci. USA, 89*(20), pp. 9895–9899.
[230] Gin E., Kirk V., and Sneyd J. (2006). A bifurcation analysis of calcium buffering. *J. Theoret. Biol., 242*(1), pp. 1–15.
[231] Li Y.-X. and Rinzel J. (1994). Equations for InsP3 receptor-mediated $[Ca^{2+}]$i oscillations derived from a detailed kinetic model: a Hodgkin–Huxley like formalism. *J. Theoret. Biol., 166*(4), pp. 461–473.
[232] Bezprozvanny I. (1994). Theoret. analysis of calcium wave propagation based on inositol (1,4,5)-trisphosphate (InsP3) receptor functional properties. *Cell Calcium, 16*(3), pp. 151–166.
[233] Dawson A.P., Lea E.J. and Irvine R.F. (2003). Kinetic model of the inositol trisphosphate receptor that shows both steady-state and quantal patterns of $Ca^{2+}$ release from intracellular stores. *Biochem. J., 370*(Pt 2),=break pp. 621–629.
[234] Kaftan E.J., Ehrlich B.E. and Watras J. (1997). Inositol 1,4,5-trisphosphate ($InsP_3$) and calcium interact to increase the dynamic range of $InsP_3$ receptor-dependent calcium signaling. *J. Gen. Physiology, 110*(5), pp. 529–538.
[235] Mak D.O., McBride S. and Foskett J.K. (2001). Regulation by $Ca^{2+}$ and inositol 1,4,5-trisphosphate ($InsP_3$) of single recombinant type 3 $InsP_3$ receptor channels. $Ca^{2+}$ activation uniquely distinguishes types 1 and 3 $InsP_3$ receptors. *J. Gen. Physiology, 117*(5), pp. 435–446.
[236] LeBeau A.P., Yule D.I., Groblewski G.E. and Sneyd J. (1999). Agonist-dependent phosphorylation of the inositol 1,4,5-trisphosphate receptor: A possible mechanism for agonist-specific calcium oscillations in pancreatic acinar cells. *J. Gen. Physiology, 113*(6), pp. 851–872.
[237] Sneyd J. and Falcke M. (2005). Models of the inositol trisphosphate receptor. *Progr. Biophys. Mol. Biol., 89*(3), pp. 207–245.
[238] Blackwell K.T. and Kotaleski J. (2003). Modeling the dynamics of second messenger pathways. In R. Kötter (Ed.), *Neuroscience Databases* (pp. 63–79). Springer US.

[239] Nakano T., Yoshimoto J. and Doya K. (2013). A model-based prediction of the calcium responses in the striatal synaptic spines depending on the timing of cortical and dopaminergic inputs and post-synaptic spikes. *Frontiers Comput. Neurosci.*, *7*(119).

[240] Sachs F., Qin F., and Palade P. (1995). Models of $Ca^{2+}$ release channel adaptation. *Science (N.Y.)*, *267*(5206), pp. 2010–2011.

[241] Keizer J. and Levine L. (1996). Ryanodine receptor adaptation and $Ca^{2+}$(-) induced $Ca^{2+}$ release-dependent $Ca^{2+}$ oscillations. *Biophysi. J.*, *71*(6), pp. 3477–3487.

[242] Saftenku E., Williams A.J. and Sitsapesan R. (2001). Markovian models of low and high activity levels of cardiac ryanodine receptors. *Biophys. J.*, *80*(6), pp. 2727–2741.

[243] Schuster S., Marhl M. and Hofer T. (2002). Modelling of simple and complex calcium oscillations: From single-cell responses to intercellular signalling. *European J. Biochem./FEBS*, *269*(5), pp. 1333–1355.

[244] Bondarenko V.E., Szigeti G.P., Bett G.C., Kim S.J. and Rasmusson R.L. (2004). Computer model of action potential of mouse ventricular myocytes. *Amer. J. Physiology: Heart Circulatory Physiol.*, *287*(3), pp. H1378–1403.

[245] Blaustein M.P. and Lederer W.J. (1999). Sodium/calcium exchange: its physiological implications. *Physiol. Rev.*, *79*(3), pp. 763–854.

[246] Zador A., Koch C. and Brown T.H. (1990). Biophysical model of a Hebbian synapse. *Proc. Natl. Acad. Sci. USA*, *87*(17), pp. 6718–6722.

[247] Matsuoka S., Sarai N., Kuratomi S., Ono, K. and Noma A. (2003). Role of individual ionic current systems in ventricular cells hypothesized by a model study. *Japanese J. Physiol.*, *53*(2), pp. 105–123.

[248] Yano K., Petersen O.H. and Tepikin A.V. (2004). Dual sensitivity of sarcoplasmic/endoplasmic $Ca^{2+}$-ATPase to cytosolic and endoplasmic reticulum $Ca^{2+}$ as a mechanism of modulating cytosolic $Ca^{2+}$ oscillations. *Biochem. J.*, *383*(Pt 2), pp. 353–360.

[249] Higgins E.R., Cannell M.B. and Sneyd J. (2006). A buffering SERCA pump in models of calcium dynamics. *Biophys. J.*, *91*(1), pp. 151–163.

[250] Shannon T.R., Ginsburg K.S. and Bers D.M. (2000). Reverse mode of the sarcoplasmic reticulum calcium pump and load-dependent cytosolic calcium decline in voltage-clamped cardiac ventricular myocytes. *Biophy. J.*, *78*(1), pp. 322–333.

[251] Blackwell K.T. (2013). Approaches and tools for modeling signaling pathways and calcium dynamics in neurons. *J. Neurosci. Methods*, *220*(2), pp. 131–140.

[252] Alves R., Antunes F. and Salvador A. (2006). Tools for kinetic modeling of biochemical networks. *Nat. Biotech.*, *24*(6), pp. 667–672.

[253] Brette R. *et al.* (2007). Simulation of networks of spiking neurons: a review of tools and strategies. *J. Comput. Neurosci*, *23*(3), pp. 349–398.

[254] Good T.A. and Murphy R.M. (1996). Effect of beta-amyloid block of the fast-inactivating K+ channel on intracellular $Ca^{2+}$ and excitability in a modeled neuron. *Proc. Natl. Acad. Sci. USA*, *93*(26), pp. 15130–15135.

[255] Wilson N.P., Gates B. and Castellanos M. (2013). Modeling the short time-scale dynamics of $\beta$-amyloid–neuron interactions. *J. Theoret. Biol.*, *331*(0), pp. 28–37.

[256] Wang S.S.S., Kazantzi V. and Good T.A. (2003). A kinetic analysis of the mechanism of $\beta$-amyloid induced G protein activation. *J. Theoret. Biol.*, *221*(2), pp. 269–278.

[257] Kidd J. and Sattelle D. (2006). The effects of amyloid peptides on A-type K+ currents of Drosophila larval cholinergic neurons: modeled actions on firing properties. *Invert. Neurosci.*, *6*(4), pp. 207–213.

[258] Morse T.M., Carnevale N.T., Mutalik P.G., Migliore M. and Shepherd G.M. (2010). Abnormal excitability of oblique dendrites implicated in early Alzheimer's: a computational study. *Frontiers in neural circuits*, *4*(16).

[259] Tiveci S., Akin A., Cakir T., Saybasili H. and Ulgen K. (2005). Modelling of calcium dynamics in brain energy metabolism and Alzheimer's disease. *Comput. Biol. Chem.*, *29*(2), pp. 151–162.

[260] Toivari E., Manninen T., Nahata A.K., Jalonen T.O. and Linne M.L. (2011). Effects of transmitters and amyloid-beta peptide on calcium signals in rat cortical astrocytes: Fura-2AM measurements and stochastic model simulations. *PloS one*, *6*(3), p. e17914.

[261] Riera J., Hatanaka R., Uchida T., Ozaki T. and Kawashima R. (2011). Quantifying the uncertainty of spontaneous $Ca^{2+}$ oscillations in astrocytes: particulars of Alzheimer's disease. *Biophys. J.*, *101*(3), pp. 554–564.

[262] De Caluwé J. and Dupont G. (2013). The progression towards Alzheimer's disease described as a bistable switch arising from the positive loop between amyloids and $Ca^{2+}$. *J. Theoret. Biol.*, *331*(0), pp. 12–18.

## Chapter 7

# Biophysics of A$\beta$ and Computational Modeling

### 7.1. Introduction

Having discussed the role of calcium in AD in Chapter 6, now we dive deeply into the roles A$\beta$ plays in the calcium related pathways. In this chapter, we take a quantitative approach to elucidate the mechanisms of A$\beta$ interactions and mathematically model them. As we have seen, A$\beta$ oligomers interact with multiple key proteins in glutamatergic synaptic transmission (Fig. 7.1). We investigate the following disturbances of A$\beta$: (1) A$\beta$ reduces glutamate uptake by glutamate transporters; (2) A$\beta$ promotes glutamate vesicle release from the presynaptic terminal and astrocytes; and (3) A$\beta$ induces the internalization of the surface expression of the glutamate receptors: AMPARs and the NMDARs. Conditions (1) and (2) may lead to an upregulation in glutamatergic signaling by increasing the glutamate availability to its receptors. This may be a reason for the $Ca^{2+}$ overload observed in Alzheimer's disease (AD) through the overactivation of NMDARs located at extrasynaptic sites that, subsequently, trigger the death pathways [1, 2]. In contrast, condition (3) may contribute to the depression of synaptic activity and a reduction in memory formation [3]. However, this reduction in the surface receptor numbers is suggested to be a neuroprotective mechanism, to some extent, in response to the glutamate-induced excitotoxicity and excessive $Ca^{2+}$ influxes from the extrasynaptic receptors [5]. These experimental observations show paradoxical

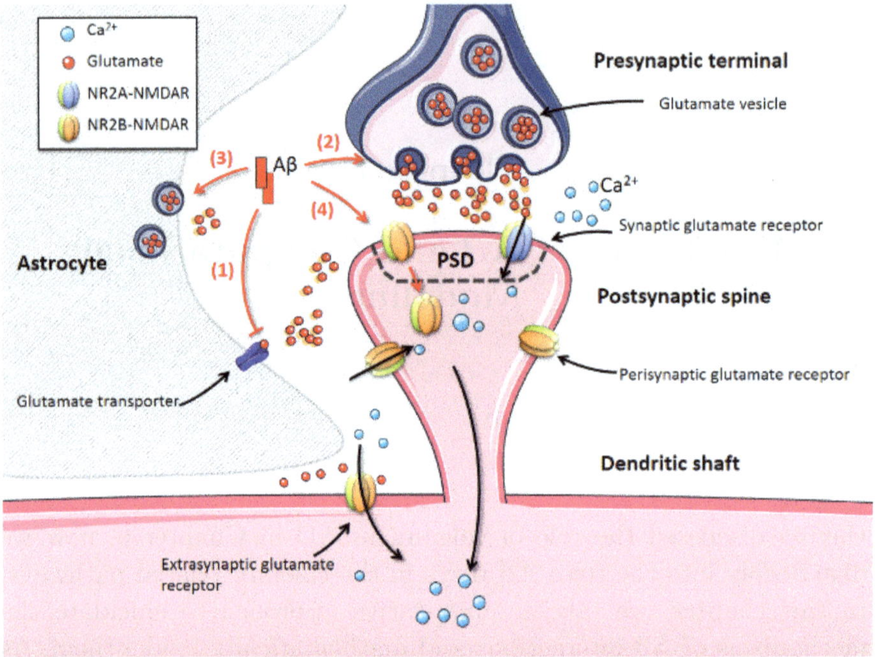


Fig. 7.1. Disturbances in glutamatergic synaptic transmission by A$\beta$ in AD. Referring to the arrow labels: ① A$\beta$ inhibits glutamate clearance by the glutamate transporters; and ② and ③ A$\beta$ promotes glutamate vesicle release from the presynaptic terminal and ambient astrocytes, respectively; and ④ A$\beta$ also mediates the internalization of surface receptors at the synaptic site. ①, ② and ③ result in an increase in extracellular glutamate concentration and, ultimately, may lead to the over activation of synaptic glutamate receptors or of receptors at distant locations from the release site. In contrast, ④ leads to a decrease in functional synaptic receptors which may depress synaptic activity. This figure is produced using Servier Medical Art (http://www.servier.com/Powerpoint-image-bank). Acknowledgment [4]: Liang, J., Kulasiri, D. and Samarasinghe, S. (2017) Computational investigation of Amyloid-$\beta$-induced location-and subunit-specific disturbances of NMDAR at hippocampal dendritic spine in Alzheimer's disease. *PLoS One 12*(8), p. e0182743.

effects of A$\beta$ on the $Ca^{2+}$ dynamics of the postsynaptic neurons. These effects lead to different interpretations of disturbances in the downstream events that are mediated by the cytosolic $Ca^{2+}$ levels.

We develop a computational model of a CA1 pyramidal dendritic spine to explore these issues both individually and globally.

This spine model is stimulated by presynaptic electrical stimulus inputs, which trigger a sequence of events that include glutamate release and transmission, glutamate receptor activation at the postsynaptic neuron at various sites, and the $Ca^{2+}$ influx via the opened receptor channels. Therefore, we can use this model to study the above-mentioned disturbances and identify their contributions to the $Ca^{2+}$ dynamic in the postsynaptic spine.

NMDAR is a key receptor that is involved in, or relates to, all events included in this model of the glutamate-mediated state transition behavior of the NMDAR and the NMDAR-dependent $Ca^{2+}$ influx. NMDAR is located at both the synaptic and extrasynaptic sites, and the current understanding is that it plays opposite physiological roles in mediating intracellular signaling and cell death pathways (see [6]). A$\beta$ shows differential disturbances on the activities of NMDARs with different subunit composition and at different locations. Therefore, we include location-specific and NR2 subunit-specific characteristics in this model to understand the A$\beta$-induced dysregulation of intracellular $Ca^{2+}$ and its relationship with the balance between synaptic and extrasynaptic NMDARs (sNMDARs and eNMDARs) in the dendritic spine of the pyramidal neurons.

Under experimental conditions, A$\beta$ is usually given at a much higher concentration than that in human brains with AD [7]. It is inappropriate to study the concentration dependence of A$\beta$ disturbances based on the results of these experiments; therefore, we mimic the disturbances of A$\beta$ by perturbing key related parameters and observing the alterations in model behavior. The perturbation in the parameters represents the degree of A$\beta$ disturbance at different stages of AD pathology.

The main contribution of the model we develop in this chapter is that it allows us to investigate different motifs of disturbance from A$\beta$ on glutamatergic signaling and NMDAR-mediated postsynaptic $Ca^{2+}$ dynamics. Specific numbers and subtypes of NMDAR at three locations are chosen, based on the literature, to demonstrate typical conditions in the CA1 pyramidal neurons. We develop the model in Section 7.2. In Section 7.3, we present the calibration and the estimation of unknown parameters with respect to the experimental findings based on the healthy condition. In Section 7.4, we discuss the model performance under healthy conditions. In Section 7.5, we use the model to mimic the A$\beta$-dependent disturbances in synaptic transmission and receptor distribution under AD conditions. In Section 7.6, we give a brief discussion and conclusions.

## 7.2. Model Overview

To model NMDAR-mediated glutamate-induced $Ca^{2+}$ dynamics in a single synaptic spine and the adjacent dendritic shaft of a CA1 pyramidal neuron, we construct a mathematical model consisting of three parts: (1) glutamate release, diffusion and uptake (Fig. 7.2(a)); (2) glutamate receptor state transition (Fig. 7.2(b)); and (3) $Ca^{2+}$ dynamics (Fig. 7.2(c)). In this section, we present the model development in detail.

### 7.2.1. *Glutamate release, diffusion and uptake*

We simulate the glutamate release from the presynaptic terminal and diffusion inside the synaptic cleft, and in the extrasynaptic space, based on the model of Rusakov and Kullmann [8]. The schematic two-dimensional profile of the glutamate diffusion model is given in Fig. 7.2. The spine head and the presynaptic terminal are modeled as two opposite hemispheres with the same radius and there is no glutamate diffusion within them. The synaptic cleft is a flat cylinder between these two hemispheres, with a height of 20 nm. We assume the volume of the spine head is 0.1 $\mu m^3$, which gives a radius of 363 nm for the synaptic cleft and the two opposite hemispheres. The extrasynaptic space is a spherically isotropic porous medium surrounded by the two hemispheric obstacles.

The glutamate vesicle release site is assumed to be a point source and is placed in the center of the presynaptic terminal surface. Each vesicle contains 100 mM glutamate, which is about 1500 glutamate molecules [9]. Single and multiple vesicular release events can be simulated by changing the total number of glutamate molecules released. The time course of a single glutamate vesicle release is

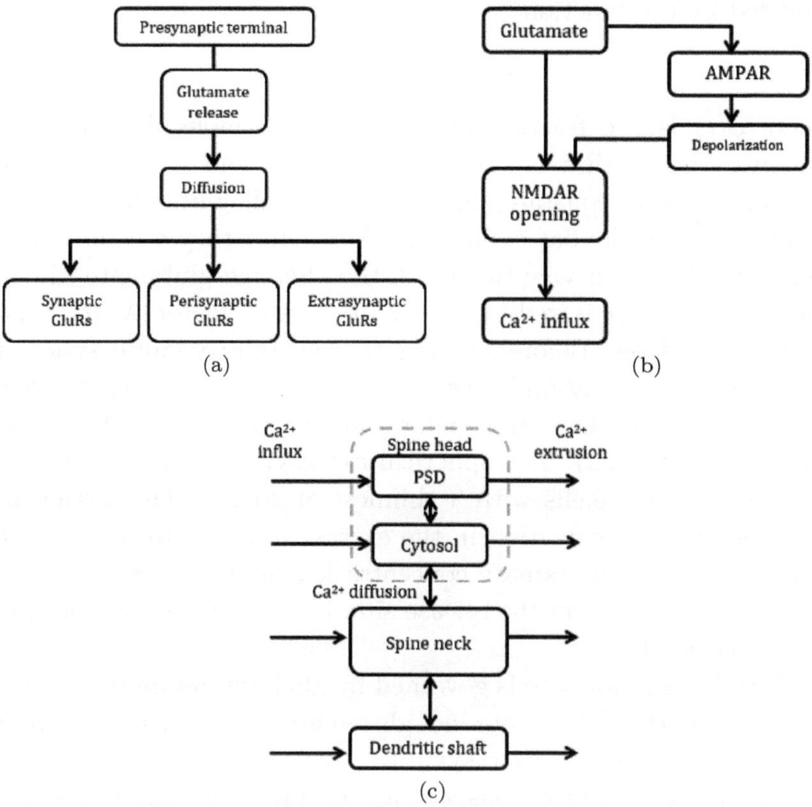


Fig. 7.2. Conceptual framework of the three parts of the model. (a) After presynaptic stimulation, glutamate is released from the presynaptic terminal into the synaptic cleft. It is then diffused across the synaptic cleft and into the extrasynaptic space. Through diffusion, glutamate can bind to the glutamate receptors (GluRs) at different locations. (b) NMDAR and AMPAR are the most common ionotropic GluRs. NMDAR is the major $Ca^{2+}$ channel; $Ca^{2+}$ influx via NMDAR requires both the binding of glutamate to NMDAR and the removal of its $Mg^{2+}$ blockage. The latter can be achieved by membrane depolarization after the activation of AMPARs. (c) A four-compartment $Ca^{2+}$ model of the dendritic spine and its adjacent dendritic shaft includes the mechansims for $Ca^{2+}$ influx, extrusion and buffering in each compartment and diffusion between the two neighboring compartments. Acknowledgment [4]: Liang, J., Kulasiri, D. and Samarasinghe, S. (2017) Computational investigation of Amyloid-$\beta$-induced location-and subunit-specific disturbances of NMDAR at hippocampal dendritic spine in Alzheimer's disease. *PLoS One 12*(8), p. e0182743.

modeled by the function

$$\phi(t) = \sigma^2 t \exp(-\sigma t), \tag{7.1}$$

where $\phi(t)$ is the fraction of all glutamate molecules in a single synaptic vesicle that is released at time $t$ and $\sigma = 39\ \text{ms}^{-1}$ is the release time constant [8]. Once released, glutamate molecules diffuse through the flat cylindrical cleft. After they escape from the cleft into the extrasynaptic space, the effective glutamate diffusion coefficient ($D_{\text{Glu}}$) is reduced by a tortuosity factor $\lambda$, from $D_{\text{Glu}}$ to $D^*_{\text{Glu}} = \frac{D_{\text{Glu}}}{\lambda^2}$. Before moving to the extrasynaptic space, the glutamate concentration in the last cylinder shell of the cleft is scaled by $\frac{1}{\alpha}$, where $\alpha$ is the extracellular volume fraction (see Appendix B for the explanation). The spherical extrasynaptic space is modeled as 50 concentric shells with a thinness of 20 nm. The resting level of glutamate concentration in the extrasynaptic space is set to 0.25 $\mu$M [10, 11]. The glutamate concentration in the last cylinder shell, which is 1.36 $\mu$m from the release site, is fixed at resting level (open boundary condition).

Uptake of glutamate is governed by glial glutamate transporters. We assume that there are no glutamate transporters inside the synaptic cleft. In the extrasynaptic space, glutamate transporters are distributed homogeneously at a concentration ($B_{\text{total}}$) of 0.5 mM [12]. A gap between the glial sheath and spine head is considered and, thus, the distribution of glutamate transporters starts at 20 nm from the edge of the dendritic cleft. A simple kinetic scheme is applied to the glutamate binding and uptake by glial transporters:

$$\text{Glu} + \text{Tr} \overset{k_{1f};k_{1b}}{\leftrightarrow} \text{Glu} - \text{Tr} \xrightarrow{k_2} \text{Tr},$$

where Glu is the glutamate, Tr is the unbound surface transporter and Glu – Tr is the glutamate-transporter complex [8]. The reactions include a rapid reversible binding between Glu and Tr with rate constants $k_{1\text{f}}$ and $k_{1\text{b}}$ and a relatively slow translocation from the surface into the glial cell at rate constant $k_2$. The values of the parameters are listed in Table 7.1. The reaction is assumed to obey mass action kinetics (see Appendix C.1 for the explanation), and the

Table 7.1. Model geometry and glutamate transmission related parameters.

| Description | Symbol | Value | References |
|---|---|---|---|
| Volume of spine head | | 0.1 $\mu m^3$ | |
| Volume of PSD | | 0.01 $\mu m^3$ | |
| Radius of PSD | $r_{PSD}$ | 150 nm | [8, 14] |
| Height of cleft | $h_{cleft}$ | 20 nm | |
| Radius of cleft | $r_{cleft}$ | 363 nm | |
| Length of dendritic neck | $l_{neck}$ | 750 nm | [15, 16] |
| Radius of dendritic neck | $r_{neck}$ | 50 nm | |
| Length of dendritic shaft | $l_{shaft}$ | 1000 nm | [15, 17] |
| Radius of dendritic shaft | $r_{shaft}$ | 500 nm | |
| Distance of the glial sheath from the synaptic cylinder surface | | 20 nm | [12] |
| Thickness of cylinder shell (cleft) | $dr_1$ | 10 nm | |
| Thickness of sphere shell (extrasynaptic space) | $dr_2$ | 20 nm | |
| Glutamate vesicle content | $Glu_0$ | 1500 molecules | [9] |
| Glutamate diffusion constant in the cleft | $D_{Glu}$ | 0.2 $\mu m^2$ $ms^{-1}$ | [8] (20–24°C) |
| Tortuosity factor | $\lambda$ | 1.34 | |
| Extracellular volume fraction | $\alpha$ | 0.12 | |
| Transporter concentration | $B_{total}$ | 0.5 mM | |
| Binding rate constant | $k_{1f}$ | 0.1 $ms^{-1}$ | |
| Unbinding rate constant | $k_{1b}$ | 5 $mM^{-1}$ $ms^{-1}$ | |
| Translocation rate | $k_2$ | 0.1 $ms^{-1}$ | |
| Resting glutamate concentration | | 0.25 $\mu M$ | [10, 11] |

model equations for transporters are in Appendix D.1. To track the glutamate concentration after release at different locations, we define the PSD site, and perisynaptic and extrasynaptic sites as follows: the surface of the PSD region and the perisynaptic zone are set from 0 to 150 nm and 365 nm away from the center of the postsynaptic surface, respectively [13] (Fig. 7.3 ① and ②). The extrasynaptic space is set beyond the outside border of the perisynaptic zone and we assume that the extrasynaptic receptors are located at the dendritic shaft. Based on the model geometry data in Table 7.1, the extrasynaptic site is about 826 nm away from the release point (Fig. 7.3 ③).

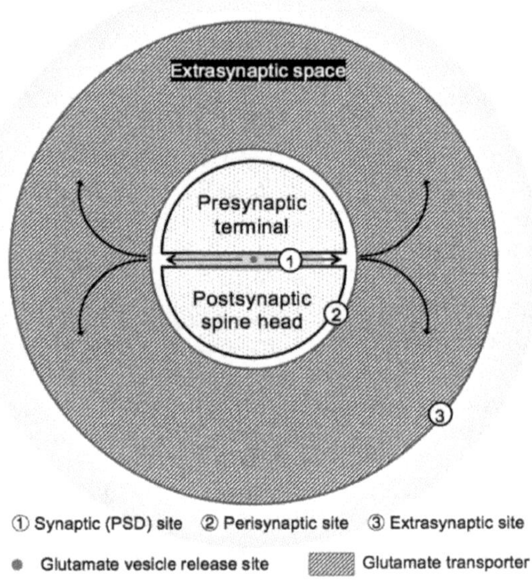


Fig. 7.3. Schematic two-dimensional profile of the glutamate diffusion model. Glutamate is released at the centre of the synaptic cleft and diffused across the synaptic cleft and into the extrasynaptic space. The synaptic cleft is assumed to be a flat cylinder with 20 thin concentric shells. The extrasynaptic space is a porous medium with spherical concentric shells. The arrows denote the diffusion direction of the glutamate. Glutamate transporters are homogenously distributed in the extrasynaptic space. Three gray areas (marked with ①, ② and ③) represent the synaptic (PSD), perisynaptic and extrasynaptic sites. The glutamate concentration at each site is used for calculating local receptor activity in the NMDAR and AMPAR models. Acknowledgment [4]: Liang, J., Kulasiri, D. and Samarasinghe, S. (2017) Computational investigation of Amyloid-$\beta$-induced location-and subunit-specific disturbances of NMDAR at hippocampal dendritic spine in Alzheimer's disease. *PLoS One 12*(8), p. e0182743.

### 7.2.2. *Activation of the ionotropic glutamate receptors: NMDARs and AMPARs*

Glutamate receptors are located at the synaptic, perisynaptic and extrasynaptic membrane surfaces and have various receptor numbers (Table 7.2). For each receptor, the local glutamate concentration it receives depends on its distance from the release site. Because the number of glutamates is much greater than the number of these receptors, we do not consider the glutamate uptake by these receptors

Table 7.2. NMDAR and AMPAR parameters.

| Receptor type | Location | Number | Reference |
|---|---|---|---|
| NR1/NR2A-NMDAR | PSD (sNR2A-NMDAR) | 12 | [19–21] |
| | Perisynaptic site | 0 | |
| | Extrasynaptic site | 0 | |
| NR1/NR2B-NMDAR | PSD (sNR2B-NMDAR) | 8 | |
| | Perisynaptic site (pNR2B-NMDAR) | 3 | |
| | Extrasynaptic site (pNR2b-NMDAR) | 8 | |
| AMPAR | PSD (sAMPAR) | 85 receptors | |
| | Extrasynaptic site (eAMPAR) | 20 receptors/$\mu m^2$ | [22] |

| Receptor type | Reaction rate constants | Value | Reference |
|---|---|---|---|
| NR2A-NMDARs (20–24°C) | $k_{\rm on}$ | 0.0316 $\mu M^{-1}$ $ms^{-1}$ | [23] |
| | $k_{\rm off}$ | 1.01 $ms^{-1}$ | |
| | $k_{\rm d1+}$ | 0.0851 $ms^{-1}$ | |
| | $k_{\rm d1-}$ | 0.0297 $ms^{-1}$ | |
| | $k_{\rm d2+}$ | 0.23 $ms^{-1}$ | |
| | $k_{\rm d2-}$ | 0.00101 $ms^{-1}$ | |
| | $k_{\rm f+}$ | 0.230 $ms^{-1}$ | |
| | $k_{\rm f-}$ | 0.178 $ms^{-1}$ | |
| | $k_{\rm s+}$ | 3.140 $ms^{-1}$ | |
| | $k_{\rm s-}$ | 0.174 $ms^{-1}$ | |
| NR2B-NMDARs (20–24°C) | $k_{\rm on}$ | 0.00283 $\mu M^{-1}$ $ms^{-1}$ | |
| | $k_{\rm off}$ | 0.0381 $ms^{-1}$ | |
| | $k_{\rm d1+}$ | 0.550 $ms^{-1}$ | |
| | $k_{\rm d1-}$ | 0.0814 $ms^{-1}$ | |
| | $k_{\rm d2+}$ | 0.112 $ms^{-1}$ | |
| | $k_{\rm d2-}$ | 0.00091$ms^{-1}$ | |
| | $k_{\rm f+}$ | 0.048 $ms^{-1}$ | |
| | $k_{\rm f-}$ | 0.23 $ms^{-1}$ | |
| | $k_{\rm s+}$ | 2.836 $ms^{-1}$ | |
| | $k_{\rm s-}$ | 0.175 $ms^{-1}$ | |

*(Continued)*

Table 7.2. (*Continued*)

| Receptor type | Reaction rate constants | Value | Reference |
|---|---|---|---|
| AMPAR (22°C) | $k_{11}$ | 0.00459 $\mu M^{-1}$ $ms^{-1}$ | [24] |
| | $k_{10}$ | 4.26 $ms^{-1}$ | |
| | $k_{21}$ | 0.0284 $\mu M^{-1}$ $ms^{-1}$ | |
| | $k_{20}$ | 3.26 $ms^{-1}$ | |
| | $k_{31}$ | 0.00127 $\mu M^{-1}$ $ms^{-1}$ | |
| | $k_{30}$ | 0.0457 $ms^{-1}$ | |
| | $\alpha$ | 4.24 $ms^{-1}$ | |
| | $\beta$ | 0.9 $ms^{-1}$ | |
| | $\alpha_1$ | 2.89 $ms^{-1}$ | |
| | $\beta_1$ | 0.0392 $ms^{-1}$ | |
| | $\alpha_2$ | 0.172 $ms^{-1}$ | |
| | $\beta_2$ | 0.000727 $ms^{-1}$ | |
| | $\alpha_3$ | 0.0177 $ms^{-1}$ | |
| | $\beta_3$ | 0.004 $ms^{-1}$ | |
| | $\alpha_4$ | 0.0168 $ms^{-1}$ | |
| | $\beta_4$ | 0.1904 $ms^{-1}$ | |

in the glutamate diffusion model. After computing the local glutamate concentrations at the location of the different receptors from the glutamate diffusion model, we input these glutamate concentrations into the receptor state kinetic models to investigate the dynamics of these receptors. We assume each receptor is independent of one another. Therefore, to get the total number of receptors opened by ligand at each time point, we first calculate the fraction of a single receptor in the open state and multiply it by the total number of receptors at each particular location. The multi-state kinetic schemes of each type of receptor are described by groups of deterministic ordinary differential equations (ODEs) with reaction rate constants. Differential equations are numerically solved using *ode15s* in Matlab. The set of ODEs is in Appendix E.

#### 7.2.2.1. *Distribution of receptor numbers*

Both AMPAR and NMDAR are not fixed on the plasma membrane. They undergo trafficking from the intracellular synthesis sites to function sites on the surface of the plasma membrane. They are inserted into the membrane by exocytosis; move between different sites via lateral diffusion on the membrane; and can be internalized by endocytosis [18]. Although the trafficking of receptors plays important roles in mediating postsynaptic activity, we have not included it in this model because it is a relatively slower process compared to the reactions in this model. Therefore, we assume the numbers of all receptors at different locations are constant during each simulation.

The NMDAR number per synapse is relatively stable, about 20, among the different sizes of synapses [13, 19]. A hippocampal slice study suggests that the ratio of eNMDARs to synaptic ones is around 1:2, with no exchange between them [21]. Moreover, only about 25% of eNMDARs are located at the perisynaptic site [6].

NMDAR consists of two NR1 subunits and two NR2 subunits. The subunit composition of NMDARs at different locations changes during postnatal development (reviewed in [25]). The ratio of NR2A to NR2B increases at the synaptic site and decreases at the extrasynaptic site, respectively, during postnatal development. In mature synapses, NR2A-NMDARs are predominant at the synaptic sites, which takes about 60% of the total synaptic NMDARs [20]. In contrast, perisynaptic NMDARs and extrasynaptic NMDARs are mainly NR1/NR2B-NMDARs (pNR2B-NMDAR and eNR2B-NMDAR). Based on the experimental observations we select numbers for each type of NMDAR at different locations to represent NMDAR distribution in a mature synapse, as listed in Table 7.2.

Unlike NMDARs, the number of AMPARs depends on the spine geometry and is positively correlated with the PSD size [13, 26]. The average density of AMPARs is suggested to be uniformly distributed at different synapses and will be of a much higher density in synaptic sites rather than in extrasynaptic sites [22]. We assume that the functional AMPARs are homogenously located in the membrane of the PSD and at the dendritic shaft with different densities, and there is no AMPAR in the rest of the membrane of the spine. The number of AMPARs per spine is critical for the generation of the temporary depolarization of the postsynaptic membrane potential after stimulation, which is also called the excitatory postsynaptic potential (EPSP). We estimate the AMPAR numbers in the PSD based on the established experimental data in Section 7.3.3. The number of AMPARs at the dendritic shaft is low and stable, and is calculated based on the membrane surface area. The receptor density in Table 7.2 is taken from [22].

### 7.2.2.2. *Kinetic model of a single NMDAR*

NMDARs are heteromeric tetramers and the activation of an NMDAR requires both the binding of glutamate molecules to the NR2 subunits and the binding of coagonists (d-serine or glycine) to the NR1 subunits [27]. In this study, we simulate the state transition of NMDAR by using a subtype specific kinetic model [23] (Fig. 7.4(a)). We assume that the two NR1 subunits are

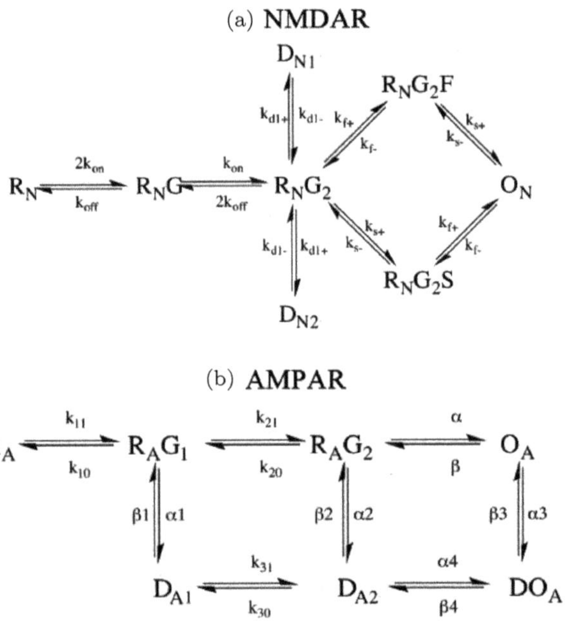


Fig. 7.4. Markov kinetic scheme of (a) NMDAR and (b) AMPAR. The kinetic rate constants are listed in Table 7.2. Acknowledgment [4]: Liang, J., Kulasiri, D. and Samarasinghe, S. (2017) Computational investigation of Amyloid-$\beta$-induced location-and subunit-specific disturbances of NMDAR at hippocampal dendritic spine in Alzheimer's disease. *PLoS One 12*(8), p. e0182743.

always occupied by co-agonists. Therefore, the activation of the receptor depends only on the glutamate concentration. In this model, NR1/NR2A-NMDAR and NR1/NR2B-NMDAR share the same activation schemes but with different reaction rate constants (Table 7.2), which allows us to investigate the effects of the different types of NMDAR. This eight-state kinetic scheme incorporates separate binding steps of two glutamate molecules, two desensitized states ($D_1$ and $D_2$) and two conformational change stages (RA2f and RA2s) before opening (O). Channel opening requires both faster and slower conformational changes in the NR1 and NR2 subunits, respectively. These two transition processes are independent and can happen in any order. The dynamics of this kinetic scheme is simulated by seven ODEs (see Appendix E.2). Therefore, by solving the ODEs, the open fraction of a single NMDAR, $Po_{\mathrm{NMDAR}}(t)$, is calculated from the fraction of NMDARs at state O and at time $t$.

#### 7.2.2.3. *Kinetic model of AMPAR*

The dynamics of a single AMPAR is simulated by a seven-state model (Fig. 7.4(b)) [24]. This model contains binding states of one and two glutamate molecules ($R_AG$ and $R_AG_2$), the open state ($O_A$), and their corresponding desensitized states ($D_1$, $D_2$, $D_0$). We apply the same simulation procedure as for the NMDAR dynamic simulation to obtain the open fraction of AMPAR, $Po_{\mathrm{AMPAR}}(t)$, by solving six ODEs (see Appendix E.3).

#### 7.2.2.4. *Differential contribution of AMPAR and NMDAR in $Ca^{2+}$ signaling*

NMDAR is the major $Ca^{2+}$ channel on the neuron membrane that contributes to the $Ca^{2+}$ influx from the extracellular space during the synaptic activation in CA1 pyramidal neurons [28]. The full opening of a NMDAR depends on both the ligand concentration and the membrane voltage. In this section, we discuss the ligand concentration-dependent property of NMDARs. Modeling of the voltage-dependent property of NMDAR is presented in Section 7.2.4.

Experimental evidence shows that AMPAR has a low $Ca^{2+}$ ion permeability [29], thus, we have not included AMPAR as a $Ca^{2+}$ channel. However, AMPAR indirectly mediated the $Ca^{2+}$ dynamics leading to membrane depolarization after activation. This will remove the $Mg^{2+}$ blockage of NMDARs, as well as activate the voltage-dependent $Ca^{2+}$ channels (VDCCs) on the membrane, and allow extracellular $Ca^{2+}$ ions to enter into the neuron.

### 7.2.3. *Membrane potential*

The removal of the $Mg^{2+}$ blockage of NMDARs and the activation of VDCC depends on the membrane depolarization after stimulation. We build an electrical model (Fig. 7.5) of a single spine and its adjacent dendritic shaft to capture the dynamics of the membrane potential. We construct the electrical model based on the morphological features, as described in Section 7.2.1.

#### 7.2.3.1. *Spine head and spine neck*

The spine head is modeled as an isopotential hemispherical compartment. Its membrane capacitance and resistance can be calculated by its surface area ($A_{\text{head}}$), the specific membrane resistance ($R_m$) and the specific capacitance ($C_m$) [30]:

$$R_{\text{head}} = \frac{R_m}{A_{\text{head}}},$$

and

$$C_{\text{head}} = C_m A_{\text{head}}.$$

Therefore, the postsynaptic membrane potential is

$$C_{\text{head}} \frac{dV_{\text{head}}}{dx} = \frac{E_{\text{rest}} - V_0}{R_{\text{head}}} - I_{syn} + I_{\text{receptor},1}, \tag{7.2}$$

where $I_{\text{receptor},1}$ represents the current of receptors on the spine head and $I_{syn}$ represents the current flow through the spine neck. Based on Ohm's law, $I_{syn}$ can be calculated as the voltage drop from the spine head ($V_{\text{head}}$) to the dendrite shaft ($V_d$) across the spine neck

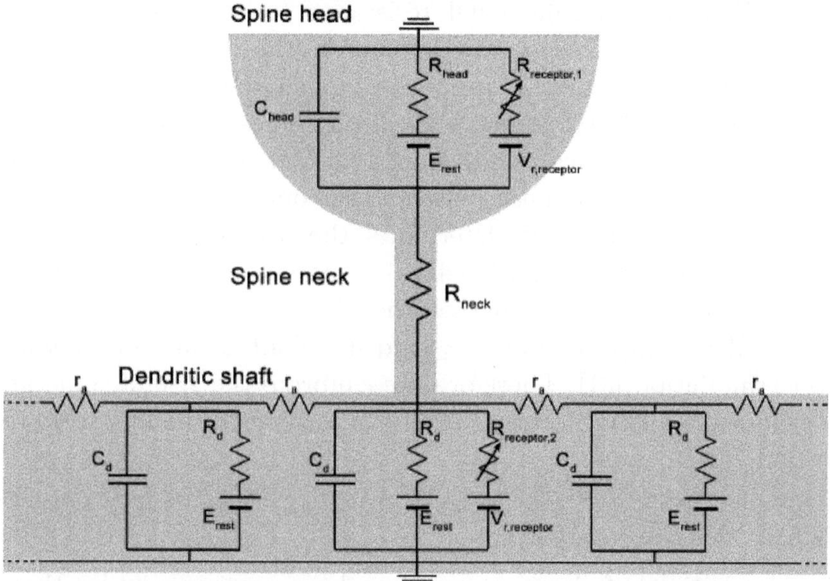


Fig. 7.5. Passive electrical model of a dendritic spine and the adjacent dendritic shaft of a CA1 pyramidal neuron. The spine head and dendritic shaft are modeled as separate compartments, with membrane capacitances, $C_{\text{head}}$ and $C_d$, and resistances, $R_{\text{head}}$ and $R_d$, respectively. $R_{\text{receptor},1}$ and $R_{\text{receptor},2}$ are the resistance of receptors in the membrane of the spine head and the dendritic shaft, respectively. The spine head and dendritic compartments are connected by the spine neck, with neck resistance, $R_{\text{neck}}$. The dendritic shaft is modeled as a series of identical cylindrical compartments. The resting potential, $E_{\text{rest}}$, is assumed to be the same in all compartments ($E_{\text{rest}} = -70$ mV). Acknowledgment [4]: Liang, J., Kulasiri, D. and Samarasinghe, S. (2017) Computational investigation of Amyloid-β-induced location-and subunit-specific disturbances of NMDAR at hippocampal dendritic spine in Alzheimer's disease. *PLoS One 12*(8), p. e0182743.

resistance ($R_{\text{neck}}$)

$$I_{syn} = \frac{V_{\text{head}} - V_d}{R_{\text{neck}}}.$$

### 7.2.3.2. *Spine neck resistance*

The spine neck is simulated as an electrical cylindrical resistor of length $l_{\text{neck}}$ and radius $r_{\text{neck}}$, therefore, its resistance is

$$R_{\text{neck}} = R_i l_{\text{neck}} \pi r_{\text{neck}}^2,$$

where $R_i$ is the specific axial resistance (or specific cytoplasmic resistivity). In the CA1 pyramidal neuron, the values of neck length and width range from 0.157 $\mu$m to 1.8 $\mu$m and 0.059 $\mu$m to 0.292 $\mu$m, respectively [16]. If we assume a specific axial resistance of $R_i = 200\Omega$ cm, the neck resistance ranges from 13.4 M$\Omega$ to 330 M$\Omega$. The large resistance of the spine neck between the compartments of the spine head and dendritic shaft determines the decrement of the EPSP amplitude from the spine head to its parent dendritic shaft after stimulation. It is suggested that the $R_{\text{neck}}$ is negatively correlated to the EPSP amplitude of the dendritic shaft membrane potential after stimulation [31]. Therefore, we calibrate the size of spine neck to generate realistic electrical results according to the literature (see Section 7.3.2).

#### 7.2.3.3. *Dendritic shaft*

The dendritic shaft is simulated as a long passive cylinder that is divided into several identical compartments [32]. Each compartment has a length of $\lambda$ and a diameter of $d$. The current can flow between neighboring compartments, which is expressed as the voltage drop between neighboring compartments across the axial resistance. Therefore, for compartment $j$, its membrane potential ($V_{d,j}$) is

$$C_{d,j}\frac{dV_{d,j}}{dt} = \frac{E_{\text{rest}} - V_{d,j}}{R_{d,j}} + \frac{V_{d,j+1} - V_{d,j}}{r_a} - \frac{V_{d,j} - V_{d,j-1}}{r_a} + I_{syn,j} + I_{\text{receptor},2}, \tag{7.3}$$

where $C_{d,j}$, and $R_{d,j}$ are the membrane capacitance and membrane resistance of the dendritic compartments and $r_a$ is the axial resistance between the neighboring dendritic compartments. They can be calculated from the specific resistances, $R_m$ and $R_a$, and the size of the compartment. Thus, Eq. (7.3) can be written as

$$\pi d\lambda C_m\frac{dV_{d,j}}{dt} = \frac{E_{\text{rest}} - V_{d,j}}{\frac{R_m}{\pi d\lambda}} + \frac{V_{d,j+1} - V_{d,j}}{\frac{R_a}{\pi\left(\frac{d}{2}\right)^2}\lambda} - \frac{V_{d,j} - V_{d,j-1}}{\frac{R_a}{\pi\left(\frac{d}{2}\right)^2}\lambda} + I_{syn,j}I_{\text{receptor},2}.$$

Multiply both sides by $\frac{R_m}{\pi d \lambda}$ and then

$$\tau_m \frac{dV_{d,j}}{dt} = E_{\text{rest}} - V_{d,j} + \frac{R_m d}{4R_a}\left(\frac{V_{d,j+1} - 2V_{d,j} + V_{d,j}}{\lambda^2}\right) + \frac{R_m}{\pi d \lambda}(I_{syn,j} + I_{\text{receptor},2}).$$

Let $\lambda = \sqrt{\frac{R_m d}{4R_i}}$ be the compartment length.

Then

$$\tau_m \frac{dV_{d,j}}{dt} = E_{\text{rest}} - V_{d,j} + (V_{d,j+1} - 2V_{d,j} + V_{d,j}) + \sqrt{\frac{4R_m R_i}{\pi^2 d^3}}(I_{syn,j} + I_{\text{receptor},2}).$$

In the last term, $I_{syn,j}$ is the total input current by the attached spine necks:

$$I_{syn,j} = \sigma I_{syn},$$

where $\sigma$ is the density of the attached spines per compartment. $I_{\text{receptor},2}$ represents the current of the receptors on the dendritic shaft. We assume $\sigma = 1$ because we focus on the stimulation on a single dendritic spine rather than the co-activation between neighboring spines.

In this study, the basal dendrite is represented as a series of cylindrical compartments with sealed ends on both sides. The sealed end boundary condition assumes that the resistance of the compartment is very high; therefore, the exit current flow via the end is negligible. We assume only one spine is attached to the middle point of the dendrite. Therefore, for $N$ dendritic compartments, the changes in voltages are

$$\tau_m \frac{dV_{d,1}}{dt} = E_{\text{rest}} - V_{d,1} + 2V_{d,2} - 2V_{d,1} + \sqrt{\frac{4R_m R_i}{\pi^2 d^3}}(I_{syn,j} + I_{\text{receptor},2}), \tag{7.4}$$

$$\tau_m \frac{dV_{d,j}}{dt} = E_{\text{rest}} - V_{d,j} + (V_{d,j+1} - 2V_{d,j} + V_{d,j}),$$
$$j = 2, \ldots, N-1, \tag{7.5}$$

and

$$\tau_m \frac{dV_{d,N}}{dt} = E_{\text{rest}} - V_{d,1} + 2V_{d,N-1} - 2V_{d,N}. \tag{7.6}$$

7.2.3.4. *Receptor current*

The total input current of the spine head and the shaft are a summation of the current of both local NMDARs and AMPARS:

$$I_{\text{receptor},1} = I_{\text{NMDAR,PSD}} + I_{\text{NMDAR,perisynaptic}} + I_{\text{AMPAR,PSD}},$$
$$I_{\text{receptor},2} = I_{\text{NMDAR},d} + I_{\text{AMPAR},d}.$$

The receptor current at a specific location $i$ ($I_{\text{NMDAR},I}$ and $I_{\text{AMPAR},I}$) is

$$I_{\text{NMDAR},i} = Po_{\text{NMDAR},i} G_{\text{NMDAR},i} B(V_i)(V_i - V_{r,\text{NMDAR}}), \tag{7.7}$$
$$I_{\text{AMPAR},i} = Po_{\text{AMPAR},i} G_{\text{AMPAR},i}(V_i - V_{r,\text{AMPAR}}), \tag{7.8}$$

where $V_{r,\text{NMDAR}}$ and $V_{r,\text{AMPAR}}$ are the reversal potentials for NMDAR and AMPAR, respectively. $Po_{\text{NMDAR},I}$ and $Po_{\text{AMPAR},I}$ are the fractions of the NMDARs and AMPARs in the open states at site $i$, respectively. $V_i$ is the membrane voltage of site $i$.

$G_{\text{NMDAR},I}$ and $G_{\text{AMPAR},I}$ are the maximum conductances for the corresponding type of receptor at site $i$. These are the conductances when all NMDARs and AMPARs are saturated:

$$G_{\text{NMDAR},i} = g_{\text{NMDAR}} * N_{\text{NMDAR},i}, \tag{7.9}$$
$$G_{\text{AMPAR},i} = g_{\text{AMPAR}} * N_{\text{AMPAR},i}, \tag{7.10}$$

where $g_{\text{NMDAR}}$ and $g_{\text{AMPAR}}$ are the single channel conductance of NMDARs and AMPARs, respectively. $N_{\text{NMDAR},i}$ and $N_{\text{AMPAR},i}$ are the total NMDAR and AMPAR numbers at site $i$.

Table 7.3. Membrane potential related parameters.

| Description | Symbol | Value | Reference |
|---|---|---|---|
| Specific resistance | $R_m$ | $30\ \mathrm{k\Omega \cdot cm^2}$ | [34] |
| Specific capacitance | $C_m$ | $1\ \mu\mathrm{F/cm^2}$ | |
| Specific cytoplasmic resistivity | $R_i$ | $100\ \mathrm{k\Omega \cdot cm}$ | |
| Membrane time constant | $\tau_E$ | $R_m \cdot C_m$ | |
| Resting membrane potential | $E_{\mathrm{rest}}$ | $-70$ mV | [35] |
| Single channel conductance of NMDAR | $g_{\mathrm{NMDAR}}$ | 45 pS (20–24°C) | [36] |
| Single channel conductance of AMPAR | $g_{\mathrm{AMPAR}}$ | 10 pS (20–24°C) | [36] |
| Reversal potentials for NMDAR and AMPAR | $V_{r,\mathrm{NMDAR}}$ $V_{r,\mathrm{AMPAR}}$ | 0 mV | [33] |
| Extracellular magnesium concentration | [Mg] | 1 mM | [33] |

In Eq. (7.6), $B(V_I)$ describes the $Mg^{2+}$ blockage of NMDARs under $V_I$ at each time step [33],

$$B(V_i) = \frac{1}{1 + \frac{[\mathrm{Mg}]}{3.57} \exp(-0.062V_i)}, \tag{7.11}$$

where $[Mg]$ is the extracellular magnesium concentration (1 mM). The values of the parameters in this section are listed in Table 7.3.

### 7.2.4. *Compartmental model of $Ca^{2+}$ dynamics*

We construct a four-compartment model to represent a pyramidal neuron dendritic spine and its adjacent dendritic shaft. The four compartments are the PSD, cytosol, spine neck and dendritic shaft (Fig. 7.6). The geometry of the spine is consistent with what is described in the diffusion model (Table 7.1). The spine head is assumed to be a hemisphere and is divided into PSD and

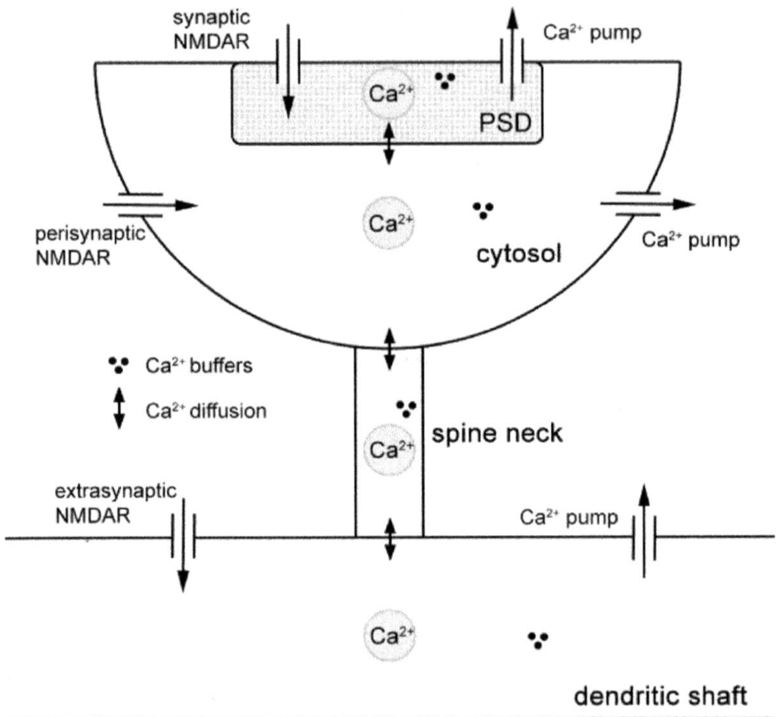


Fig. 7.6. Schematic diagram of $Ca^{2+}$ dynamics at the dendritic spine head and its adjacent dendritic shaft, for a CA1 pyramidal neuron. The spine head is divided into two compartments, the PSD and cytosol. A thin long spine neck links the spine head to the dendritic shaft, which allows $Ca^{2+}$ to diffuse from the spine head to the dendrite. $Ca^{2+}$ enters the PSD, cytosol and the dendritic shaft via NMDARs and is extruded by $Ca^{2+}$ pumps in all compartments. $Ca^{2+}$ buffers are distributed homogenously within each compartment. Diffusion of $Ca^{2+}$ between compartments is also considered. This schematic diagram does not represent the actual scale of the model. Acknowledgment [4]: Liang, J., Kulasiri, D. and Samarasinghe, S. (2017) Computational investigation of Amyloid-$\beta$-induced location-and subunit-specific disturbances of NMDAR at hippocampal dendritic spine in Alzheimer's disease. *PLoS One 12*(8), p. e0182743.

cytosol compartments. The PSD compartment is a cylinder attached to the postsynaptic membrane and the rest of the spine head is in the cytosol compartment. We assume the volume of the spine head is 0.1 $\mu m^3$ (radius = 0.363 $\mu$m) and the PSD takes 10% of the total volume of the spine head [37, 38]. Therefore, the compartment volumes of the PSD and cytosol are 0.01 and 0.09 $\mu m^3$, respectively. The radius of the PSD is 150 $\mu m$ and the height of its compartment is calculated from its volume, which is about 0.142 $\mu m$. The spine neck is represented as a long thin cylinder, which is coaxial with the spine head. The size of the spine neck is discussed in Section 7.3.2. The dendritic shaft is another cylinder that is attached to the bottom of the spine neck with a radius of 0.5 $\mu m$ and a length of 1 $\mu m$.

Extracellular $Ca^{2+}$ ions entering each compartment result from the local membrane $Ca^{2+}$ current that is then buffered by various $Ca^{2+}$ buffer proteins extruded via membrane $Ca^{2+}$ pumps and diffused from the spine head to the dendrite. We assume the molecules are well-mixed in each compartment and only diffusion of $Ca^{2+}$ between the compartments is considered. The $Ca^{2+}$ dynamics in PSD, spine head, spine neck and dendritic shaft are governed by

$$\frac{d[Ca^{2+}]_j}{dt} = J_{I_{Ca},j} + J_{\text{mem},j} + J_{\text{diffusion},j} - J_{\text{buffer},j}, \tag{7.12}$$

where $j$ indicates the compartment index ($j = 1$, PSD; 2, cytosol; 3, spine neck; 4, dendritic shaft). The values of the parameters are in Table 7.4.

Table 7.4. Parameters for the spine compartment model.

| Description | Symbol | Value | References |
|---|---|---|---|
| Pump affinity (PMCA) | $K_{M,1}$ | 0.2 $\mu$M | [39–41] |
| Turnover rate (PMCA) | $V_{max,1}$ | 100 $ms^{-1}$(37°C) | [39–41] |
| Pump affinity (NCX) | $K_{M,2}$ | 20 Mm | [39–41] |
| Turnover rate (NCX) | $V_{max,2}$ | 1000 $ms^{-1}$(37°C) | [39–41] |
| $Ca^{2+}$ diffusion coefficient | $D_{Ca}$ | 0.220 $\mu m^2 ms^{-1}$ (20–24°C) | [42] |
| Forward rate constant of mobile buffer | $k_{bf}$ | 0.176 $\mu M^{-1} ms^{-1}$ | MCMC estimated |
| Backward rate constant of mobile buffer | $k_{bb}$ | 0.624 $ms^{-1}$ | MCMC estimated |
| Total buffer concentration in compartment $j$ | $[B]_{\text{total},j}$ | 108.78 $\mu$M (spine head) 116.97 $\mu$M (dendritic shaft) | MCMC estimated |

### 7.2.4.1. $Ca^{2+}$ *influx by* $Ca^{2+}$ *current*

In Eq. (7.12), $J_{I_{Ca},j}$ is the $Ca^{2+}$ influx by the $Ca^{2+}$ current, mediated by the membrane ionotropic receptors (NMDAR in this study). The $Ca^{2+}$ current at the spine head and dendritic shaft is assumed to be a fixed fraction of the total local input current. The fraction of the $Ca^{2+}$ current in the total cation current through NMDARs is about 10% [35] whereas, through AMPARs, is about 0.6% [43]. Therefore, the $Ca^{2+}$ influx into the spine head and dendritic shaft is mainly mediated by local NMDARs. We only consider the contribution of AMPARs to the depolarization of the synaptic membrane and ignore the $Ca^{2+}$ influx via AMPAR in this study. The $Ca^{2+}$ influx via NMDAR is given by

$$J_{\mathrm{NMDAR},i} = -\frac{I_{\mathrm{NMDAR},i}}{Z_{\mathrm{Ca}} F \mathrm{Vol}_j}, \tag{7.13}$$

where $I_{\mathrm{NMDAR},i}$ is the NMDAR current calculated in Eq. (7.7), $f_{\mathrm{Ca}} =$ 10% is the fraction of $Ca^{2+}$ current carried by NMDAR [35], $F =$ 96485.3°C $\mathrm{Mol}^{-1}$ is Faraday's constant, $Z_{\mathrm{Ca}} = 2$ is the valence of $Ca^{2+}$ ions and $\mathrm{Vol}_j$ is the volume of compartment $j$. $J_{I_{\mathrm{Ca}},j}$ is not valid when $j = 3$.

### 7.2.4.2. *VDCCs*

VDCCs are located at both the spine head and the dendritic shaft. We calculate the $Ca^{2+}$ current by VDCCs as

$$I_{\mathrm{VDCC},i} = G_{\mathrm{VDCC},i} m h (V_i - V_r^{\mathrm{Ca}^{2+}}),$$

where $G_{\mathrm{VDCC},i}$ and $V_i$ are the maximum conductance of VDCC and membrane potential at site $I$, respectively. $G_{\mathrm{VDCC},i}$ is estimated in Section 7.3.2. $V_r^{\mathrm{Ca}^{2+}}$ is the membrane reversal potential of the $Ca^{2+}$ ions:

$$V_r^{\mathrm{Ca}^{2+}} = \frac{RT}{Z_{\mathrm{Ca}} F} \ln \frac{\mathrm{Ca}_o}{\mathrm{Ca}_i}, \tag{7.14}$$

where $R$ is the ideal gas constant, $T$ is the absolute temperature in Kelvin, $Z_{ca}$ is the valence of $Ca^{2+}$ and $F$ is the Faraday constant. $\mathrm{Ca}_o$ represents the extracellular $Ca^{2+}$ concentration ($\mathrm{Ca}_o = 100\ \mu$M) and $\mathrm{Ca}_i$ represents the cytosolic $\mathrm{Ca}^{2+}$ concentration in compartment $i$, respectively. $m$ and $h$ represent the activation and deactivation of VDCC, and they are calculated as

$$\frac{dm}{dt} = (1-m)\alpha_m - m\beta_m,$$
$$\frac{dh}{dt} = (1-h)\alpha_h - h\beta_h,$$

where

$$\alpha_m = \frac{8.5}{1+\exp\left(-\frac{V_i-8}{12.5}\right)},$$
$$\beta_m = \frac{35}{1+\exp\left(\frac{V_i+74}{14.5}\right)},$$
$$\alpha_h = \frac{0.0015}{1+\exp\left(\frac{V_i+29}{8}\right)},$$
$$\beta_h = \frac{0.0055}{1+\exp\left(-\frac{V_i+23}{8}\right)}.$$

### 7.2.4.3. *$Ca^{2+}$ pumps and leakage*

We assume there is no ER in the dendritic spine, therefore, $\mathrm{Ca}^{2+}$ extrusion from the cytosol of CA1 pyramidal neurons is mainly mediated by two types of membrane pump: the plasma membrane $\mathrm{Ca}^{2+}$ ATPase (PMCA) and the sodium $\mathrm{Ca}^{2+}$ exchanger (NCX). PMCA has about a 10-fold higher affinity for $\mathrm{Ca}^{2+}$ but a lower turnover rate than the NCX. We follow the model of Schiegg, Gerstner, Ritz and van Hemmen [35] to simulate $\mathrm{Ca}^{2+}$ extrusion using first-order Michaelis–Menten kinetics (see Appendix C.4 for the explanation). They are modeled as

$$J_{\mathrm{mem},j} = -\sum_n \frac{A_j}{V_j} Ps_n V_{max,n} \frac{[\mathrm{Ca}^{2+}]_j}{[\mathrm{Ca}^{2+}]_j + K_{d,n}} + \frac{A_j}{V_j} J_{\mathrm{leak},j}, \qquad (7.15)$$

where the maximum pumping velocity of pump $n, (n=1, \text{PMCA}; 2, \text{NCX})$ at compartment $j$ is calculated by the multiplication of the maximum turnover rates of $Ca^{2+}$ of pump $n$, $V_{max,n}$ is surface density, $Ps_n$, and the ratio of surface area to volume of compartment $j$, $\frac{A_j}{V_j}$. $K_{d,n}$ is the dissociation constant for pump $n$. $J_{\text{leak},j}$ is the $Ca^{2+}$ leakage flux via the membrane, which balances the $Ca^{2+}$ concentration at the resting level. The densities of pumps at different locations are difficult to be precisely estimated from experimental data; therefore, we estimate the values of them in Section 7.3.4 using a Markov chain Monte Carlo (MCMC) (see Appendix F.1) from the experimental data [39].

#### 7.2.4.4. *Diffusion*

We simulate the $Ca^{2+}$ diffusion between compartments using Fick's first law [32]. The concentration change of $Ca^{2+}$ in compartment $j$ due to diffusion is

$$J_{\text{diffusion},j} = -\frac{D_{\text{Ca}}}{V_j}\left[\left(\frac{A}{\delta}\right)_{j,j-1}([\text{Ca}^{2+}]_j - [\text{Ca}^{2+}]_{j-1}) - \left(\frac{A}{\delta}\right)_{j+1,j}([\text{Ca}^{2+}]_{j+1} - [\text{Ca}^{2+}]_j)\right], \tag{7.16}$$

where $D_{\text{Ca}}$ is the diffusion coefficient of $Ca^{2+}$, and $V_j$ is the volume of compartment $j$. $\left(\frac{A}{\delta}\right)_{p,q} = \frac{A_{pq}}{\Delta l_{pq}}$ describes the movement of $Ca^{2+}$ ions from compartment $p$ to $q$, where $A_{pq}$ is the cross-sectional area of the two compartments and $\Delta l_{pq}$ is the distance between the midpoints of compartments $p$ and $q$.

#### 7.2.4.5. *Buffering*

In each compartment, we include a general form of an endogenous immobile $Ca^{2+}$ buffer protein and two mobile buffer proteins: calbindin (CaD) and calmodulin (CaM). The buffers' capacity for $Ca^{2+}$ decides the shape of the decay in the cytosolic $Ca^{2+}$ concentration after reaching a peak. The last part in Eq. (7.12) is $Ca^{2+}$ buffering by all types of endogenous buffers. The concentration of immobile buffer protein in each compartment is estimated in Section 7.3.2. The simulation details of immobile $Ca^{2+}$ buffer, CaD and CaM are in Appendix D.2.

## 7.3. Parameter Calibration and Estimation

### 7.3.1. *Temperature Coefficient ($Q_{10}$) corrections*

Since most reaction and diffusion rates are affected by the environmental temperature, it is necessary to correct the values of the rate constants in regard to their temperature sensitivity. The temperature coefficient ($Q_{10}$) measures the degree of temperature dependence when increasing the temperature by 10°C. It is calculated according to

$$Q_{10} = \left(\frac{R_2}{R_1}\right)^{\frac{10}{T_2 - T_1}},$$

where $R_1$ and $R_2$ are the rate constants at temperatures $T_1$ and $T_2$ ($T_1 < T_2$), respectively. Therefore, the corrected value, $R_{\text{sim}}$, of a parameter with original value, $R_o$, from the original temperature $T_o$ to simulation temperature $T_{\text{sim}}$ is

$$R_{\text{sim}} = \begin{cases} R_o \times Q_{10}^{\frac{T_{\text{sim}} - T_o}{10}}, & T_{\text{sim}} > T_o, \\ \dfrac{R_o}{Q_{10}^{\frac{T_o - T_{\text{sim}}}{10}}}, & T_{\text{sim}} < T_o. \end{cases}$$

The values of $Q_{10}$ for each type of parameter are listed in Table 7.5. The simulation temperature used in this chapter is 34°C.

### 7.3.2. *Resistance*

We explore the dependency of the spine neck size on the $EPSP_{\text{spine}}$ and the ratio of EPSP at the spine head to the dendrite shaft ($EPSP_{\text{spine}}/EPSP_{\text{shaft}}$). We vary the length and width of the spine neck in the physiological range [16] and investigate its effects on the $EPSP_{\text{spine}}$ and the $EPSP_{\text{spine}}/EPSP_{\text{shaft}}$ (Fig. 7.7). The results are consistent with the experimental observation that as the increment in spine neck length increases/decreases, the $EPSE_{\text{spine}}$ decrement of

Table 7.5. $Q_{10}$ values for biological processes.

| Biological processes | $Q_{10}$ |
|---|---|
| Diffusion | 1.3 |
| Glutamate transporter kinetics | 3 |
| AMPAR kinetics | 2.4 |
| AMPAR conductance | 1.5 |
| NMDAR kinetics | 3 |
| NMDAR conductance | 1.6 |
| Pump kinetics | 3 |
| Buffer kinetics | 2.15 |
| VDCC gating kinetics | 3 |
| VDCC conductance | 1.5 |

the EPSP amplitude from the spine head to its parent dendritic shaft increases linearly. The longer spine neck results in a larger resistance and allows more EPSP to stay in the spine head with a higher $EPSP_{peak}$. In contrast, the spine head EPSP decrement decreases exponentially with the enlargement in the spine width.

### 7.3.3. *Synaptic AMPAR number and spine*

Based on experimental observation, we assume that the EPSP amplitude at the synaptic site is lower than 5 mV after a single synaptic stimulation [44, 45]. The EPSP is mainly AMPAR-dependent and partially NMDAR-dependent. Therefore, we adjust the number of AMPARs to generate the expected EPSP amplitude. The range of AMPAR numbers for testing is calculated based on the spine head size and the density of AMPAR [22]. It has been reported that the AMPAR density in synaptic site ranges from 874 to 946 receptor per $\mu m^2$ in single synapses in the immature cerebellum. The density range is much wider in the adult cerebellum, 48 to 1210 receptors per $\mu m^2$ [46]. The area of the PSD in this model is 0.07 $\mu m^2$ ($r = 0.15$ $\mu m$), which gives a range of AMPAR from 3 to 86. Figure 7.8 shows the number of AMPARs in the PSD vs. the $EPSP_{peak}$ at the spine head. The EPSP amplitude increases linearly with the increasing AMPAR density. The simulation is based on $R_{neck} = 157$ M$\Omega$, representing the resistance of a medium sized

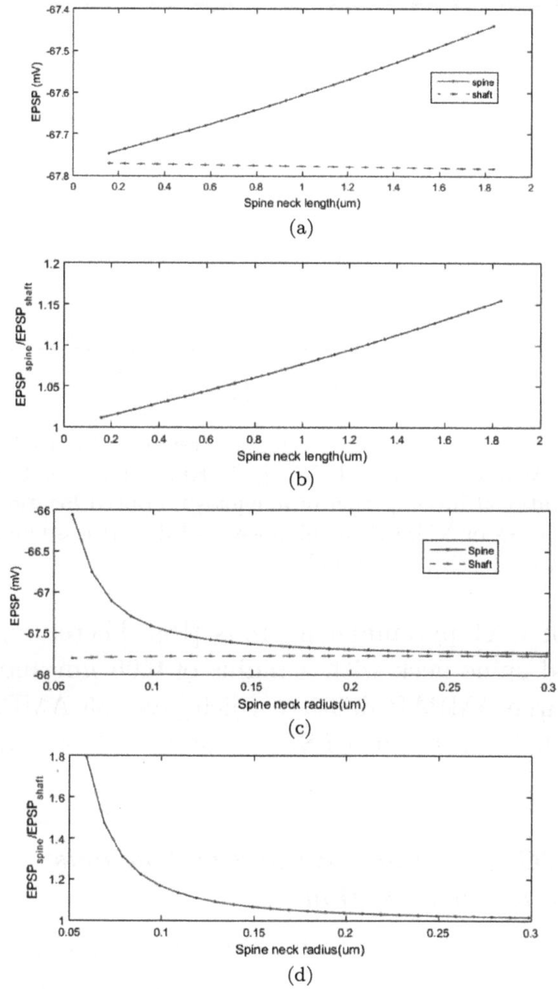


Fig. 7.7. Effect of spine neck size on the EPSP amplitude in the spine head and the spine-shaft EPSP ratio. Images (a) and (b) are simulated by varying the spine neck length from 0.157 to 1.8 $\mu$m while fixing the radius to 0.086 $\mu$m. Images (c) and (d) are simulated by varying the spine neck width from 0.059 to 0.292 $\mu$m while fixing the length to 0.689 $\mu$m. The model is simulated with a medium AMPAR density in PSD (AMPAR = 42). Acknowledgment [4]: Liang, J., Kulasiri, D. and Samarasinghe, S. (2017) Computational investigation of Amyloid-$\beta$-induced location-and subunit-specific disturbances of NMDAR at hippocampal dendritic spine in Alzheimer's disease. *PLoS One 12*(8), p. e0182743.

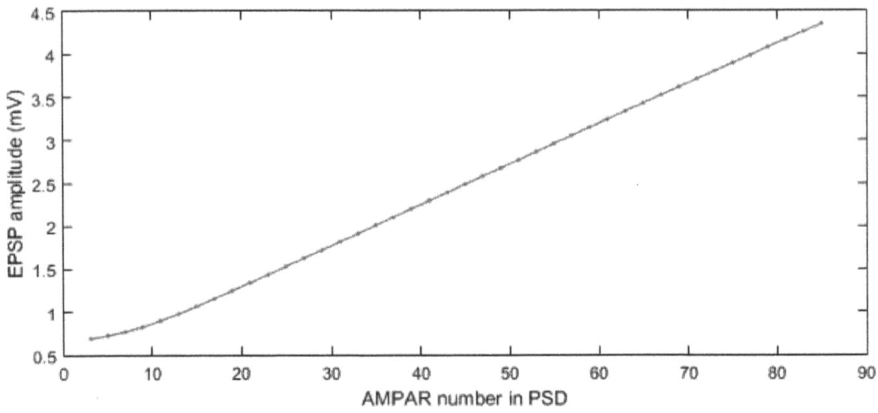


Fig. 7.8. AMPAR numbers in PSD positively relate to the EPSP amplitude in the spine head. Acknowledgment [4]: Liang, J., Kulasiri, D. and Samarasinghe, S. (2017) Computational investigation of amyloid-$\beta$-induced location-and subunit-specific disturbances of NMDAR at hippocampal dendritic spine in Alzheimer's disease. *PLoS One 12*(8), p. e0182743.

spine neck of CA1 pyramidal neurons [16]. Therefore, we chose a medium sized spine neck with a radius of 0.05 $\mu$m and a length of 0.75 and a large AMPAR density, which gives 85 AMPARs in PSD in this model. It causes an $EPSP_{spine}$ of 4.5 mV and an $EPSP_{shaft}$ of 1.7 mV.

### 7.3.4. *VDCC, pump density and endogenous protein concentration*

The pump densities for VDCC, PMCA and NCX and the concentration, forward/backward reaction rates of the endogenous immobile buffer and its concentration in the dendritic and spine locations are estimated according to the experimental observations in Sabatini *et al.* [39]. They estimate the $Ca^{2+}$ dynamics at the dendritic spines of the CA1 pyramidal neurons and their parent dendrites by a single bAP in the absence of exogenous buffer ($Ca^{2+}$ indicator) and the washout of mobile buffers. The results correspond to spines and small dendrites with surface-to-volume ratios of 4–20 $\mu m^{-1}$ and 1–4 $\mu m^{-1}$, respectively. The geometry of our model lies well in these ranges (spine: $\mu m^{-1}$; dendrite: $\mu m^{-1}$). We choose four pieces of target data from the experiment by Sabatini, Oertner and Svoboda [39]: the

Table 7.6. Values of target experimental data, means of MCMC samples and with the parameter set with the lowest mean absolute percentage error (MAPE).

| Target experimental data | Target value | Mean of MCMC samples | Set with the lowest MAPE |
|---|---|---|---|
| $\Delta[\mathrm{Ca}]_{\mathrm{bAP}}$ in spine ($\mu$M) | 1.7 ($\pm$0.6) | 1.200 | 1.668 |
| $\Delta[\mathrm{Ca}]_{\mathrm{bAP}}$ in dendrite ($\mu$M) | 1.5 ($\pm$0.5) | 1.248 | 1.471 |
| $\tau_{\mathrm{decay}}$ in spine (ms) | 12 ($\pm$4) | 12.8772 | 11.8580 |
| $\tau_{\mathrm{decay}}$ in dendrite (ms) | 15 ($\pm$5) | 16.4846 | 14.9430 |
| $\frac{[\mathrm{Ca}]_{\mathrm{PMCA}}}{[\mathrm{Ca}]_{\mathrm{NCX}}}$ in spine | 1.070 | 1.0716 | 0.9213 |
| $\frac{[\mathrm{Ca}]_{\mathrm{PMCA}}}{[\mathrm{Ca}]_{\mathrm{NCX}}}$ in dendrite | 0.825 | 0.8064 | 0.5409 |
| MAPE | | 0.1097 | 0.0896 |

amplitudes of $Ca^{2+}$ transient ($\Delta[\mathrm{Ca}]_{\mathrm{bAP}}$) in response to a single bAP in the spine and dendrite, and the corresponding decay time constant $\tau_{\mathrm{decay}}$ (Table 7.6). All temperature-dependent rate constants are adjusted according to the temperature their experiments are conducted at 34°C (a near-physiological temperature).

The bAP is simulated by injecting potential into the spine head and dendrite compartment of the passive electrical model to generate amplitudes of bAPs 66.4 and 66.7 mV, respectively [47]. The simulation details of bAP stimulation is in Chapter 2. This leads to $Ca^{2+}$ influx via VDCC at the spine head or the dendritic shaft. According to the experimental conditions in the experiment by Sabatini, Oertner and Svoboda [39], we set the glutamate concentration at the resting level (0.5 mM) to eliminate the NMDAR-dependent $Ca^{2+}$ transient. Therefore, VDCCs provide the only $Ca^{2+}$ source for the elevation of cytosolic $Ca^{2+}$ level during bAP. Under this condition, the mobile buffers are significantly washed out. Therefore, we set the concentration of CaD and CaM in all compartments as zero.

The clearance of cytosolic $Ca^{2+}$ is mediated by the endogenous protein and membrane pumps. The clearance ability can be evaluated by measuring the $Ca^{2+}$ decay time constant $\tau_{\mathrm{decay}}$. We calculated the value of $\tau_{\mathrm{decay}}$ according to the experiment of Sabatini, Oertner and Svoboda [39] by fitting a single exponential to the $Ca^{2+}$ concentration after reaching the peak value and until $t = 200$ ms. The contribution of PMCA and NCX on $Ca^{2+}$ extrusion in response to a single bAP in the dendritic spine is reported in Scheuss *et al.* (2006). We use the ratio of the fraction of $Ca^{2+}$ removed by PMCA to NCX $\left(\frac{[\mathrm{Ca}]_{\mathrm{PMCA}}}{[\mathrm{Ca}]_{\mathrm{NCX}}}\right)$ in the spine and dendrites from their research as an additional target data for parameter estimation (Table 7.6).

We estimate values of these parameters using the MCMC method [48, 49] (see Appendix F.1). We use the summation of the mean absolute percentage error (MAPE) to measure the goodness of the fit:

$$M = \sum_{n=1}^{8} \left| \frac{x_n - x_n^t}{x_n} \right|,$$

where $x_n$ and $x_n^t$ are the stimulated values and target values, respectively. Based on this equation, the MCMC converges to a set of parameter values with a minimum summation of the MAPE. The ranges of the parameters are selected from the established experimental literature.

The range of VDCC density is calculated from the $\Delta[\mathrm{Ca}]_{\mathrm{bAP}}$ and $\tau_{\mathrm{decay}}$ in [39], which gives the numbers of $Ca^{2+}$ ions entered via VDCC at compartments, the spine head and the dendritic shaft as about 1490–3116 and 13243–26486, respectively.

The means of MCMC samples and the parameter set with the lowest MAPE are listed in the last two columns in Table 7.7.

In comparing the simulation results using the mean of MCMC samples and parameter set with the lowest MAPE, the latter shows that $\Delta[\mathrm{Ca}]_{\mathrm{bAP}}$ in spine and dendrite are closer to the mean of the experimental data and show differences in the amplitudes in $Ca^{2+}$ transients between these two locations (Table 7.6). Therefore, we choose the parameters with lowest MAPE for our model.

## 7.4. Model Performance under Healthy Conditions

In this section, we simulate the healthy condition/control condition using the values of parameters estimated in Section 7.3.

Table 7.7. Parameter ranges for MCMC and the estimated values by MCMC.

| Parameter | Range for MCMC | Mean of MCMC samples | Set with the lowest MAPE |
|---|---|---|---|
| VDCC density in spine head | 2.23–4.66 $\mu m^{-2}$ | 3.577 $\mu m^{-2}$ | 3.498 $\mu m^{-2}$ |
| VDCC density in shaft | 7.29–14.57 $\mu m^{-2}$ | 11.106 $\mu m^{-2}$ | 9.855 $\mu m^{-2}$ |
| PMCA density in spine head | 200–2000 $\mu m^{-2}$ | 359.851 $\mu m^{-2}$ | 236.554 $\mu m^{-2}$ |
| PMCA density in shaft | 200–2000 $\mu m^{-2}$ | 577.689 $\mu m^{-2}$ | 296.785 $\mu m^{-2}$ |
| NCX density in spine head | 10–700 $\mu m^{-2}$ | 0.480 $\mu m^{-2}$ | 258.991 $\mu m^{-2}$ |
| NCX density in shaft | 100–1000 $\mu m^{-2}$ | 733.058 $\mu m^{-2}$ | 551.426 $\mu m^{-2}$ |
| Immobile buffer concentration in spine head | 50–200 $\mu M$ | 138.069 $\mu M$ | 108.782 $\mu M$ |
| Immobile buffer concentration in shaft | 50–200 $\mu M$ | 142.134 $\mu M$ | 116.966 $\mu M$ |
| Forward rate constant of immobile buffer | 0.05–1 $\mu M^{-1} ms^{-1}$ | 0.205 $\mu M^{-1} ms^{-1}$ | 0.176 $\mu M^{-1} ms^{-1}$ |
| Backward rate constant of immobile buffer | 0.05–1 $ms^{-1}$ | 0.569 $ms^{-1}$ | 0.624 $ms^{-1}$ |

We investigate the model performance in response to presynaptic stimulation (EPSP stimuli; see Chapter 2 for the detailed explanation) from different aspects: the glutamate profile, the open fraction of receptors at different locations and, consequently, the $Ca^{2+}$ dynamics in the spine head and its parent dendritic shaft.

### 7.4.1. *Glutamate profile and receptor activity*

After release from the vesicle, the glutamates diffuse rapidly from the synaptic cleft. This gives a peak concentration of 0.7 mM and a sharp decrease to the resting level of less than 1 ms (Fig. 7.9(a) left panel). The maximum fractions of receptors opened are 0.37 (sNR2A-NMDAR), 0.044 (sNR2B-NMDAR) and 0.25 (sAMPAR), respectively (Fig. 7.9(b) top panel). A single EPSP stimulus is too weak to activate receptors at the perisynaptic and extrasynaptic sites (Fig. 7.9(b) bottom panel) because of the large extrasynaptic volume in comparison with the synaptic cleft, and the long distance between them and the release site. Even for the presynaptic NMDARs, the local glutamate concentration peak is 33 $\mu$M (Fig. 7.9(a) middle panel), which gives a maximum open fraction lower than 0.0025 (Fig. 7.9(b) bottom left panel).

### 7.4.2. *Postsynaptic membrane depolarization*

A single EPSP stimulus leads to maximums of 4.5 mV and 1.7 mV depolarization at the spine head and the dendritic shaft, respectively (Fig. 7.10). The depolarization levels are not strong enough to open more VDCCs than under the resting conditions. The $Ca^{2+}$ influx depends mostly on the opening of NMDARs by depolarization and the glutamates diffused from the release sites.

### 7.4.3. *Mobile buffer proteins*

In Section 7.3, we removed the mobile buffers, CaD and CaM, by setting their concentration to zero to satisfy the experimental conditions for parameter estimation. Before starting further simulation, we need to return them back to their standard levels (Table 7.4). In response to a single vesicle release event from the presynaptic terminal (a single EPSP stimulus), the $Ca^{2+}$ peak amplitude in the spine head is lower ($\Delta[\mathrm{Ca}]_{\mathrm{EPSP}} = 0.5\,\mu\mathrm{M}$) with a slower decay time in the presence of CaD and CaM than in the absence ($\Delta[\mathrm{Ca}]_{\mathrm{EPSP}} = 1.8\,\mu\mathrm{M}$) (Fig. 7.11). A large fraction of free $Ca^{2+}$ ions are removed by the mobile buffers. The value of $\Delta[\mathrm{Ca}]_{\mathrm{EPSP}}$ lies in the range estimated under experimental conditions [32]. A single EPSP stimulus limits

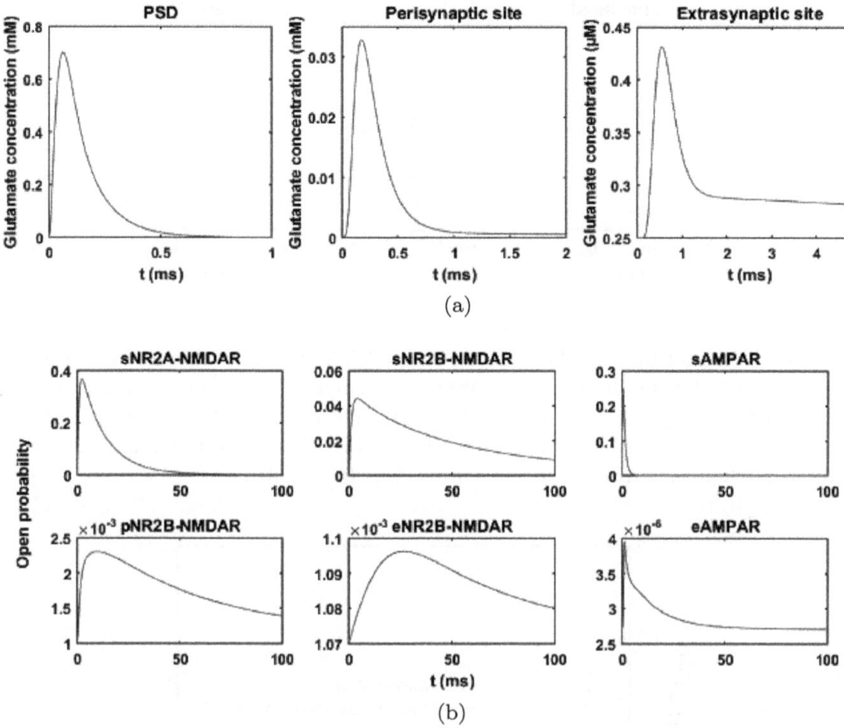


Fig. 7.9. (a) Glutamate concentration and (b) receptor response at different locations in response to a single EPSP stimulus under the healthy condition. Acknowledgment [4]: Liang, J., Kulasiri, D. and Samarasinghe, S. (2017) Computational investigation of Amyloid-$\beta$-induced location-and subunit-specific disturbances of NMDAR at hippocampal dendritic spine in Alzheimer's disease. *PLoS One 12*(8), p. e0182743.

the effect on cytosolic $Ca^{2+}$ to the spine head only. The change in $Ca^{2+}$ levels in the dendritic shaft is negligible.

### 7.4.4. *Multipulse stimulation*

We next simulate the model in response to different stimulation patterns: 1 s stimulation in LFS (10 Hz) and HFS (100 Hz) (see Chapter 2 for the detailed explanation). In both HFS and LFS conditions, glutamate transients at each stimulation pulse are well separated from each other because of the rapid diffusion. The peak

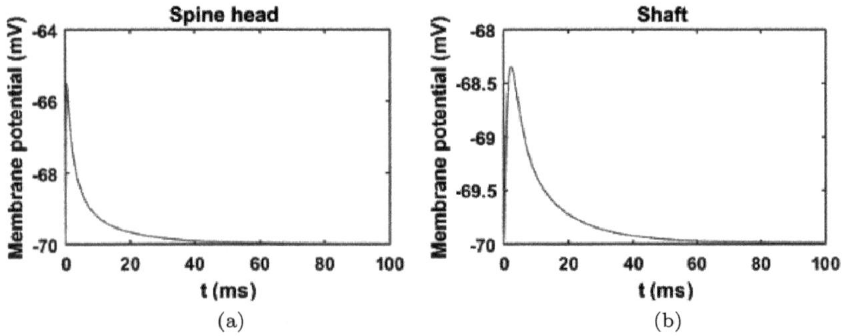


Fig. 7.10. Membrane depolarization by a single EPSP stimulus at (a) the spine head and (b) the dendritic shaft. Acknowledgment [4]: Liang, J., Kulasiri, D., and Samarasinghe, S. (2017) Computational investigation of Amyloid-$\beta$-induced location-and subunit-specific disturbances of NMDAR at hippocampal dendritic spine in Alzheimer's disease. *PLoS One 12*(8), p. e0182743.

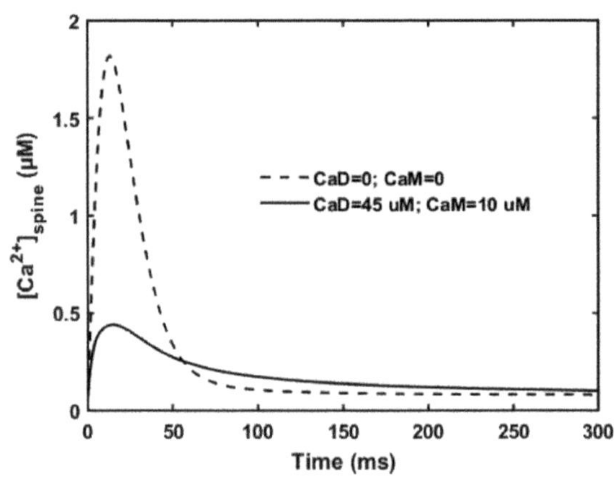


Fig. 7.11. $Ca^{2+}$ transient in the spine head in response to a single-pulse presynaptic stimulation in the presence and absence of mobile buffers. Acknowledgment [4]: Liang, J., Kulasiri, D. and Samarasinghe, S. (2017) Computational investigation of Amyloid-$\beta$-induced location-and subunit-specific disturbances of NMDAR at hippocampal dendritic spine in Alzheimer's disease. *PLoS One 12*(8), p. e0182743.

concentrations induced by each stimulation are at the same level, with the only exception being at the extrasynaptic site. Under HFS, the peak glutamate concentration at the extrasynaptic site increases as more stimulation pulses arrive and approach a maximum

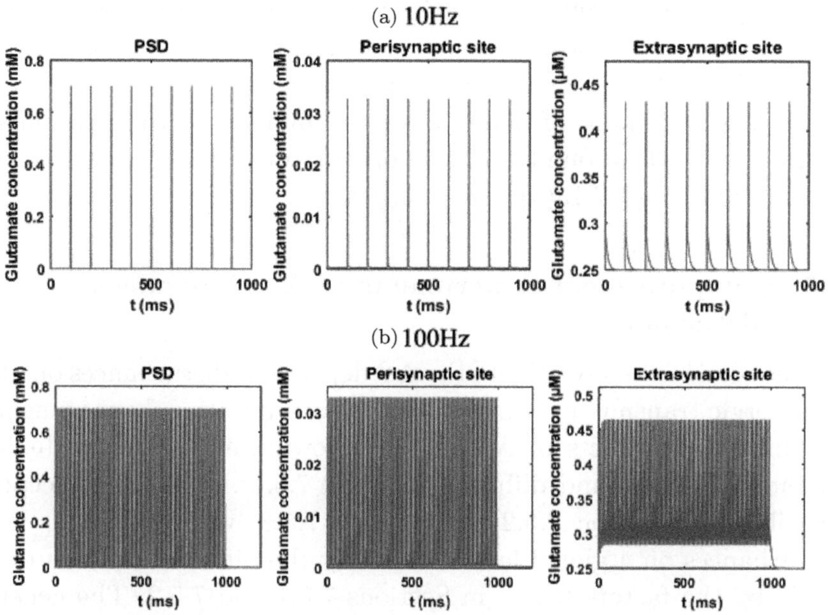


Fig. 7.12. Glutamate concentration at different locations in response to (a) 1 s of LFS (10 Hz) and (b) 1 s of HFS (100 Hz). Acknowledgment [4]: Liang, J., Kulasiri, D. and Samarasinghe, S. (2017) Computational investigation of Amyloid-β-induced location-and subunit-specific disturbances of NMDAR at hippocampal dendritic spine in Alzheimer's disease. *PLoS One 12*(8), p. e0182743.

concentration of around 0.47 $\mu$M in the HFS condition (Fig. 7.12(b) right panel).

The maximum fraction of NMDARs in PSD increases in response to the first few pulses and then decreases to a lower level under both LFS and HFS (Figs. 7.13(a) and 7.13(b)). The sNR2B-NMDAR shows a faster decrease in amplitude than under LFS and sNR2A-NMDAR. The decrease results in the fast desensitization of NR2B-NMDAR.

In the perisynaptic and extrasynaptic sites, a higher fraction of NMDARs open as more pulses arrive at the synapse. The maximum fraction and the increase in the peak fraction are higher in HFS than in LFS, indicating that the temporal summation of HFS allows more glutamate to accumulate after escaping from the synaptic cleft. This leads to a higher open fraction of glutamate receptors; however, it causes a significant increase in receptor desensitization because of the prolonged exposure to higher levels of glutamate than in the resting level. Consequently, the $Ca^{2+}$ transient in the spine head rises to a maximum of 3 $\mu$M after the first seven to eight pulses and decreases to around 1 $\mu$M before 400 ms after the start of the stimulation (Figs. 7.13(a) and 7.13(b)).

## 7.5. Computational Experiments of A$\beta$-Dependent Disturbances

In this section, we investigate the A$\beta$-dependent disturbances on glutamatergic transmission through various computational experiments. We mimic the effects of A$\beta$ on the following aspects individually: glutamate release and diffusion (Section 7.5.1), glutamate receptor distribution (Section 7.5.2). In Section 7.5.3, we investigate these disturbances on a global level to identify the different contributions made by the factors tested in Sections 7.5.1 and 7.5.2. The general procedure in the experiments is as follows:

Step 1. We first set a reasonable range for the factor to be tested and pick one value at a time while keeping the other parameters at the standard levels.

Step 2. We run the model for long enough without any stimulation input to reach a steady state, the values of which are used as the initial value for the experiment.

Step 3. For each set of parameters, three types of presynaptic stimulation patterns are used as inputs of the model: (1) a single EPSP stimulus (1 pulse), (2) low-frequency stimulation (LHS) at 10 Hz, and (3) high-frequency stimulation (HFS) at 100 Hz (see Chapter 2 for the detailed explanation). All stimulation patterns last for 1 s, thus, there are 1, 10 and 100 pulses for each stimulation pattern, respectively. Simulation stops when the system returns to the resting levels (initial state).

The major outputs we are collecting after the simulation are as follows:

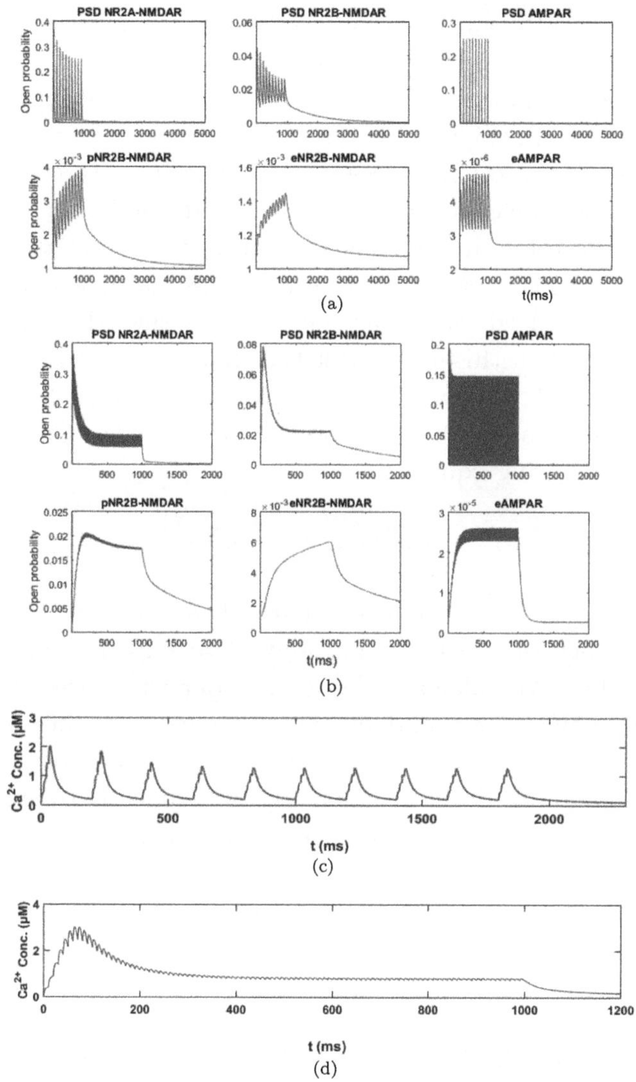


Fig. 7.13. States transition of the receptors and $Ca^{2+}$ transient in the spine head in response to 1 s of LFS (10 Hz) and 1 s of HFS (100 Hz). States transition of the receptors by type and location in response to LFS (a) and HFS (b). The $Ca^{2+}$ ions transient in the spine head are induced by LFS (c) and HFS (d). Acknowledgment [4]: Liang, J., Kulasiri, D. and Samarasinghe, S. (2017) Computational investigation of Amyloid-β-induced location-and subunit-specific disturbances of NMDAR at hippocampal dendritic spine in Alzheimer's disease. *PLoS One 12*(8), p. e0182743.

(1) Peak concentration of glutamate ($[\mathrm{Glu}]_{\mathrm{peak}}$) at the synaptic site, the perisynaptic site and the extrasynaptic site.
(2) The additional time for each receptor staying in the open/desensitized/bound states. We calculate the area under the curve of the fraction of receptors staying in a state against time and deduct the effects of the background fraction at resting level to get the total time in this state from the stimulation applied. The difference between this total time and the one produced from a standard value parameter set is the additional time of the receptor in this state after the change in testing factor.
(3) Additional $\mathrm{Ca}^{2+}$ ions entered by each type of NMDAR. The $\mathrm{Ca}^{2+}$ current from a NMDAR can be transferred to the $\mathrm{Ca}^{2+}$ ion influx by the following equation:

$$\mathrm{Ca}^{2+}\mathrm{ion} = -\frac{I_{\mathrm{Ca}^{2+}}(t)}{Z_{\mathrm{Ca}}C},$$

where $I_{\mathrm{Ca}^{2+}}(t)$ is the $\mathrm{Ca}^{2+}$ current at time $t$, $Z_{\mathrm{Ca}} = 2$ is the valence of the $\mathrm{Ca}^{2+}$ ion and $C = 6.24\mathrm{e}18$, where $1\ \mathrm{C} = 1\ \mathrm{A} \times 1\ \mathrm{s}$. We calculate the area under the curve of $\mathrm{Ca}^{2+}$ ion influx against time and deduct the effects of the background influx rate at resting level to get the total $\mathrm{Ca}^{2+}$ ions entered by the stimulation applied. Then, the difference between the total $\mathrm{Ca}^{2+}$ ions numbers and the one produced with a standard value parameter set is the additional number of $\mathrm{Ca}^{2+}$ ions entered through this type of receptor by the change in the factor tested.
(4) Peak amplitude of intracellular $\mathrm{Ca}^{2+}$ transient ($[\mathrm{Ca}^{2+}]_{\mathrm{peak}}$) and peak amplitude of membrane depolarization ($\mathrm{V}_{\mathrm{peak}}$) in the spine head and dendritic shaft, respectively.
(5) Dynamics of the intracellular $\mathrm{Ca}^{2+}$and membrane potential in the spine head and dendritic shaft, respectively.

### 7.5.1. *Aβ-dependent disturbance on glutamate transmission*

Experimental evidence suggests that the presence of Aβ may lead to an increase in the extrasynaptic glutamate levels by promoting the release of glutamate vesicles from the presynaptic terminal [50],

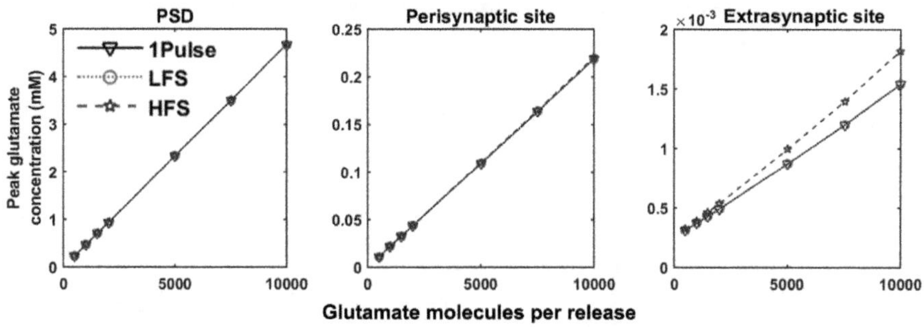


Fig. 7.14. Maximum amplitudes of glutamate concentration at different locations with various amounts of glutamate released in response to three types of stimulation patterns. Acknowledgment [4]: Liang, J., Kulasiri, D., and Samarasinghe, S. (2017) Computational investigation of Amyloid-β-induced location-and subunit-specific disturbances of NMDAR at hippocampal dendritic spine in Alzheimer's disease. *PLoS One 12*(8), p. e0182743.

inhibiting the glutamate uptake by glial glutamate transporters [51, 52], or inducing glutamate release from ambient astrocytes [11]. In this section, we investigate these disturbances one by one.

7.5.1.1. *Increase in presynaptic release of glutamate*

In this model, the number of vesicles per release is represented by the total number of glutamate molecules in the model. We vary the number of glutamates per release from 500 to 10000 and keep other parameters at the standard values. Because of the fast diffusion of glutamate in synaptic cleft and the extrasynaptic space, HFS shows no or minor (at extrasynaptic site) additive impacts on the $[\text{Glu}]_{\text{peak}}$. The $[\text{Glu}]_{\text{peak}}$ after a single pulse stimulation, LFS and HFS in three locations all linearly increase with the number of glutamate molecules released (Fig. 7.14).

Multiple vesicle releases have little effect on the transition states of sNR2A-NMDAR (Fig. 7.15) and $Ca^{2+}$ influx via it (Fig. 7.16(a)). For sNR2B-NMDAR, multiple vesicle releases lead to increases in its additional open time, additional desensitized time and additional bound time, however, these effects on its states transition are limited to conditions of a single-pulse stimulation and LFS (Fig. 7.15) and lead to more $Ca^{2+}$ influx under these conditions (Fig. 7.16(a)).

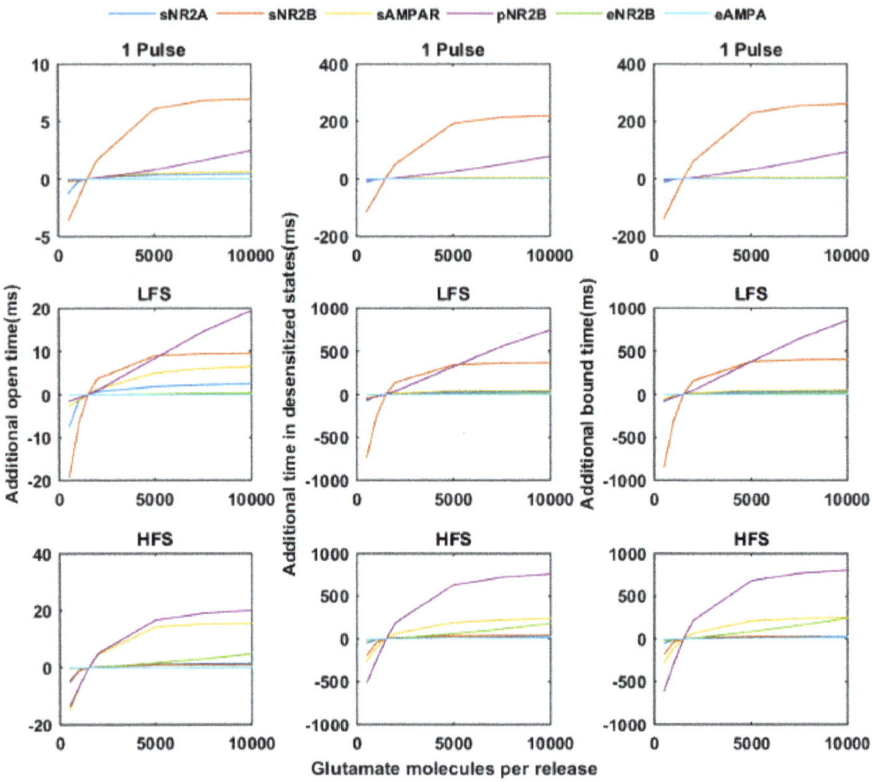


Fig. 7.15. Additional times in the open and desensitized states per receptor in response to various numbers of glutamate releases in comparison with the amount under control conditions (1500 molecules per release). sNR2A: synaptic NR2A-NMDAR; sNR2B: synaptic NR2A; sAMPAR: synaptic AMPAR; pNR2B: perisynaptic NR2B-NMDAR; eNR2B: extrasynaptic NR2B-NMDAR; eAMPAR: extrasynaptic AMPAR. Acknowledgment [4]: Liang, J., Kulasiri, D., and Samarasinghe, S. (2017) Computational investigation of Amyloid-$\beta$-induced location-and subunit-specific disturbances of NMDAR at hippocampal dendritic spine in Alzheimer's disease. *PLoS One 12*(8), p. e0182743.

In contrast, the pNR2B-NMDAR shows an increase in additional open time, desensitized time, bound time and entered $Ca^{2+}$ ions in response to all three tested stimulation protocols (Figs. 7.15 and 7.16(a)). All these increases lead to a higher $[Ca^{2+}]_{peak}$ in the spine head with increasing numbers of glutamate releases (Fig. 7.16(b)). Especially under HFS, the $Ca^{2+}$ level rises to a peak of about

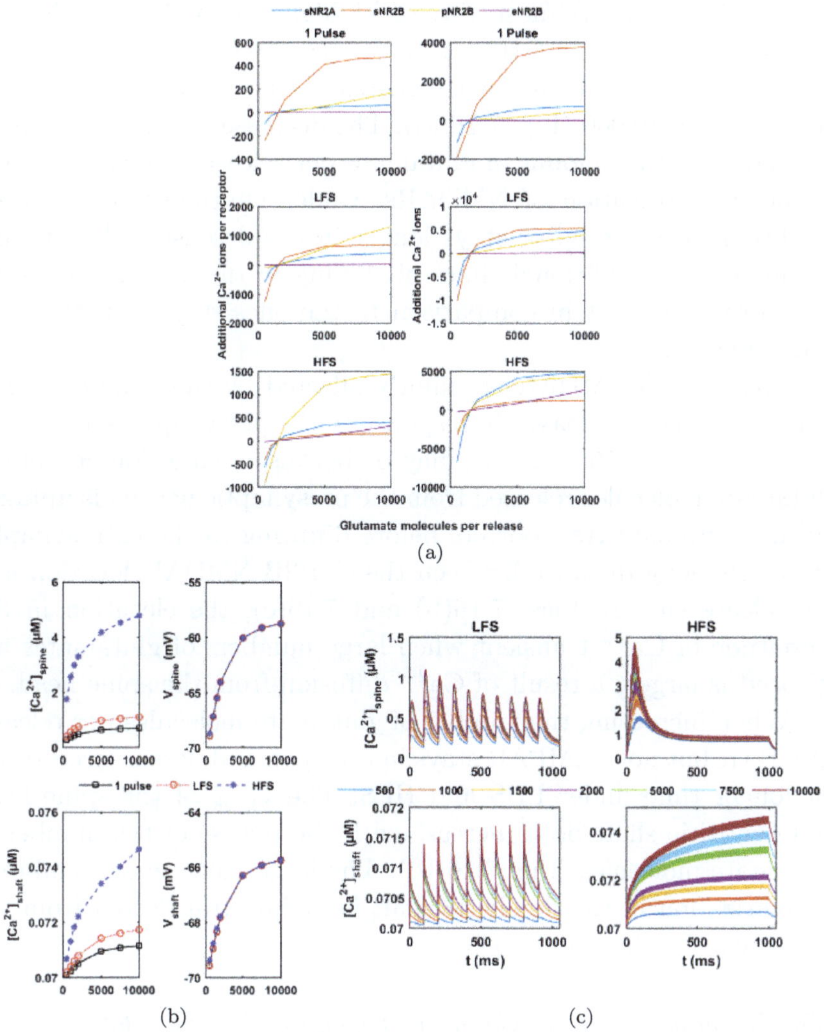


Fig. 7.16. (a) Additional numbers of $Ca^{2+}$ ions by NMDARs in response to various numbers of glutamate molecules released compared with the numbers under the standard condition (1500 molecules per release). (b and c) $Ca^{2+}$ ions and the voltage responses in the spine head and dendritic shaft in response to a single-pulse stimulation, LFS and HFS by various numbers of glutamate molecules released. sNR2A: synaptic NR2A-NMDAR; sNR2B: synaptic NR2A; pNR2B: perisynaptic NR2B-NMDAR; eNR2B: extrasynaptic NR2B-NMDAR. Acknowledgment [4]: Liang, J., Kulasiri, D. and Samarasinghe, S. (2017) Computational investigation of Amyloid-β-induced location-and subunit-specific disturbances of NMDAR at hippocampal dendritic spine in Alzheimer's disease. *PLoS One 12*(8), p. e0182743.

4.8 $\mu$M in 55 ms (almost three-fold higher than it is under the healthy condition), but then rapidly decreases to less than 1 $\mu$M after 35 pulses (350 ms), when increases the glutamate molecules per release to 10,000 (Fig. 7.16(c)). The decrease in rate is positively correlated to the amount of glutamate per release, which is because of the desensitization of NMDARs. Under multiple vesicle release conditions, each receptor stays longer in the desensitized state and the increase in additional times of staying in the desensitized state are more significant in comparison to the ones staying in the open state (Fig. 7.15).

The eNR2B-NMDAR is minorly affected by the number of glutamate molecules released in response to the presynaptic stimulation (Figs. 7.15 and 7.16(a)). This may be because a large fraction of the glutamate molecules released from the presynaptic neuron is uptaken by the glutamate transporters before diffusing to the extrasynaptic site or the long distance between the eNR2B-NMDAR location and the release site. In Figs. 7.16(b) and 7.16(c), the elevation in the amplitude of $Ca^{2+}$ transient when large numbers of glutamates are released is largely a result of $Ca^{2+}$ diffusion from the spine head.

When increasing the number of glutamate molecules per release, sAMPAR but not eAMPAR stays in the open and desensitized states for longer time under LHS and HFS. The $V_{\text{peak}}$ of the spine head and dendritic shaft both increase with the release of the number of glutamate molecules (Fig. 7.16(b)). The large increase in membrane depolarization in the dendritic shaft is mainly transferred from the spine head by the spine neck.

#### 7.5.1.2. *Inhibition in uptaking extrasynaptic glutamate by the glutamate transporter*

Experimental evidence suggests that A$\beta$ may inhibit extrasynaptic glutamate uptake by reducing the number of glutamate transporters or their activities (Carney *et al.* 1995; Wang *et al.* 1996; Beitz *et al.* 1997, 2004; Augusto *et al.* 2008). We examine if this down-regulation contributes to the abnormal opening of NMDARs by reducing the concentration of total glutamate transporter concentration ($B_{total}$)

from the standard value of 0.5 mM (0% reduction) to 0 mM (100% reduction) and then running the model under different stimulation patterns.

The results show that there is no effect on the $[\mathrm{Glu}]_{\mathrm{peak}}$ at the postsynaptic site from reducing $B_{\mathrm{total}}$ whereas, when $B_{\mathrm{total}} = 0$ mM (100% reduction), the $[\mathrm{Glu}]_{\mathrm{peak}}$ at the perisynaptic and extrasynaptic sites increases by about 3 $\mu$M and 1.5 $\mu$M (Fig. 7.17(a)), respectively, with slower decay to baseline (Fig. 7.17(b)). There are relatively minor effects on the transition state of the synaptic receptors in comparison with ones of pNR2B-NMDAR and eNR2B-NMDAR, especially under HFS (Fig. 7.18(a)). These changes result in no obvious effects on $Ca^{2+}$ dynamics in the spine head, but increases in the peaks of $Ca^{2+}$ transient in the dendritic shaft (Fig. 7.18(b)).

We then increase the glutamate number per release from 1500 to 5000 to investigate if there are any significant changes in multi-vesicle release events by $B_{\mathrm{total}}$. The results are consistent with the single-vesicle release experiment (Fig. 7.19), suggesting that the down-regulation in glutamate uptake by glutamate transporters does not directly affect the postsynaptic spine in response to the presynaptic stimulations. Instead, it promotes $Ca^{2+}$ influx via extrasynaptic receptors in the dendritic shaft, especially under HFS.

#### 7.5.1.3. *Increase in the resting glutamate levels in the extrasynaptic space*

Elevation in the resting level of extrasynaptic glutamate ($[\mathrm{Glu}]_{\mathrm{rest}}$) has been observed in the hippocampus of AD transgenic mice that overexpress human APP [11]. They reported that the resting level of 0.8 $\mu$M in 12-month-old mice is in contrast to 0.25 $\mu$M in the control ones, and with an even higher level of 3.3 $\mu$M in 22 to 24-month-old mice. To examine how the $[\mathrm{Glu}]_{\mathrm{rest}}$ affects the system, we vary the $[\mathrm{Glu}]_{\mathrm{rest}}$ from 0.01 $\mu$M to 100 $\mu$M and run the model for long enough to calculate the fraction of NR2A-NMDAR, NR2B-NMDAR and AMPAR in the open and desensitized bound states, respectively. Both NR2A- and NR2B-NMDAR reach the maximum open fractions (8% and 2%, respectively) when glutamate increases

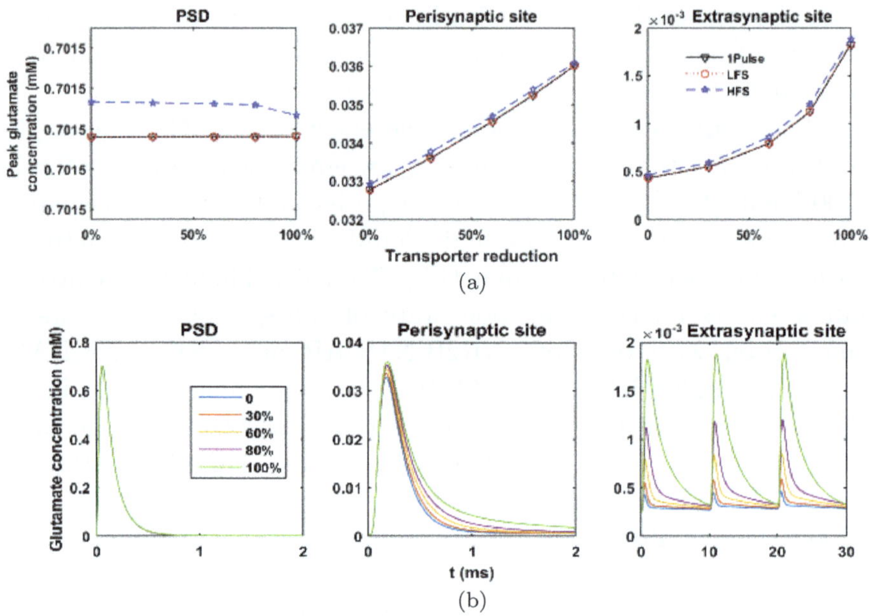


Fig. 7.17. (a) Maximum amplitude and (b) time course of glutamate concentration at different locations with the reduction in transporter levels in response to three types of stimulation patterns. The x-axis denotes the degree of reduction in total glutamate transporter concentration ($B_{\text{total}}$) from the standard value, 0.5 mM (0: no reduction; 1: fully reduction and $B_{\text{total}} = 0$). Acknowledgment [4]: Liang, J., Kulasiri, D., Samarasinghe, and S. (2017) Computational investigation of Amyloid-$\beta$-induced location-and subunit-specific disturbances of NMDAR at hippocampal dendritic spine in Alzheimer's disease. *PLoS One 12*(8), p. e0182743.

to about 2 $\mu$M (Fig. 7.20(a)). In contrast, the fraction of NR2A-NMDAR and NR2B-NMDAR in the desensitized state approaches a maximum of 81% at the same time (Fig. 7.20(b)). Only about 10% of receptors are not bound to glutamate (Fig. 7.20(c)). This may reduce the sensitivity of the postsynaptic neurons in response to stimulation.

The background opening of NR2B-NMDAR by $[\text{Glu}]_{\text{rest}}$ causes a persistent inward current and $Ca^{2+}$ influx. When increasing the glutamate concentration to 10 $\mu$M, the background $Ca^{2+}$ influx reaches the maximum of 580 (NR2A-NMDAR) and 160 (NR2B-NMDAR) $Ca^{2+}$ ions per second, which is 5–10 $Ca^{2+}$ ions per second under control conditions (Fig. 7.20d).

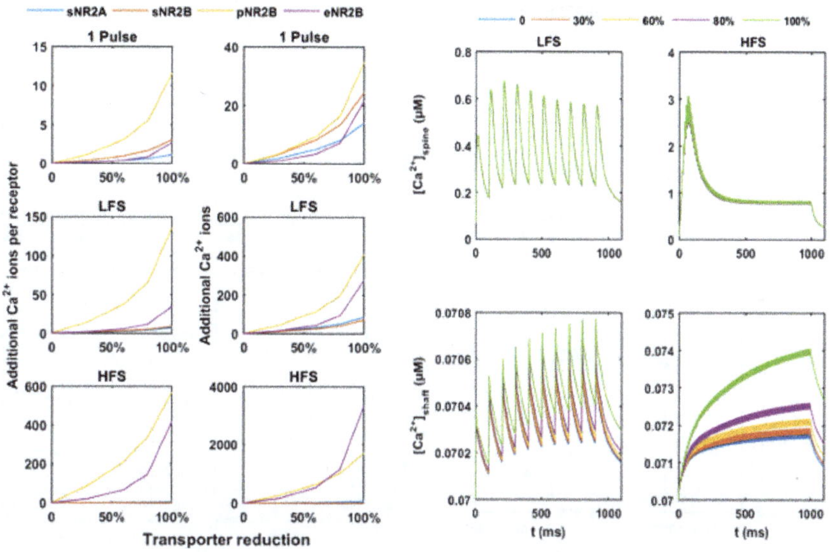


(a) (b)

Fig. 7.18. (a) Additional numbers of $Ca^{2+}$ ions by NMDARs and (b) $Ca^{2+}$ dynamics in the spine head and dendritic shaft with the reduction in transporter levels in response to three types of stimulation patterns. The glutamate number per release is 1500. sNR2A: synaptic NR2A-NMDAR; sNR2B: synaptic NR2A; pNR2B: presynaptic NR2B-NMDAR; eNR2B: extrasynaptic NR2B-NMDAR. Acknowledgment [4]: Liang, J., Kulasiri, D. and Samarasinghe, S. (2017) Computational investigation of Amyloid-$\beta$-induced location-and subunit-specific disturbances of NMDAR at hippocampal dendritic spine in Alzheimer's disease. *PLoS One 12*(8), p. e0182743.

The open fraction of AMPAR is less affected by the elevation of glutamate concentration while its desensitized fraction increases gradually and when the glutamate reaches a high level greater than 90 $\mu$M, it is almost fully desensitized (Fig. 7.20(b)).

In the experiment of presynaptic stimulation, the elevation in resting extrasynaptic glutamate concentration increases the desensitization time of all receptors to different extents while shows neglectable effects on their open time (Fig. 7.21(a)). In the spine head, sNR2A-NMDAR, sNR2B-NMDAR and pNR2B-NMDAR are most affected, in terms of several milliseconds lower open time in comparison to the control conditions. Together with the increase in desensitization time, especially for pNR2B-NMDAR, this leads

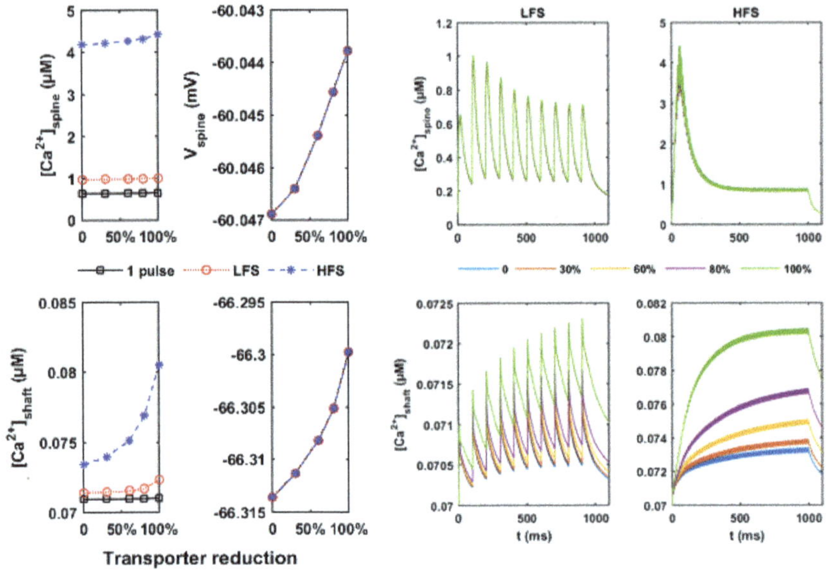


(a) (b)

Fig. 7.19. (a) Amplitudes of $Ca^{2+}$ and voltage response and (b) $Ca^{2+}$ dynamics in the spine head and dendritic shaft with the reduction in transporter levels in response to three types of stimulation patterns. The glutamate number per release is 5000. Acknowledgment [4]: Liang, J., Kulasiri, D. and Samarasinghe, S. (2017) Computational investigation of Amyloid-$\beta$-induced location-and subunit-specific disturbances of NMDAR at hippocampal dendritic spine in Alzheimer's disease. *PLoS One 12*(8), p. e0182743.

to fewer $Ca^{2+}$ ions entering (Fig. 7.21(b)) and decreases in the $[Ca^{2+}]_{peak}$ in the spine head (Fig. 7.21(c)). For the eNR2B-NMDAR, it stays in the states for a longer time by the elevation in $[Glu]_{rest}$ (Fig. 7.21(a)). The open time of the eNR2B-NMDAR and the number of $Ca^{2+}$ ions entering via it are not affected. In addition, the high level of $[Glu]_{rest}$ results in a high background open fraction of eNR2B-NMDAR and small increases in response to the presynaptic stimulations (Fig. 7.22).

#### 7.5.1.4. *Non-synaptic release of glutamate*

A$\beta$ has been found to induce glutamate release from astrocytes in AD transgenic mice [11]. The astrocyte stays close to the dendritic spine,

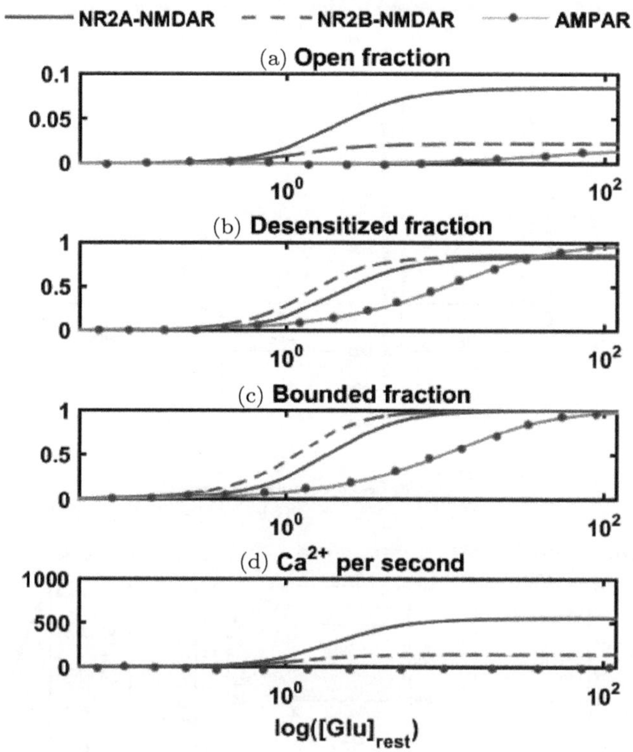


Fig. 7.20. Effects of the resting level of extrasynaptic glutamate concentration $[Glu]_{rest}$ on the resting states of NR2A-NMDAR, NR2B-NMDAR and AMPAR. $[Glu]_{rest}$ increases from 0.01 $\mu$M to 100 $\mu$M. Under control conditions $[Glu]_{rest}$ is 0.25 $\mu$M. Acknowledgment [4]: Liang, J., Kulasiri, D. and Samarasinghe, S. (2017) Computational investigation of Amyloid-$\beta$-induced location-and subunit-specific disturbances of NMDAR at hippocampal dendritic spine in Alzheimer's disease. *PLoS One 12*(8), p. e0182743.

which can potentially induce the extrasynaptic glutamate receptors. We simulate the astrocytic release of glutamate by applying a brief pulse of 1 mM glutamate to the extrasynaptic sites where eNR2B-NMDAR and eAMPAR are located. The lengths of the brief pulses are in the range of 1 ms to 20 ms. We assume the [Glu] in the synaptic cleft and presynaptic site will not be affected by this stimulation. The stimulation causes large amounts of $Ca^{2+}$ to enter (Fig. 7.23(b)), which is increased with the length of the stimulation pulse. The $\Delta$ $[Ca^{2+}]$ in the dendritic shaft ranges from 13 to 17 nM and is not

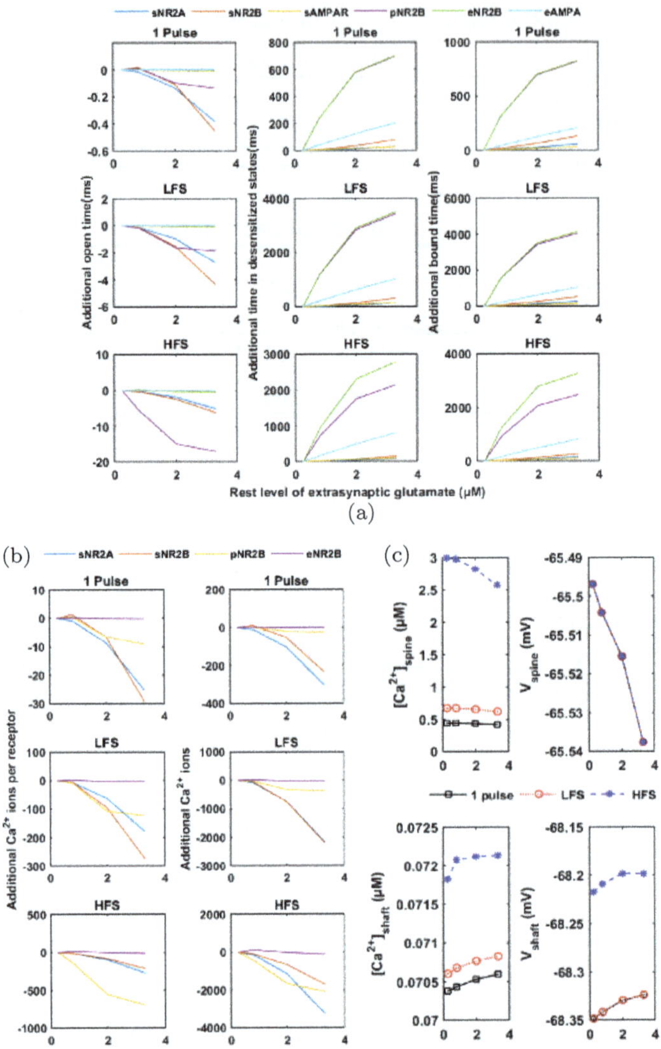


Fig. 7.21. (a) Additional times in the open and desensitization states per receptor and (b) additional numbers of $Ca^{2+}$ ions by NMDARs in response to presynaptic stimulation (a single-pulse stimulation, LFS and HFS) under the conditions of the increased resting level of the extrasynaptic glutamate concentration ($[Glu]_{rest}$) in comparison with the amount under standard conditions. (c) Amplitudes of $Ca^{2+}$ and voltage response in the spine head and dendritic shaft at various $[Glu]_{rest}$. sNR2A: synaptic NR2A-NMDAR; sNR2B: synaptic NR2A; sAMPAR: synaptic AMPAR; pNR2B: perisynaptic NR2B-NMDAR; eNR2B: extrasynaptic NR2B-NMDAR; eAMPAR: extrasynaptic AMPAR.

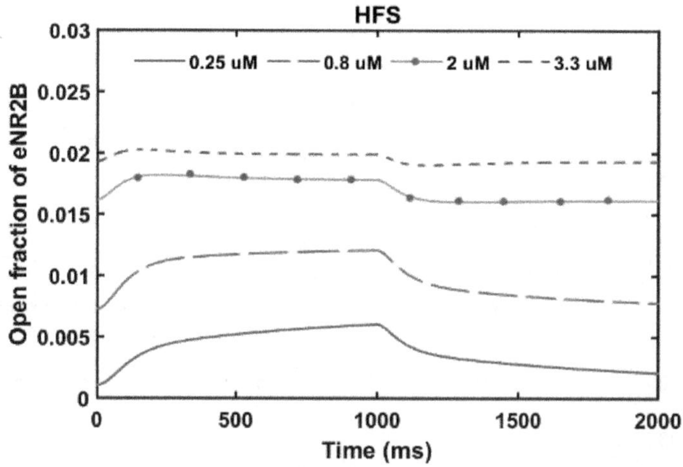


Fig. 7.22. The fraction of eNR2B-NMDAR in the open state during simulation under HFS. The resting synaptic glutamate concentrations are 0.25 $\mu$M in the control condition and 0.8 $\mu$M, 2 $\mu$M and 3.3 $\mu$M, representing the different stages of the disease. Acknowledgment [4]: Liang, J., Kulasiri, D. and Samarasinghe, S. (2017) Computational investigation of Amyloid-$\beta$-induced location-and subunit-specific disturbances of NMDAR at hippocampal dendritic spine in Alzheimer's disease. *PLoS One 12*(8), p. e0182743.

significantly increased by the longer stimulation (Fig. 7.23(c)). This is because the volume of the dendritic shaft is about 0.79 $\mu m^3$ in this model, about eight-fold the spine head volume, whereas the low number of eNR2B-NMDA is similar to the dendritic ones. The $Ca^{2+}$ ions entered are largely diluted by the volume, therefore the $[Ca^{2+}]_{peak}$ is much less than in the spine head in response to the presynaptic stimulation. Moreover, the eAMPAR activated by the astrocytic glutamate release creates a 4–6 mV depolarization (Fig. 7.23(d)). Because of the lower density of AMPAR in the extrasynaptic site (20 receptor/$\mu m^2$) than in the synaptic location (up to 1000 receptor/$\mu m^2$), even when increasing the stimulation time to 20 ms, it still fails to create a larger depolarization to activate other voltage-dependent $Ca^{2+}$ channels on the dendritic shaft membrane.

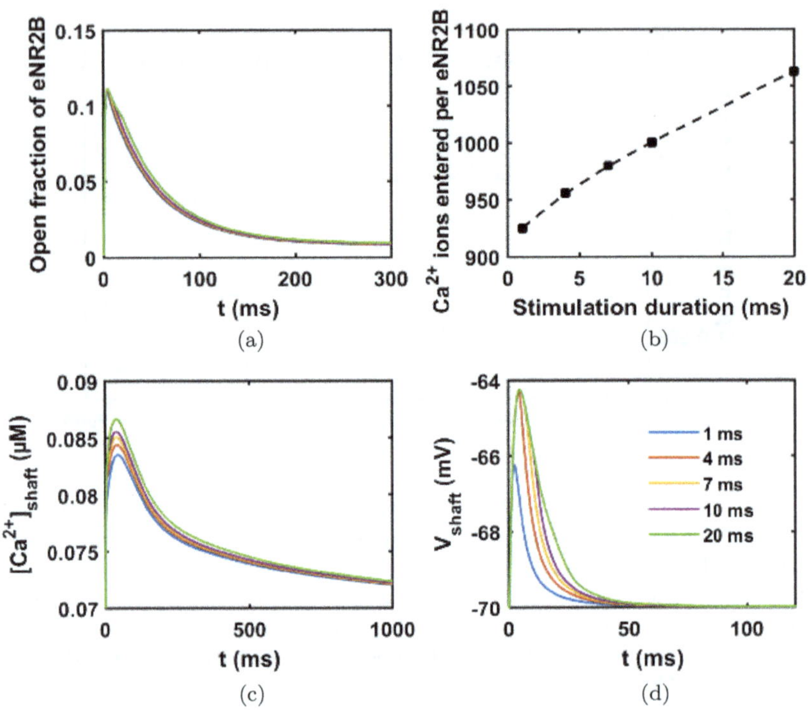


Fig. 7.23. Effects of eNR2B-NMDAR activation by the astrocytic release of glutamate. In response a pulse of 1 mM glutamate with various time durations (1 ms, 4 ms, 7 ms, 10 ms and 20 ms). The fraction of eNR2B-NMDAR opens (a) and allows $Ca^{2+}$ ions to enter the dendritic shaft (b). The corresponding $Ca^{2+}$ transient and membrane depolarizations are shown in (c) and (d), respectively. Acknowledgment [4]: Liang, J., Kulasiri, D. and Samarasinghe, S. (2017) Computational investigation of Amyloid-$\beta$-induced location-and subunit-specific disturbances of NMDAR at hippocampal dendritic spine in Alzheimer's disease. *PLoS One 12*(8), p. e0182743.

### 7.5.2. *Receptors*

#### 7.5.2.1. *Reduction in surface expression of NMDAR and AMPAR*

To investigate the inhibition of A$\beta$ on the membrane surface expression of different types of receptors, we decrease the number of each receptor from the standard value to zero to mimic the degree of reduction in their surface expression.

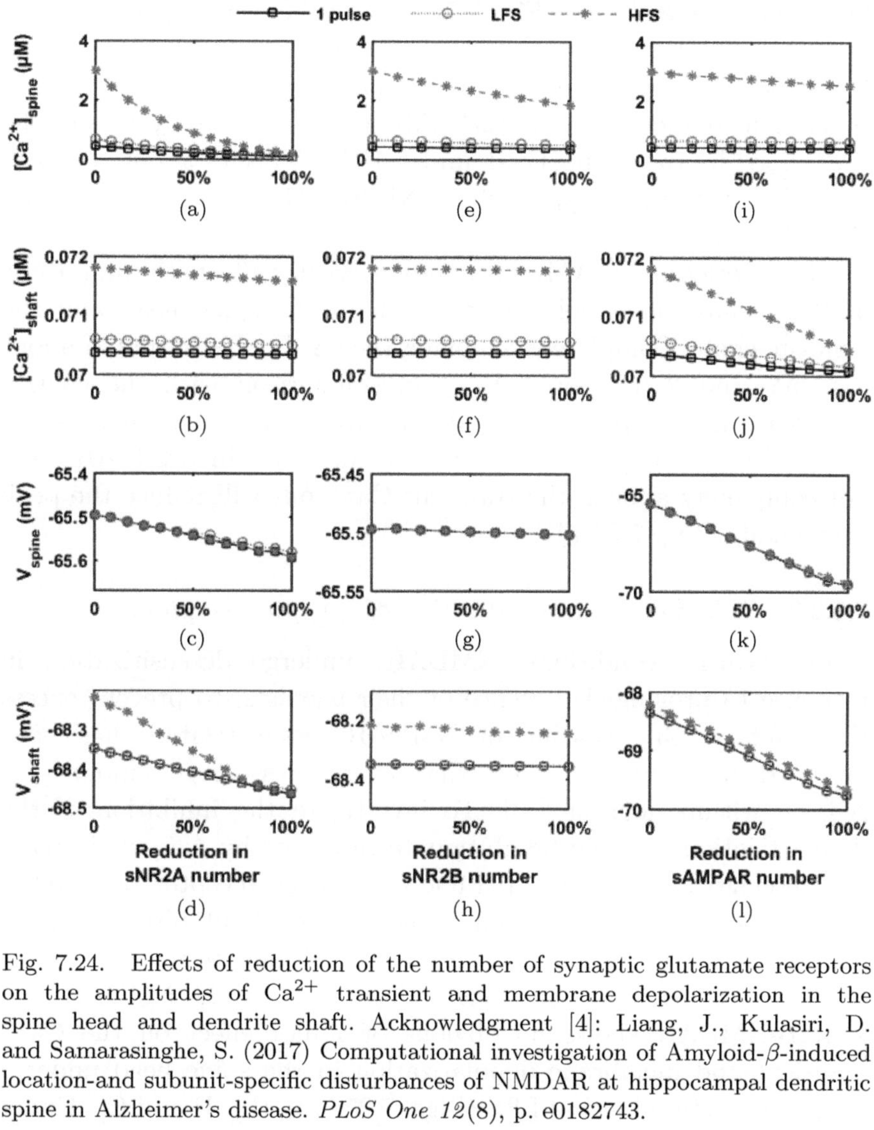


Fig. 7.24. Effects of reduction of the number of synaptic glutamate receptors on the amplitudes of $Ca^{2+}$ transient and membrane depolarization in the spine head and dendrite shaft. Acknowledgment [4]: Liang, J., Kulasiri, D. and Samarasinghe, S. (2017) Computational investigation of Amyloid-β-induced location-and subunit-specific disturbances of NMDAR at hippocampal dendritic spine in Alzheimer's disease. *PLoS One 12*(8), p. e0182743.

The reduction in sNR2A-NMDAR numbers shows negligible effects on the $Ca^{2+}$ transient amplitude under a single-pulse stimulus and LFS whereas, under HFS, the peak gradually decreases from 3 $\mu$M to nearly 0 $\mu$M with the increase in the reduction level (Fig. 7.24(a)). It also shows minor effects on the depolarization of the postsynaptic membrane (Fig. 7.24(c,d)). When setting the sNR2A-NMDAR number to zero, it causes about a 0.1 mV decrease in the depolarization amplitude in comparison to the control conditions. The reduction in sNR2B-NMDAR numbers only affects the $[Ca^{2+}]_{peak}$ in the spine head under HFS. This is about a 1 $\mu$M reduction in the $[Ca^{2+}]_{peak}$ when sNR2B-NMDAR is fully removed (Fig. 7.24(e)).

The reduction in sAMPAR numbers greatly affects the amplitude of the membrane depolarization both in the spine head and the dendrite shaft. Completely removing the sAMPARs results in about a 5 mV and 2 mV lower $V_{peak}$ in comparison with the control conditions in the spine head and dendritic shaft membrane, respectively (Fig. 7.24(l)). In addition, the decrease in sAMPARs will not completely abolish the transient $Ca^{2+}$ but will reduce the peak fractionally (Fig. 7.24(i,j)).

### 7.5.2.2. *Reduction in desensitization of synaptic receptors*

Under healthy conditions, NMDARs undergo desensitization in response to prolonged presence of their agonists, to prevent excess $Ca^{2+}$ influx [53]. Slowing of NMDAR desensitization has been observed in transgenic mice that overexpress large amounts of $A\beta_{1-42}$ within neurons [54]. To investigate the inhibition of the desensitization of synaptic glutamate receptors by $A\beta$, we decrease the desensitization related parameters of each receptor by 0–100% of the standard value to mimic the degree of inhibition in their desensitization in response to presynaptic glutamate release.

The inhibition of the desensitization of sNR2A-NMDAR and sNR2B-NMDAR shows no obvious or minor effect on the $Ca^{2+}$ transient and membrane depolarization in the spine head under a single-pulse stimulus and LFS (Figs. 7.25A(a)–(h), B and C). Under HFS, the peak of $Ca^{2+}$ transient increases significantly from 2.9 $\mu$M to 17.9 $\mu$M with the increase in the inhibition level of the sNR2A-NMDAR desensitization (Fig. 7.25A(a)). When the desensitization of sNR2A-NMDAR is fully inhibited, the cytosolic $Ca^{2+}$ concentration in the spine head maintains at a high level after reaching a peak

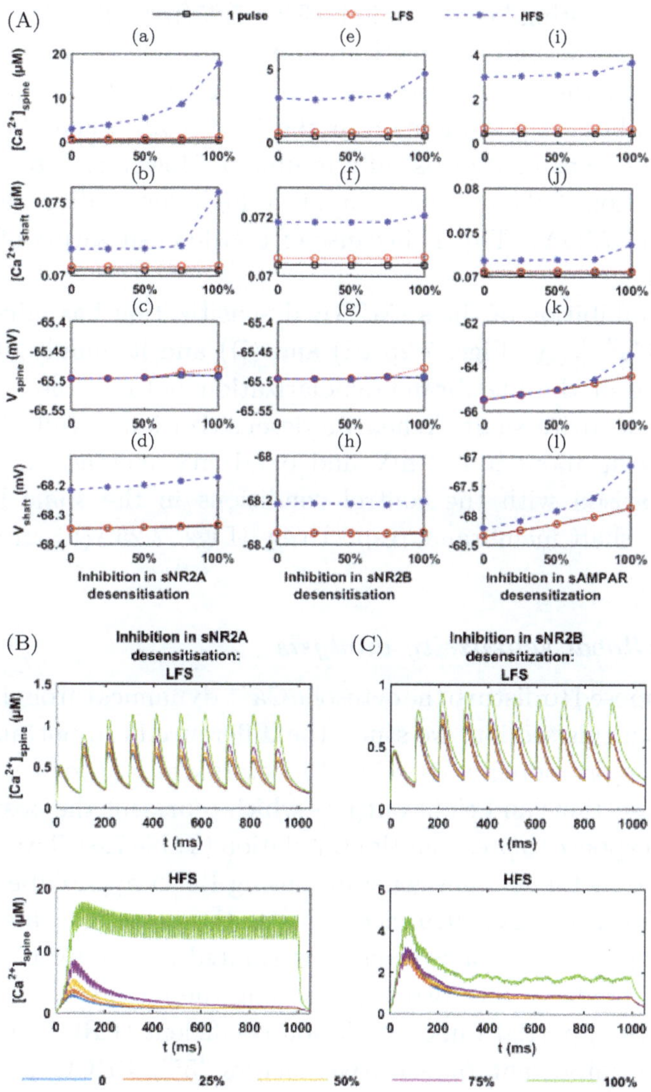


Fig. 7.25. (A) Effects of inhibition of the desensitization of synaptic glutamate receptors on the amplitudes of $Ca^{2+}$ transient and membrane depolarization in the spine head and dendrite shaft. The $Ca^{2+}$ time course in the spine head under LFS and HFS, with the desensitization inhibition of sNR2A-NMDAR (B) and sNR2B-NMDAR (C), respectively. Acknowledgment [4]: Liang, J., Kulasiri, D. and Samarasinghe, S. (2017) Computational investigation of Amyloid-$\beta$-induced location-and subunit-specific disturbances of NMDAR at hippocampal dendritic spine in Alzheimer's disease. *PLoS One 12*(8), p. e0182743.

through the stimulation period (Fig. 7.25(b)). The inhibition of the sNR2B-NMDAR desensitization also positively affects the $Ca^{2+}$ transient in the spine head under HFS (Figs. 7.25A(e) and C). It leads to about 2 $\mu$M increase in the $[Ca^{2+}]_{peak}$ when the sNR2B-NMDAR desensitization is fully inhibited. However, cytosolic $Ca^{2+}$ concentration fails to maintain at a high level after reaching a peak (Fig. 7.25A). This is because of the desensitization of sNR2A-NMDAR.

The inhibition of the sAMPAR desensitization has minor effects on the $[Ca^{2+}]_{peak}$ (Figs. 7.25A(i) and (j)) and it mainly affects the amplitude of the membrane depolarization both in the spine head and the dendrite shaft. When its desensitization is fully inhibited, it results in about a 1–2 mV and 0.5–1 mV increase in the $V_{peak}$ in comparison with the control conditions in the spine head and dendritic shaft membrane, respectively (Figs. 7.25A(k) and (l)).

### 7.5.3. *Global sensitivity analysis*

A$\beta$ is proposed to disturb the cytosolic $Ca^{2+}$ dynamics through multiple targets. In this section, we study the difference in importance of the various factors globally. We select 15 factors which have been tested in previous sections and eight outputs which represent the postsynaptic neuron response to presynaptic stimulation (Table 7.8). Two thousand perturbed factor sets are generated using Latin hypercube sampling (LHS) (McKay *et al.*, 1979) according to the range in Table 7.8. LHS makes sure each factor is evenly distributed in the given range and effectively reduces the correlation among the factors.

We use a partial rank correlation coefficient (PRCC) to identify the most important or sensitive factors [55]. PRCC is useful to measure the nonlinear but monotonic relationships between outputs and parameters (see Appendix F.2 for the explanation). Therefore, the factor ranges are chosen according to Sections 7.5.2 and 7.5.2, to ensure there is not a non-monotonic relationship between the selected factors and outputs. In each realization, we run the model with one set from the factor matrix generated by LHS while keeping other parameters at the standard values. Both the LFS and HFS conditions are simulated.

Table 7.8. Fifteen factors and eight outputs selected for PRCC.

| Factor | Biological meaning | Standard value | PRCC ranges |
|---|---|---|---|
| (A) *Factors and their biological meaning, standard values and ranges for PRCC* | | | |
| G0 | Glutamate number per release | 1500 | 500–10000 |
| Grest | Rest extrasynaptic glutamate concentration | 0.25 $\mu$M | 0–1 $\mu$M |
| TP | Glutamate transporter concentration | 0.5 $\mu$M | 0–1 $\mu$M |
| Ds2A | Inhibition on desensitization of sNR2A-NMDAR | 0 | 0–1 |
| Ds2B | Inhibition on desensitization of sNR2B-NMDAR | 0 | 0–1 |
| DsAMPAR | Inhibition on desensitization of sAMPAR | 0 | 0–1 |
| Dp2B | Inhibition on desensitization of pNR2B-NMDAR | 0 | 0–1 |
| De2B | Inhibition on desensitization of eNR2B-NMDAR | 0 | 0–1 |
| DeAMPAR | Inhibition on desensitization of eAMPAR | 0 | 0–1 |
| sNR2A | sNR2A-NMDAR number | 12 | 6–18 |
| sNR2B | sNR2B-NMDAR number | 8 | 4–12 |
| AMPAR | sAMPAR number | 85 | 43–130 |
| pNR2B | pNR2B-NMDAR number | 3 | 1–5 |
| eNR2B | eNR2B-NMDAR number | 8 | 4–12 |
| eAMPAR | eAMPAR number | $20/\mu m^2$ | 15–30 $\mu m^2$ |

| Output | Biological meaning |
|---|---|
| (B) *Outputs and their biological meaning for PRCC* | |
| $Ca^{2+}$ by sNR2A | $Ca^{2+}$ ions entered through of sNR2A-NMDAR |
| $Ca^{2+}$ by sNR2B | $Ca^{2+}$ ions entered through of sNR2B-NMDAR |
| $Ca^{2+}$ by pNR2B | $Ca^{2+}$ ions entered through of pNR2B-NMDAR |
| $Ca^{2+}$ by eNR2B | $Ca^{2+}$ ions entered through of eNR2B-NMDAR |
| Peak $Ca^{2+}$@spine | Peak concentration of $Ca^{2+}$ transient in spine head |
| Peak $Ca^{2+}$@shaft | Peak concentration of $Ca^{2+}$ transient in dendritic shaft |
| Peak V@spine | Peak membrane potential in spine head |
| Peak V@shaft | Peak membrane potential in dendritic shaft |

We calculate the PRCC value and the corresponding $p$-value for each factor against each output using a PRCC Matlab toolbox [55]. PRCC results if its corresponding $p$-value is greater than 0.05, which fails to reject the null hypothesis that no relationship exists between the factor and the output. The PRCC results are shown in Fig. 7.26. Generally, we assume there is a correlation between the factor and the output when the absolute PRCC value is greater than 0.5. Based on this, we modified the results and, therefore, in Fig. 7.26, the white color denotes that there is no correlation between corresponding factor and output.

From the point of view of parameter contributions (read Fig. 7.26 in a vertical direction), all outputs, except the $Ca^{2+}$ ion number entering through sNR2B-NMDAR, are very sensitive to the amount of presynaptic released glutamate. The correlation levels decrease under HFS in comparison with under LFS, with the only exception being that the $Ca^{2+}$ ion numbers entered through eNR2B-NMDAR show the highest correlation (PRCC = 0.95) with the glutamate release number under both conditions. The resting extrasynaptic glutamate level positively contributes to the $Ca^{2+}$ influx through eNR2B-NMDAR under LFS but not under HFS. The resting extrasynaptic glutamate level negatively contributes to the $Ca^{2+}$ influx through pNR2B-NMDAR under HFS, suggesting an elevation in $[Glu]_{rest}$ will cause a stronger desensitization of pNR2B-NMDAR in response to HFS rather than LFS. Only sNR2A-NMDAR is sensitive to sAMPAR, which implies that only sNR2A-NMDAR, but not NR2A-NMDAR, is sensitive to membrane depolarization under LFS and HFS. The transporter concentration, desensitization of eNR2B-NMDAR and the eAMPAR show no relationship with any output, which suggests they are of less importance in the postsynaptic response to presynaptic stimulation.

From the point of view of output sensitivity (read Fig. 7.26 in a horizontal direction), the $Ca^{2+}$ transient in the spine head is most sensitive to sNR2A-NMDAR followed by sNR2B-NMDAR. Under HFS, the importance of sNR2B-NMDAR decreases, which suggests a relatively lower activity of sNR2B-NMDAR in HFS than under LFS. The AMPAR number becomes more important in

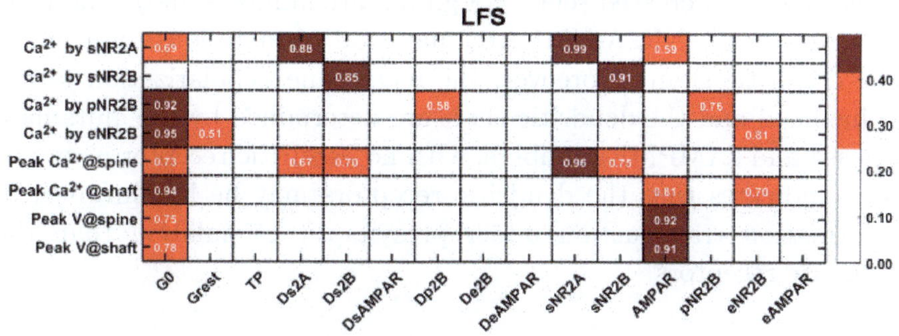


(a)

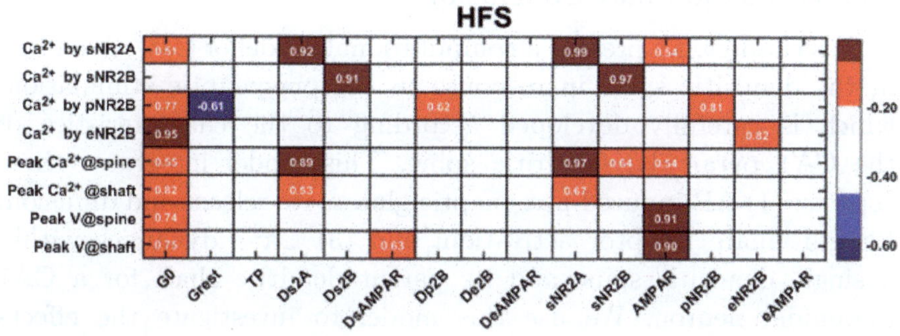


(b)

Fig. 7.26. Heat maps of PRCC results for the model in response to (a) LFS and (b) HFS. The PRCC values for 15 factors against eight outputs are represented by colors, with the corresponding PRCC values written in white. Red and blue denote the positive and negative correlations, respectively. Only PRCCs greater than 0.5 and with $p$-value $< 0.05$ are shown in the figures. The white color means there is no relationship between the corresponding factor and the output. Acknowledgment [4]: Liang, J., Kulasiri, D. and Samarasinghe, S. (2017) Computational investigation of Amyloid-$\beta$-induced location-and subunit-specific disturbances of NMDAR at hippocampal dendritic spine in Alzheimer's disease. *PLoS One 12*(8), p. e0182743.

comparison to LFS, indicating that depolarization by HFS brings more $Ca^{2+}$ ions into the spine head. The $Ca^{2+}$ transient in the dendritic shaft is mostly sensitive to the amount of glutamate released, while eNR2B is only important under LFS. Surprisingly, it is also sensitive to the sNR2A-NMDAR number. This suggests that under HFS, large amounts of $Ca^{2+}$ ions influx from NR2A, either by the increased release of glutamate numbers and/or activity or numbers of sNR2A-NMDAR, may lead to more ions diffusing to the dendritic shaft. Moreover, the membrane depolarization in the spine head and the dendritic shaft are determined by the glutamate release and sAMPMR numbers, with no dendritic receptor involved. This indicates that the dendritic receptors may be less involved in information transduction under presynaptic stimulation than the synaptic receptors.

## 7.6. Discussion and Conclusions

In this chapter, we present a computational model of $Ca^{2+}$ dynamics in the dendritic spine in response to the presynaptic stimulation, which is carefully developed according to the characteristics of the CA1 pyramidal dendritic spine. This model integrates three aspects: (1) AP-induced presynaptic glutamate release and diffusion; (2) glutamate receptor activation; and (3) $Ca^{2+}$ dynamics within a single dendritic spine and its parent dendrite shaft for a CA1 pyramidal neuron. We use this model to investigate the effects of A$\beta$-dependent disturbances on activation patterns of NMDARs with different subunit composition and at different locations and NMDAR-mediated $Ca^{2+}$ dynamics in the dendritic spine and its parent dendrite shaft.

Our model shows that the increase in the amount of glutamate released leads to different levels of increases in the availability of glutamate to receptors at each tested location. It predicts that increased glutamate availability in the synaptic cleft and perisynaptic zone after presynaptic stimulation leads to higher $Ca^{2+}$ responses in the spine head. Global sensitivity analysis suggests a great sensitivity of the postsynaptic response to the glutamate amount released during presynaptic stimulation. This confirms that the A$\beta$-induced increase in synaptic glutamate release may play a major role in the over-excitation and $Ca^{2+}$ overload of postsynaptic neurons. However, the spill-over effect of glutamate from the synaptic cleft shows much lower effects on activation patterns of extrasynaptic NMDARs than perisynaptic NMDARs. This is because of the fast diffusion of glutamate and the uptake by glial glutamate transporters before it reaches to eNMDAR. Therefore, it suggests that A$\beta$-induced multiple vesicle releases from the presynaptic terminal alone may not be sufficient to cause an overactivation of the extrasynaptic receptors.

In contrast, the results show that the inhibition of the ability to uptake extrasynaptic glutamate only affects peri- and extrasynaptic receptors in our model experiments. Furthermore, on a longer time scale, this inhibition may cause an accumulation of the glutamate released from the presynaptic neurons and astrocytes in the extrasynaptic space. Therefore, it may contribute to a gradual increase in the resting level of extrasynaptic glutamate.

We have shown that elevation in the resting extrasynaptic glutamate concentration reduces the sensitivity of the postsynaptic neurons to the presynaptic signals, as a result of increased background opening under conditions of rest. The large background opening of eNMDARs will cause an especially high level of net $Ca^{2+}$ ion influx into the dendritic shaft. As well, the A$\beta$-induced astrocytic glutamate release also lead to a high level of $Ca^{2+}$ ion influx in the absence of presynaptic stimulation. These findings are consistent with experimental observations, which show excitotoxicity results from A$\beta$-induced over activation of the extrasynaptic NMDAR, but not the synaptic NMDAR [56], which may also, in turn, promote A$\beta$ production [57]. Even though these abnormal $Ca^{2+}$ influxes have failed to induce large amplitudes of $Ca^{2+}$ transient in the dendritic shaft, due to the dilution of its relatively large volume, it may still induce downstream pathways by affecting proteins located close to the receptors. Over a long time span, this may contribute to the $Ca^{2+}$ overload and neuronal death in AD [58].

The experimental observations suggest that A$\beta$ induces the internalization of surface glutamate receptors, especially sNR2B-NMDAR, which may affect the synaptic transmission [18, 59]. Our simulation shows that sNR2B-NMDAR contributes less to the synaptic $Ca^{2+}$ transient compared to sNR2A-NMDAR. Even when removing all sNR2B-NMDAR, the sNR2A-NMDAR can still lead to considerable $Ca^{2+}$ transients. This seems to be in conflict with the experimental observations. In fact, besides its role as a $Ca^{2+}$

channel, sNR2B-NMDAR binds to $Ca^{2+}$/calmodulin-dependent protein kinase II (CaMKII), a critical protein in LTP formation and, therefore, sNR2B-NMDAR are involved in downstream pathways mediating synapse strength and plasticity [60]. Reducing the functional surface of sNR2B-NMDAR may affect the accessibility of CaMKII which, consequently, disturbs information transmission and memory formation. We discuss this further by combining the current model with a CaMKII transition model in Chapter 9.

## Appendix B: Glutamate Diffusion Model

### B.1. Fick's First Law

Adolf Fick was the first one to describe the diffusion process quantitatively [61]. Fick's first law states that the diffusion flux is proportional to the concentration gradient, and the rate of concentration change at a point in space is proportional to the second derivative of concentration with space. For example, Fick's law in one dimension is:

$$J(x,t) = -D\frac{\partial C(x,t)}{\partial x}, \tag{B.1}$$

where $J(x,t)$ is the diffusion flux (the amount of a substance crossing through a unit area per unit time), $D$ is the diffusion coefficient, $C(x,t)$ is the concentration of substance at position $x$ at time $t$. The negative sign before $D$ denotes the diffusion is from a higher concentration to a lower one.

Consider the diffusion space as separated into a series of compartments. In each compartment, the substance is well mixed. Therefore, its concentration in compartment $i$ is changed by the diffusion between neighbor compartments $i-1$ and $i+1$:

$$\begin{aligned}\frac{dC_i}{dt} &= \frac{J_{i,i-1} + J_{i,i+1}}{V_i}\\ &= -\frac{D}{V_i\left[\left(\frac{A}{\delta}\right)_{i,i-1}(C_i - C_{i-1}) - \left(\frac{A}{\delta}\right)_{i+1,i}(C_{i+1} - C_i)\right]},\end{aligned} \tag{B.2}$$

where $\left(\frac{A}{\delta}\right)_{p,q} = \frac{A_{pq}}{\Delta l_{pq}}$ describes the movement of $Ca^{2+}$ ions from compartment $p$ to $q$, where $A_{pq}$ is the cross-sectional area of the two compartments and $\Delta l_{pq}$ is the distance between the midpoints of compartments $p$ and $q$.

### B.1.1. *Diffusion inside the Synaptic Cleft*

The synaptic cleft is divided into $N$ concentric cylindrical rings, with a thinness of $\Delta R = 20$ nm. Therefore, the dynamics of glutamate in the $i$th ring is

$$\frac{d[\mathrm{Glu}_i]}{dt} = -\frac{D_{\mathrm{Glu}}}{V_i}\left[\left(\frac{A}{\delta}\right)_{i,i-1}([\mathrm{Glu}]_i - [\mathrm{Glu}]_{i-1}) - \left(\frac{A}{\delta}\right)_{i+1,i}([\mathrm{Glu}]_{i+1} - [\mathrm{Glu}]_i)\right], \tag{B.3}$$

where $V_i = h\pi[\Delta RI]^2 - h\pi[\Delta R(i-1)]^2$ and $(A/\delta)_{p,q} = (2h\pi\Delta Rq)/\Delta R$.

### B.1.2. *Diffusion in the extrasynaptic space*

Diffusion in the extrasynaptic space is modified from the model of Rusakov [62]. The coordinate origin is set at the center of synaptic cleft, the space compartments run in the radial direction with step $\Delta R_2$ and in the tangential direction with an angular step $\Delta\theta$. Each ring-shaped $(i,j)$th compartment has four interfaces with adjacent compartments. In the tangential direction, the interface area is $S_T(i,j) = 2\pi i\Delta R_2 \sin\theta(j)\Delta\theta$. In the radial direction, the interface area is $S_R(i,j) = 2\pi(i\Delta R_2)^2[\cos\theta(j) - \cos\theta(j-1)]$. The volume is $V(i,j) = 1/2[S_T(i,j)+S_T(i-1,j)]\Delta R_2$. In our simulation, $\Delta R_2 = 20$ nm and $\Delta\theta = \pi/9$.

The dynamic of glutamate concentration in compartment $(i,j)$ is

$$\frac{d[\mathrm{Glu}_{i,j}]}{dt} = -\frac{D^*_{\mathrm{Glu}}}{V_{i,j}}\left(\sum J_R(i,j)S_R + \sum J_T(i,j)S_T\right), \tag{B.4}$$

where $D^*_{\text{Glu}}$ is the effective diffusion coefficient in the extrasynaptic space, $\sum J_R(i,j)$ and $\sum J_T(i,j)$ are the summation of fluxes through radial and tangential interfaces, respectively. The diffusion flux between two adjacent compartment $n$ and $n+1$ can be calculated calculate them using the Fick's first law (Eq. (B.2)) and obtain:

$$J_{n,n+1} = \frac{1}{\delta_{n,n+1}}([\text{Glu}]_n - [\text{Glu}]_{n+1}),$$

where $\delta_{n,n+1}$ is the spatial distance between compartment $n$ and $n+1$.

## Appendix C: Mathematical Expressions for Rate Laws

### C.1. Mass Action-Based Models

The law of mass action is a fundamental law of a chemical reaction, which defines that the rates of a chemical reaction at constant temperature are proportional to the product of the reactant concentrations [63]. A single step reaction that follows mass action kinetics is called elementary reaction. Based on the number of reactant species, there are unimolecular, bimolecular and termolecular reactions. For example, the following simple bimolecular elementary reaction:

$$A + B \xrightarrow{k} C, \tag{C.1}$$

where $A$ and $B$ are the reactants, $C$ is the product and $k$ is the reaction rate constant or proportionality rate. Based on the law of mass action, the rate of this reaction, $V$, is proportional to $[A][B]$, and therefore,

$$V = k[A][B], \tag{C.2}$$

where $[A]$ and $[B]$ represent the concentrations of reactant $A$ and $B$, respectively. (In the following section, we use $[x]$ to denote the concentration of $x$.) The concentration changes of $A$, $B$ and $C$ over time can be described by a set of differential equations (ODEs):

$$\frac{d[A]}{dt} = -k[A][B] = -V, \tag{C.3}$$

$$\frac{d[B]}{dt} = -k[A][B] = -V, \tag{C.4}$$

$$\frac{d[C]}{dt} = k[A][B] = V. \tag{C.5}$$

An elementary reaction can be reversible, such as the following reaction:

$$A + B \overset{k_1;k_{-1}}{\longleftrightarrow} C \tag{C.6}$$

can be treated as two simultaneous reactions: one forward and one backward reaction, where $k_1$ and $k_{-1}$ denote the forward and backward reaction rate constants, respectively. According to the law of mass action, the forward reaction rate, $V_1$, is

$$V_1 = k_1[A][B], \tag{C.7}$$

and the backward reaction rate, $V_{-1}$, is

$$V_{-1} = k_{-1}[C]. \tag{C.8}$$

The concentration changes of all species in this system over time can be described with three ODEs:

$$\frac{d[A]}{dt} = -k_1[A][B] + k_{-1}[C] = -V_1 + V_{-1}, \tag{C.9}$$

$$\frac{d[B]}{dt} = -k_1[A][B] + k_{-1}[C] = -V_1 + V_{-1}, \tag{C.10}$$

$$\frac{d[C]}{dt} = k_1[A][B] - k_{-1}[C] = V_1 - V_{-1}. \tag{C.11}$$

Many complex reactions contain multiple steps, and each step is an elementary reaction. For example, the following chain reaction

$$A \xrightarrow{k_1} B \xrightarrow{k_2} C \tag{C.12}$$

consists of two elementary reaction: $A \xrightarrow{k_1} B$ and $B \xrightarrow{k_2} C$, with reaction rate constant $k_1$ and $k_2$, respectively. In the first reaction step, the reactant $A$ converts into an intermediate, $B$, and the reaction rate, $V_1$ is

$$V_1 = k_1[A]. \tag{C.13}$$

In the second reaction step, the intermediate $B$ converts into the final product $C$, and the reaction rate, $V_2$, is

$$V_2 = k_2[B]. \tag{C.14}$$

Therefore, the ODEs for the concentration changes of all species in this system over time are

$$\frac{d[A]}{dt} = -k_1[A] = -V_1, \tag{C.15}$$

$$\frac{d[B]}{dt} = k_1[A] - k_2[B] = V_1 - V_2, \tag{C.16}$$

$$\frac{d[C]}{dt} = k_2[B] = V_2. \tag{C.17}$$

## C.2. Michaelis–Menten Model

### C.2.1. *Equilibrium approximation*

The Michaelis–Menten model is a classic model for enzyme kinetics [64]. In enzyme-catalyzed reactions:

$$E + S \overset{k_1;k_{-1}}{\longleftrightarrow} ES \overset{k_2}{\longrightarrow} E + P, \tag{C.18}$$

a substrate, $S$, reversibly binds to an enzyme, $E$, to form a substrate-enzyme complex, $ES$, with the forward reaction rate constant, $k_1$, and the backward reaction rate constant, $k_{-1}$. Then, $ES$ irreversibly converts into the production, $P$, with rate constant $k_2$ and releases $E$ at the same time. Therefore, based on the law of mass action, the concentration changes of all species in this system can be described by four ODEs:

$$\frac{d\,[S]}{dt} = -k_1\,[E]\,[S] + k_{-1}[ES], \tag{C.19}$$

$$\frac{d\,[E]}{dt} = -k_1\,[E]\,[S] + (k)_{-1} + k_2[ES], \tag{C.20}$$

$$\frac{d\,[ES]}{dt} = k_1\,[E]\,[S] - (k)_{-1} + k_2[ES], \tag{C.21}$$

$$\frac{d\,[P]}{dt} = k_2[ES]. \tag{C.22}$$

The total concentration of the enzyme ($E_0$) is a constant, where

$$[E_0] = [E] + [ES]\,. \tag{C.23}$$

The original assumption by Michaelis and Menten [64] was that the substrate-enzyme complex, ES, reaches its equilibrium instantly ($k_1\,[E]\,[S] = k_{-1}\,[ES]$). Therefore, the concentration of the substrate-enzyme complex is given by

$$[ES] = \frac{k_1}{k_{-1}}\,[E]\,[S]\,. \tag{C.24}$$

According to the Eq. (C.23), Eq. (C.24) is modified to

$$[ES] = \frac{k_1}{k_{-1}}\,[S]\,([E_0] - [ES])\,. \tag{C.25}$$

Let $K_d = \frac{k_{-1}}{k_1}$ represent the dissociation constant of the ES complex. Substitute $K_d = \frac{k_{-1}}{k_1}$ into Eq. (C.25) and rearrange it to

$$[ES] = \frac{[S]\,[E_0]}{K_d + [S]}. \tag{C.26}$$

Therefore, the reaction rate of the final production $P$ (Eq. (B.22)), $V$, is given by

$$V = \frac{d\,[P]}{dt} = k_2\,[ES] = k_2\frac{[S]\,[E_0]}{K_d + [S]}. \tag{C.27}$$

Let $V_{\max} = k_2\,[E_0]$sent the maximum reaction rate the system can reach at a saturating substrate concentration. Then, Eq. (C.27) becomes

$$V = \frac{V_{\max}[S]}{K_d + [S]}. \tag{C.28}$$

When [S] $= K_d$, $V = V_{\max}/2$ equilibrium approximation is valid when the reversible reaction is much faster than the irreversible reaction

$k_{-1} \gg k_2$, and thus, the reaction rate of the production $P$ is limited by the irreversible reaction limits.

### C.2.2. *Quasi-steady-state approximation*

Briggs and Haldane [65] suggested an alternative approximation, the quasi-steady-state approximation, which assumes the concentration of the intermediate substrate-enzyme complex ES does not change during the production of P. Therefore, $\frac{d[ES]}{dt} = 0$ and rearrange the Eq. (C.21) to

$$[ES] = \frac{k_1}{k_{-1} + k_2} [E][S]. \tag{C.29}$$

Based on the conservation of the enzyme in Eq. (C.23), Eq. (C.29) is modified to

$$[ES] = \frac{[S][E_0]}{K_M + [S]}. \tag{C.30}$$

where $K_M = \frac{k_{-1}+k_2}{k_1}$ is called Michaelis–Menten constant.

Substitute Eq. (C.30) into Eq. (C.22), the reaction rate of the final production of $P$, $V$, is given by

$$V = \frac{d[P]}{dt} = k_2 [ES] = k_2 \frac{[S][E_0]}{K_M + [S]} = \frac{V_{\max[S]}}{K_M + [S]}. \tag{C.31}$$

The quasi-steady-state approximation is valid when $[E_0] \ll [S_0]$, where $S_0$ is the total substrate concentration ($[S]_0 = [S] + [ES]$). Equation (C.31) is very similar to Eq. (C.28), with the only difference that the equilibrium approximation uses $K_M$ (Eq. (C.31)) whereas the equilibrium approximation uses $K_{\max}$(C.28)). Notably, when $k_{-1} \gg k_2$, $K_M \approx K_{\max}$ these two approximation will produce similar results.

## C.3. Hill Model

The Hill model is a type of model that describes the cooperative binding of ligands to receptors [66]. It is widely used as the mathematical formula to quantify the binding between ligands and receptors by expressing the fraction of the bound receptor as a function of the ligand concentration [67]. The Hill model provides a way to describe the dependence between chemical substances in modeling the cooperative activation or inhibition in complex biochemical systems.

Consider a receptor protein consists of $n$ subunits and each subunit contains one binding site for a ligand. The binding of the ligands to the receptor can be separated into n reversible elementary reaction steps as follows:

$$L + R + \overset{k_1;k_{-1}}{\longleftrightarrow} LR, \tag{C.32}$$

$$L + LR \overset{k_2;k_{-2}}{\longleftrightarrow} L_2R, \tag{C.33}$$

$$\vdots$$

$$L + L_{n-1}R \overset{k_n;k_{-n}}{\longleftrightarrow} dL_nR\ , \tag{C.34}$$

where $L$ is the ligand, $R$ is the free receptor, $L_nR$ is the $n$-ligand bound receptor, and $k_n$ and $k_{-n}$ are the forward and backward reaction for the $n$-th elementary reaction steps, respectively. Besides, the rate of ions entering through this receptor is proportional to the concentration of the fully bound receptor, $L_nR$, with a rate constant of $k_{ion}$. Let $[x]$ be the concentration of $x$. The dynamics of all species in this system can be governed by a set of ODEs:

$$\frac{d[R]}{dt} = -k_1[R][L] + k_{-1}[LR], \tag{C.35}$$

$$\frac{d[LR]}{dt} = k_1[R][L] - k_{-1}[LR] - k_2[LR][L] + k_{-2}[L_2R], \tag{C.36}$$

$$\frac{d[L_2R]}{dt} = k_2[LR][L] - k_{-2}[L_2R] - k_3[L_2R][L] + k_{-3}[L_3R], \tag{C.37}$$

$$\vdots$$

$$\frac{d[L_{n-1}R]}{dt} = k_{n-1}\left[L_{(n-2)}R\right][L] - k_{-(n-1)}[L_{n-1}R] - k_n[L_{n-1}R][L] + k_{-n}[L_nR], \tag{C.38}$$

$$\frac{d\left[L_nR\right]}{dt} = k_n\left[L_{n-1}R\right]\left[L\right] - k_{-n}\left[L_nR\right], \tag{C.39}$$

$$\frac{d\left[L\right]}{dt} = -\left(k_1\left[R\right] + k_2\left[LR\right] + k_3\left[L_2R\right] + \cdots + k_n\left[L_{n-1}R\right]\right)\left[L\right] + \left(k_{-1}\left[LR\right] + k_{-2}\left[L_2R\right] + k_{-3}\left[L_3R\right] + k_{-n}\left[L_nR\right]\right). \tag{C.40}$$

The ion flux rate through this receptor is $f = k_{ion}\left[L_nR\right]$. Assuming there is no change in the concentrations of all binding states ($LR$, $L_2R, \ldots, L_nR$) over the timescale of $f$, thus, Eqs. (B.36)–(B.39) become zeros and we obtain:

$$\left[LR\right] = \frac{\left[R\right]\left[L\right]}{K_1}, \tag{C.41}$$

$$\left[L_2R\right] = \frac{\left[R\right]\left[L\right]^2}{K_1K_2}, \tag{C.42}$$

$$\vdots$$

$$\left[L_nR\right] = \frac{\left[R\right]\left[L\right]^n}{K_1K_2\cdots K_n}, \tag{C.43}$$

where $K_1 = \frac{k_{-1}}{k_1}$, $K_2 = \frac{k_{-2}}{k_2}, \ldots$ and $K_n = \frac{k_{-n}}{k_n}$.

Let $R_0$ be the total concentration of the receptor, $[R]_0 = [R] + [LR] + [L_2R] + \cdots + [L_nR]$ based on the law of conservation of mass. Therefore, we can calculate the fraction of receptor that has been fully bound by $n$ ligands:

$$\begin{aligned}\frac{[L_nR]}{[([R_0])]_0} &= \frac{[L_nR]}{[R] + [LR] + [L_2R] + \cdots + [L_nR]} \\ &= \frac{\frac{[R][L]^n}{K_1,K_2,\ldots,K_n}}{[R] + \frac{[R][L]}{K_1} + \frac{[R][L]^2}{K_1K_2} + \cdots + \frac{[R][L]^n}{K_1,K_2,\ldots,K_n}} \\ &= \frac{[L]^n}{K_1,K_2,\ldots,K_n + K_2\ldots K_n[L] + K_3\ldots K_n[L]^2 + \cdots + [L]^n}.\end{aligned} \tag{C.44}$$

We assume the first ligand binding is slow, but once one site is occupied, binding to the remaining sites is fast and the binding affinity for the remaining sites are changed by the first binding and ligands can bind to them simultaneously. This can be modeled as $k_1 \to 0$, $k_2 \to k_n \to \infty$ and $k_1, k_2, \ldots, k_n$ is fixed. Therefore, $K_1 \to \infty$, $K_2, \ldots, K_n \to 0$ and $K_1, K_2, \ldots, K_n$ is a constant. Let $K_a^n = K_1, K_2, \ldots, K_n$, which describes the ligand concentration producing half occupation of all binding sites. Equation (C.44) can be simplified to

$$\frac{[L_nR]}{[R]_0} = \frac{[L]^n}{K_a^n + [L]^n}. \tag{C.45}$$

Equation (C.45) is the classic form of the Hill equation, and $n$ is called the Hill coefficient. The Hill equation can also be written as

$$\frac{[L_nR]}{[R]_0} = \frac{[L]^n}{K_d + [L]^n}, \tag{C.46}$$

where $K_d = K_a^n$ denotes the apparent dissociation constant which can be derived from the law of mass action.

Rearrange Eq. (C.46), the concentration of $L_nR$ becomes a function of the ligand concentration:

$$[L_nR] = \frac{[R]_0[L]^n}{K_a^n + [L]^n}. \tag{C.47}$$

Therefore, the ion flux rate through this receptor is $f = k_{\text{ion}}[L_nR]$:

$$f = \frac{V_{\max}[L]^n}{K_a^n + [L]^n}, \tag{C.48}$$

where $V_{\max} = k_{\text{ion}}[R]_0$ is the maximum flux rate when all receptors are full bound by ligands.

The modified Hill model (Eq. (C.48)) describes a sigmoidal relationship between the flux rate $f$ and the ligand concentration $[L]$ (Fig. C.1). The curve of $f$ versus $[L]$ has an "S" shape (sigmoid curve). The Hill coefficient $n$ determines the steepness of the sigmoid curve. When $n = 1$, the Hill model becomes the Michaelis–Menten model, which shows no cooperativity binding. This is valid when each ligand binding site is independent from each other and the binding

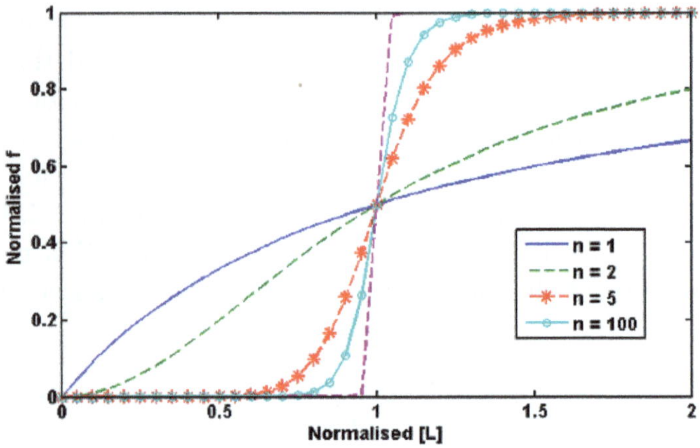


Fig. C.1. The normalized flux rate from the Hill model. The flux rate is normalized to the $V_{\max}$ while ligand concentration $[L_n]$ is normalized to the $K_a$. Acknowledgment [4]: Liang, J., Kulasiri, D. and Samarasinghe, S. (2017) Computational investigation of Amyloid-$\beta$-induced location-and subunit-specific disturbances of NMDAR at hippocampal dendritic spine in Alzheimer's disease. *PLoS One 12*(8), p. e0182743.

affinity for a ligand to a binding site is not affected if the receptor is already bound by other ligands. When $n > 1$, the Hill model describes a positively cooperative binding process that once a ligand is bound to the receptor, the affinity for the remaining ligands increases. On the contrary, when $n < 1$, the Hill model describes a negatively cooperative binding that the binding affinity decreases once a ligand is bound to the receptor.

## C.4. Application of the Michaelis–Menten Model to Simulation of $Ca^{2+}$ Extrusion Mechanisms

The intracellular $Ca^{2+}$ ions are removed from the cytosol by membrane exchangers or pumps, which can be assumed to have two steps: (1) intracellular $Ca^{2+}$ binds to the $Ca^{2+}$ pump and (2) the $Ca^{2+}$ pump or exchanger releases the bound $Ca^{2+}$ to the extracellular space:

$$\mathrm{Ca}_i^{2+} + P \overset{k_1;k_{-1}}{\longleftrightarrow} \mathrm{CaP} \xrightarrow{k_2} \mathrm{Ca}_0^{2+} + P, \tag{C.49}$$

where $Ca_i^{2+}$ and $Ca_o^{2+}$ denotes intracellular and extracellular $Ca^{2+}$, respectively. $P$ and CaP denote the free and intermediate $Ca^{2+}$-bound $Ca^{2+}$ pumps or exchangers, respectively. The $Ca^{2+}$ pumps in the membrane of internal $Ca^{2+}$ stores, such as the ER or mitochondrion, can be represented in a similar way. The only difference is that $Ca_o^{2+}$ represents the $Ca^{2+}$ in internal $Ca^{2+}$ storage. The $Ca^{2+}$ binding step is reversible, with forward rate constant, $k_1$, and backward rate constant, $k_{-1}$, respectively. The release step, with release time constant $k_2$, is irreversible when assuming the $Ca_0^{2+}$ is a constant. $Ca^{2+}$ normally binds to the pump with a high affinity, and the release process is relatively much slower than the $Ca^{2+}$ binding process $((k_{-1})k_2)$. Therefore, the $Ca^{2+}$ flux via the $Ca^{2+}$ pump or exchanger can be described by the Michaelis–Menten model:

$$J_{\text{pump}} = V_{\text{pump}} \frac{[\text{Ca}^{2+}]_i}{K_{\text{pump}} + [\text{Ca}^{2+}]_i}, \tag{C.50}$$

where $J_{\text{pump}}$ denotes the pump $Ca^{2+}$ flux (in units of concentration per unit time), $V_{\text{pump}}$ denotes the maximum pump velocity, $K_{\text{pump}}$ is the Michaelis–Menten constant, which denotes the $Ca^{2+}$ concentration with half $V_{\text{pump}}$, and $[\text{Ca}^{2+}]_i$ denotes the intracellular $Ca^{2+}$ concentration. $V_{\text{pump}}$ can be calculated from the maximum velocity and the total pump number:

$$V_{\text{pump}} = k_2 Ps \frac{A}{V},$$

where $Ps$ is the surface density of the pump (in units of molecule number per unit area of membrane), $A$ is the surface area of membrane and $V$ is the volume of the cytosol. According to Eq. (C.30), $K_{\text{pump}} = (k_{-1} + k_2)/k_1$, in the unit of concentration.

In some literature, the SERCA pump is simulated using a Hill model with a Hill coefficient greater than 1 (often using 2) [68]. The SERCA pump in the smooth ER membrane binds to two $Ca^{2+}$ ions for each ATP molecule, and can be modeled as two sequentially binding steps:

$$\text{Ca}_i^{2+} + P \overset{k_1;k_{-1}}{\longleftrightarrow} \text{Ca}P \overset{k_{r1}}{\longrightarrow} \text{Ca}_{\text{ER}}^{2+} + P, \tag{C.51}$$

$$\text{Ca}_i^{2+} + \text{Ca}P \overset{k_2;k_{-2}}{\longleftrightarrow} \text{Ca}_2P \overset{k_{r2}}{\longrightarrow} \text{Ca}_{\text{ER}}^{2+} + \text{Ca}P, \tag{C.52}$$

where $Ca^{2+}_{ER}$ denotes the $Ca^{2+}$ in the ER, $k_1$ and $k_2$ are the forward rate constants and $k_{-1}$ and $k_{-2}$ are the backward rate constants, of the each reveserble $Ca^{2+}$ binding step, respectively. $k_{r1}$ and $k_{r2}$ are the release time constant. The total concentration of SERCA, $P_0$, is a constant that $P_0 = [P] + [Ca_2P]$. Assume the concentration of CaP and $Ca_2P$ reach to the equilibriums instantly, then the $Ca^{2+}$ flux via the SERCA pump can be described by the Hill model:

$$J_{\text{pump}} = V_{\text{pump}} \frac{[\text{Ca}^{2+}]_i^n}{K_{\text{pump}}^n + [\text{Ca}^{2+}]_i^n}, \tag{C.53}$$

where the Hill coefficient $n = 2$ [68].

## Appendix D: ODEs for Tr, CaD and CaM

### D.1. Gluatmate Transporter (Tr)

Glutamate uptake by glial transporters can be represented through the following simple kinetic scheme [8]:

$$\text{Glu} + \text{Tr} \overset{k_{1f};k_{1b}}{\longleftrightarrow} \text{Glu} - \text{Tr} \overset{k_2}{\longrightarrow} \text{Tr},$$

where Glu is the glutamate, Tr is the unbound surface transporter and Glu–Tr is the glutamate-transporter complex (see Section 3.1.1 for the explanation). The dynamics of all species in the model can be described by three ODEs (Appendix C.1):

$$\frac{d[\text{Glu}]}{dt} = -k_{1f}[\text{Glu}][\text{Tr}] + k_{1b}([\text{Glu}] - [\text{Tr}]), \tag{D.1}$$

$$\frac{d[\text{Tr}]}{dt} = -k_{1f}[\text{Glu}][\text{Tr}] + (k_{1b} + k_2)([\text{Glu}] - [\text{Tr}]), \tag{D.2}$$

$$\frac{d[\text{Glu} - \text{Tr}]}{dt} = k_{1f}[\text{Glu}][\text{Tr}] - (k_{1b} + k_2)([\text{Glu}] - [\text{Tr}]). \tag{D.3}$$

According to the law of conservation of mass, $[\text{Tr}_{\text{total}}] = [\text{Tr}] + [\text{Glu} - \text{Tr}]$, $[\text{Tr}]_{\text{total}}$ is the total concentration of the transporter, therefore Eqs. (D.1)–(D.3) can be simplified to two ODEs:

$$\frac{d[\text{Glu}]}{dt} = -k_{1f}[\text{Glu}][\text{Tr}] + k_{1b}([\text{Tr}_{\text{total}}] - [\text{Tr}]), \tag{D.4}$$

$$\frac{d[\text{Tr}]}{dt} = -k_{1f}[\text{Glu}][\text{Tr}] + (k)_{1b} + k_2([\text{Tr}_{\text{total}}] - [\text{Tr}]). \tag{D.5}$$

## D.2. Buffers

Dynamics of the immobile buffer and CaD are governed by the same ODE: where $k_{bf}$ and $k_{bb}$ are the forward and backward buffer rate constants,

$$\frac{d[B]_j}{dt} = -k_{bf}[\text{Ca}^{2+}]_j[B]_j + k_{bb}([B]_{\text{total}} - [B]_j), \tag{D.6}$$

respectively, and $[B]_{\text{total},j}$ is the total concentration of immobile buffer or CaD in compartment $j$.

One CaM can bind four $Ca^{2+}$ ions:

$$\text{Ca}^{2+} + \text{CaM} \overset{k_1;k_{-1}}{\longleftrightarrow} \text{CaCaM},$$

$$\text{Ca}^{2+} + \text{CaCaM} \overset{k_2;k_{-2}}{\longleftrightarrow} \text{Ca}_2\text{CaM},$$

$$\text{Ca}^{2+} + \text{Ca}_2\text{CaM} \overset{k_3;k_{-3}}{\longleftrightarrow} \text{Ca}_3\text{CaM},$$

$$\text{Ca}^{2+} + \text{Ca}_3\text{CaM} \overset{k_4;k_{-4}}{\longleftrightarrow} \text{Ca}_4\text{CaM}.$$

The dynamics of all species in the model can be described by four ODEs (Appendix C.1):

$$\frac{d[\text{CaCaM}]}{dt} = k_1[\text{Ca}^{2+}][\text{CaM}] - k_{-1}[\text{CaCaM}] - k_2[\text{Ca}^{2+}][\text{CaCaM}] + k_{-2}[\text{Ca}_2\text{CaM}], \tag{D.7}$$

$$\frac{d[\text{Ca}_2\text{CaM}]}{dt} = k_2[\text{Ca}^{2+}][\text{CaCaM}] - k_{-2}[\text{Ca}_2\text{CaM}] - k_3[\text{Ca}^{2+}][\text{Ca}_2\text{CaM}] + k_{-3}[\text{Ca}_3\text{CaM}], \tag{D.8}$$

$$\frac{d[\mathrm{Ca_3CaM}]}{dt} = k_3[\mathrm{Ca}^{2+}][\mathrm{Ca_2CaM}] - k_{-3}[\mathrm{Ca_3CaM}]$$
$$- k_4[\mathrm{Ca}^{2+}][\mathrm{Ca_3CaM}] + k_{-4}[\mathrm{Ca_4CaM}], \quad \text{(D.9)}$$
$$\frac{d[\mathrm{Ca_4CaM}]}{dt} = k_4[\mathrm{Ca}^{2+}][\mathrm{Ca_3CaM}] - k_{-4}[\mathrm{Ca_4CaM}], \quad \text{(D.10)}$$

where $[\mathrm{CaM}] = [\mathrm{CaM}]_{\mathrm{total}} - [\mathrm{CaCaM}] - [\mathrm{Ca_2CaM}] - [\mathrm{Ca_3CaM}] - [\mathrm{Ca_4CaM}]$ ($[\mathrm{CaM}]_{\mathrm{total}}$ is the total concentration of CaM).

# Appendix E: Markov State Models of $Ca^{2+}$ Channels

## E.1. Markov Kinetic Models for a Single Ligand-Gated Ion Channel

Most voltage- and ligand-gated ion channels go through multiple conformational changes during gating [69]. The Markov kinetic model is widely used to capture the transition dynamics between different states and the ligand dependence of voltage- and ligand-gated ion channels [70]. The Markov kinetic model consists of a series of discrete states, such as open, close, desensitized and ligand-bound states, and describes the transition pathways among these states with specific rate constants, dependent or independent of voltage and/or ligand, accordingly. The state transition in the Markov kinetic model is memoryless, only depends on the state the channels in, not on time or its previous transitions.

The rate constants can be estimated by fitting experimental single-channel data to proposed Markov kinetic schemes. The experimental single-channel data by single-channel recording techniques describes the dwell times single channels spent in open or closed states. The frequency distribution of dwell-times can be described as a sum of exponential decay components by fitting the dwell-time histograms to multiple exponential functions [71]. Fit the dwell-time distributions to the Markov kinetic model to estimate values of each rate constant [72].

The deterministic interpretation of the Markov kinetic scheme can be represented as a set of ODEs [73]. For example, the following simple four-state model of a ligand-gated channel is

$$C + L \overset{k_1;k_{-1}}{\longleftrightarrow} LC_1 \overset{k_2;k_{-2}}{\longleftrightarrow} CL_2 \overset{k_3;k_{-3}}{\longleftrightarrow} O, \tag{E.1}$$

where $C$, $LC_1$, and $LC_2$ represent the close state and two ligand-bound close state, $L$ denotes the ligand and $O$ denotes the open state. $k_1$ and $k_{-1}$ are the forward and backward rate constants for the ligand binding step, respectively. $k_2$ and $k_{-2}$ are rate constants for the state transition between $LC_1$, and $LC_2$. $k_3$ and $k_{-3}$ are the rate constants of opening and the open state inactivation back to $LC_2$, respectively.

Let $[x]$ represents the fraction of a single channel in $x$ state. The dynamics can be described by a set of ODEs as following:

$$\frac{d[L]}{dt} = -k_1[C][L] + k_{-1}[LC_1], \tag{E.2}$$

$$\frac{d[C]}{dt} = -k_1[C][L] + k_{-1}[LC_1], \tag{E.3}$$

$$\frac{d[\mathrm{LC}_1]}{dt} = k_1[C][L] - k_{-1}[LC_1] - k_2[LC_1] + k_{-2}[LC_2], \tag{E.4}$$

$$\frac{d[\mathrm{LC}_2]}{dt} = k_2[LC_1] - k_{-2}[LC_2] - k_3[LC_2] + k_{-3}[O], \tag{E.5}$$

$$\frac{d[O]}{dt} = k_3[LC_2] - k_{-3}[O]. \tag{E.6}$$

Because of $[C]+[LC_1]+[LC_2]+[O] = 1$, the fraction of channels in state $C$ is

$$[C] = 1 - [LC_1] - [LC_2] - [O]. \tag{E.7}$$

Therefore, Eq. (E.2) can be removed and Eq. (E.3) becomes:

$$\frac{d[LC_1]}{dt} = k_1[L](1 - [LC_1] - [LC_2] - [O]) - k_{-1}[LC_1] - k_2[LC_1] + k_{-2}[LC_2]. \tag{E.8}$$

When the total concentration of the ligand is much greater than the number of channels, $[L]$ can be approximated to be unaffected by the binding step. Therefore, the dynamics of the four-state Markov model in Eq. (E.1) is governed by three ODEs: Eqs. (E.4), (E.5), and (E.7).

## E.2. Eight-State NMDAR Model

The Markov kinetic scheme of NMDAR in Fig. 7.4(a) contains eight states [23]. Let $[X]$ denote the fraction of a single NMDAR in state $X$, and, thus, $[R_N] + [R]_N G + [R]_N G_2 + [R]_N G_2 F + [R]_N G_2 S + [D]_1 + [D]_2 + [O]_N = 1$. We assume the glutamate concentration is not affected by the binding processes. The parameters of the reaction rates are summarized in Table 7.2. The dynamics of this Markov model are governed by seven ODEs:

$$\frac{d[R_N G]}{dt} = 2k_{\text{on}}[R_N][G] - k_{\text{off}}[R_N G] - k_{\text{on}}[R_N G][G] + 2k_{\text{off}}[R_N G_2], \tag{E.9}$$

$$\frac{d[R_N G_2]}{dt} = k_{\text{on}}[R_N G][G] - 2k_{\text{off}}[R_N G_2] - k_{d1+}[R_N G_2] + k_{d1-}[D_1] - k_{d2+}[R_N G_2] + k_{d2-}[D_2] - k_{f+}[R_N G_2] + k_{f-}[R_N G_2 F] - k_{s+}[R_N G_2] + k_{s-}[R_N G_2 S], \tag{E.10}$$

$$\frac{d[R_N G_2 F]}{dt} = k_{f+}[R_N G_2] - k_{f-}[R_N G_2 F] - k_{s+}[R_N G_2 F] + k_{s-}[O_N], \tag{E.11}$$

$$\frac{d[R_N G_2 S]}{dt} = k_{s+}[R_N G_2] - k_{s-}[R_N G_2 S] - k_{f+}[R_N G_2 S] + k_{f-}[O_N], \tag{E.12}$$

$$\frac{d[D_1]}{dt} = k_{d1+}[R_N G_2] - k_{d1-}[D_1], \tag{E.13}$$

$$\frac{d[D_2]}{dt} = k_{d2+}[R_N G_2] - k_{d2-}[D_2], \tag{E.14}$$

$$\frac{d[O_N]}{dt} = k_{s+}[R_N G_2 F] - k_{s-}[O_N] + k_{f+}[R_N G_2 S] - k_{f-}[O_N], \tag{E.15}$$

where $[R_N] = 1 - [R]_N G - [R]_N G_2 - [R]_N G_2 F - [R]_N G_2 S - [D]_1 - [D]_2 - [O]_N$. The fraction of a single NMDAR in the open state at time $t$ is $Po_{\text{NMDAR}}(t) = [O]_N(t)$.

## E.3. Seven-State AMPAR Model

The Markov kinetic scheme of AMPAR in Fig. 7.4(b) contains seven states [24]. Let $[X]$ denote the fraction of a single AMPAR in state $X$, and, thus, $[R_A] + [R_A G_1] + [R_A G_2] + [D]_1 + [D]_2 + [DO]_A + [O]_A = 1$.

We assume the glutamate concentration is not affected by the binding processes. The parameters of the reaction rates are summarized in Table 7.2. The dynamics of this Markov model are governed by six ODEs:

$$\frac{d[R_A G_1]}{dt} = k_{11}[R_A][G] - k_{10}[R_A G_1] - k_{21}[R_A G_1][G] + k_{20}[R_A G_2] - \alpha_1[R_A G_1] + \beta_1[D]_1, \quad \text{(E.16)}$$

$$\frac{d[R_A G_2]}{dt} = k_{21}[R_A G_1][G] - k_{20}[R_A G_2] - \alpha[R_A G_2] + \beta[O_A] - \alpha_2[R_A G_2] + \beta_2[D]_2, \quad \text{(E.17)}$$

$$\frac{d[D_1]}{dt} = \alpha_1[R_A G_1] - \beta_1[D]_1 - k_{31}[D_1][G] + k_{30}[D_2], \quad \text{(E.18)}$$

$$\frac{d[D_2]}{dt} = k_{31}[D_1][G] - k_{30}[D_2] + \alpha_2[R_A G_2] - \beta_2[D]_2 - \alpha_4[D_2] + \beta_4[DO]_A, \quad \text{(E.19)}$$

$$\frac{d[DO_A]}{dt} = \alpha_4[D_2] - \beta_4[DO]_A + \alpha_3[O_A] - \beta_3[DO]_A, \quad \text{(E.20)}$$

$$\frac{d[O_A]}{dt} = \alpha[R_A G_2] - \beta[O_A] - \alpha_3[O_A] + \beta_3[DO]_A, \quad \text{(E.21)}$$

where $[R_A] = 1 - [R_A G_1] - [R_A G_2] - [D]_1 - [D]_2 - [DO]_A - [O]_A$. The fraction of a single AMPAR in the open state at time $t$ is $Po_{\text{AMPAR}}(t) = [O_A](t)$.

# Appendix F: MCMC and PRCC

## F.1. MCMC for Parameter Estimation

Markov chain Monte Carlo (MCMC) is a general sampling strategy. It is always used in solving problems with high-dimensional spaces, such as parameter estimation in a multi-dimensional system [74]. In this research, we estimate the values of parameters using an MCMC Matlab toolboxes downloaded from https://mjlaine.github.io/mcmcstat/.

A Markov chain is a discrete random process that consists of a finite number of states. The probability of being at a state only depends on the previous states. Suppose a Markov chain is a sequence of random variables $(X_0, X_1, \ldots, X_N)$ at $t = 0, 1, \ldots, n$, respectively, with a state space of $X = \{x_1, x_2, \ldots, x_s\}$. Therefore, the probability of transfer from one state $X_n$ at time n to next state $X_{n+1}$ at time $n + 1$ is

$$\begin{aligned} p(X_{n+1} = x_j | X_n = x_{i_n}, X_{n-1} = x_{i_{n-1}}, \ldots, X_0 = x_{i_0}) \\ = p(X_{n+1} = x_j | X_n = x_{i_n}), \quad (n \in N, i_n, j \in s), \end{aligned}$$

where $p(x_j|x_i)$ denotes the probability of moving from state $i$ to $j$ at current time step. A transition matrix, $T$, describes the transition probabilities across the state space as

$$T = \begin{pmatrix} p(x_1|x_1) & p(x_1|x_2) & \cdots & p(x_1|x_j) & \cdots & p(x_1|x_s) \\ p(x_2|x_1) & p(x_2|x_2) & \cdots & p(x_2|x_j) & \cdots & p(x_2|x_s) \\ & & & \cdots & & \\ p(x_i|x_1) & p(x_i|x_2) & \cdots & p(x_i|x_j) & \cdots & p(x_i|x_s) \\ & & & \cdots & & \\ p(x_s|x_1) & p(x_s|x_2) & \cdots & p(x_s|x_j) & \cdots & p(x_s|x_s) \end{pmatrix}.$$

The total of the transition probabilities from a state $x_i$ to the rest states in $X$ is 1, therefore,

$$\sum_{i=1}^{s} p(x_j|x_i) = 1.$$

For example, a simple Markov chain consists a state space of three discrete states ($X = \{x_1, x_2, x_3\}$) and its transition matrix is

$$T = \begin{pmatrix} 0.5 & 0 & 0.5 \\ 0 & 0.4 & 0.6 \\ 0.1 & 0 & 0.9 \end{pmatrix}.$$

The probability distribution of states at time $n$ is $P_n$, and

$$P_{n+1} = P_n T = (P_{n-1}T)T = \cdots = P_0 T^n,$$

where $P_0$ is the initial distribution. When $n$ is large enough,

$$\lim_{n\to\infty} (T)^n = \begin{pmatrix} 0.1667 & 0 & 0.8333 \\ 0.1667 & 0 & 0.8333 \\ 0.1667 & 0 & 0.8333 \end{pmatrix}.$$

Therefore, the chain converges to an invariant or stationary distribution, (0.1667, 0, 0.08333) regardless of the initial distribution $P_0$.

When an aperiodic Markov chain starts from any state and all the rest states of the state space are reachable within finite steps, the chain will converge to an invariant or stationary distribution [74]. The MCMC is developed based on this convergence property of the Markov chain. The invariant or stationary distribution must have the following detailed balance:

$$P(x_i)T_{ij} = P(x_j)T_{ji}, \quad (i, j = 1, \ldots, s),$$

where $T_{ij} = p(x_i|x_j)$. Summing both sides of the above equation over $x_j$, thus

$$P(x_i) = \sum p(x_j)T_{ji}.$$

When we use MCMC to estimate parameter values, if we let the constructed Markov chain converge on the optimal distribution of parameters, then the optimal distribution of parameters can be directly generated from the converged chain.

The Metropolis–Hastings algorithm is one of the most general MCMC methods [75, 76]. For a given aperiodic and irreducible Markov chain, $MC_1$ with a discrete parameters space $X = \{x_1, x_2, \ldots, x_s\}$. Its transition matrix $Q$ is

$$Q = q(X_{n+1} = x_j | X_n = x_i) = Q_{ij}, \quad (i, j = 1, \ldots, s).$$

Suppose the optimal distribution of parameters is $P(X)$, and $MC_1$ does not converge to $P(X)$:

$$P(x_i)Q_{ij} \neq P(x_j)Q_{ji}, \quad (i, j = 1, \ldots, s).$$

If $MC_1$ can be transformed by a function $A$ into $MC_2$ and $MC_2$ satisfies the detailed balance condition and converges to $P(X)$:

$$P(x_i)Q_{ij}A(x_i, x_j) = P(x_j)Q_{ji}A(x_j, x_i), \quad (i, j = 1, \ldots, s).$$

Therefore, the simplest form of $A$ satisfies $A(x_i, x_j) = P(x_j)Q_{ji}$ and $A(x_j, x_i) = P(x_i)Q_{ij}$. The function $A$ is called the acceptance probability and it describes the probability of accepting that the transition happens in $MC_1$ also occurs in $MC_2$.

Then the transition matrix of $MC_2$ is

$$Q' = q'(X_{n+1} = x_j | X_n = x_i) = Q'_{ij} = Q_{ij}Q_{ji}P(x_j), \quad (i, j = 1, \ldots, s).$$

If $A$ is very small, it may take a long time for $MC_2$ to explore the whole parameter space to converge to $P(X)$. In that case, multiply $A$ with a factor $k$ where

$$kA(x_i, x_j) \leq 1, kA(x_j, x_i) \leq \text{ and } k \geq 1.$$

Then we obtain a Markov chain $MC_3$ which can converge on $P(X)$ faster than $MC_2$:

$$P(x_i)Q_{ij}kA(x_i, x_j) = P(x_j)Q_{ji}kA(x_j, x_i), \quad (i, j = 1, \ldots, s).$$

When $kA(x_i, x_j) = 1$ or $kA(x_j, x_i) = 1$, the convergence of $MC_3$ reaches the maximum speed, and *therefore*, the maximum of $kA(x_i, x_j)$ is $\min\left[1, \frac{A(x)_{i}, x_j}{A(x_j, x_i)}\right]$.

With a given *initial* state $x^0$ ($x^0 \in X$), at iteration $t$ ($t = 1, 2, \ldots, T$), the Metropolis–Hastings algorithm follows the following steps:

(1) Draw $y$ from $X$ and generate a random number, $u$, from uniform distribution, $U(0, 1)$;
(2) If $u < \min\left[1, \frac{A(x)_{i,x_j}}{A(x_j, x_i)}\right]$, $x^t = y$;

Otherwise, $x^t = x^{t-1}$.

## F.2. PRCC for Global Parameter Sensitivity

The partial ranking correlation coefficient (PRCC) is one of many popular global sensitivity analysis methods, which quantifies the sensitivity of the model output with respect to the variation of selected parameters. PRCC works well for nonlinear but monotonic relationships between outputs and inputs, therefore, it is important to check the monotonicity of the model parameters against the outputs [55].

There is a number of sampling algorithms, such as Latin hypercube sampling (LHS) [77], which provides the model with a large number of parameter sets, varying within given ranges. For a given model, the correlation coefficient (CC) between a particular parameter, $x_j (j = 1, 2, \ldots, n)$ and output, $y$, is calculated:

$$r_{x_j y} = \frac{\sum_{i=1}^{N}(x_{ij} - \bar{x})(y_i - \bar{y})}{\sqrt{\sum_{i=1}^{N}(x_{ij} - \bar{x})^2 \sum_{i=1}^{N}(y_i - \bar{y})^2}},$$

where $N$ is the total number of sets of parameters. $x_{ij}$, $\bar{x}$ and $\bar{y}$ are the values of the $i$th sample of $x_j$, the mean of samples $x_j$ and the mean of outputs, respectively. The absolute value of $r_{x_j y}$ which is close to 1 indicates a strong correlation, whereas $|r_{x_j y}|$ which is close to 0 indicates a poor correlation. The sign of $r_{x_j y}$ means negative correlation ($-$) or positive correlation ($+$). A $p$-value shows the significance level of its corresponding $r_{x_j y}$. Normally when $p$-value $<0.05$, it rejects the null hypothesis that $x_j$ and $y$ is uncorrelated.

We generate a sample parameter set simultaneously according to certain percentage around their reference values. Therefore, CC results may be affected by the variation of parameters and the reference values. The partial correlation coefficient (PCC) removes the linear effects on output $y$ of all parameters other than $x_j$ and then quantifies the relation between $x_j$ and $y$. The PCC between $x_j$ and $y$ is the CC between $(x_j - \hat{x}_j)$ and $(y - \hat{y})$, where $\hat{x}_j$ and $\hat{y}$ are calculated by following regression model:

$$\hat{x}_j = e_1 + \sum_{i=1, i \neq j}^{n} \alpha_i x_i,$$

$$\hat{y} = e_2 + \sum_{i=1, i \neq j}^{n} \beta_i x_i,$$

where $e_1$ and $e_2$ are error terms, $\alpha_i$ and $\beta_i$ are the regression coefficients for $x_i$ and $y$, respectively.

In the PRCC analysis, the parameter $x_j$ and output $y$ are first rank transformed before calculating $\hat{x}_j$ and $\hat{y}$ according to the linear regression models. The rank-transformed data is ranks (integers) which are transferred from real values. Therefore, PRCC can eliminate the effect by the variation of parameters as well as by the magnitude of the reference values on CC. The PRCC analysis in this research is using Matlab functions downloaded from http://malthus.micro.med.umich.edu/lab/usadata/.

## References

[1] Li S. *et al.* (2009). Soluble oligomers of amyloid $\beta$ protein facilitate hippocampal long-term depression by disrupting neuronal glutamate uptake. *Neuron*, *62*(6), pp. 788–801.

[2] Hardingham G.E. and Bading H. (2010). Synaptic versus extrasynaptic NMDA receptor signalling: implications for neurodegenerative disorders. *Nat. Rev. Neurosci.*, *11*(10), pp. 682–696.

[3] Mota S.I., Ferreira I.L. and Rego A.C. (2014). Dysfunctional synapse in Alzheimer's disease — A focus on NMDA receptors. *Neuropharmacology*, *76*(Pt A), pp. 16–26.

[4] Liang J., Kulasiri D. and Samarasinghe S. (2017). Computational investigation of Amyloid-$\beta$-induced location-and subunit-specific disturbances of NMDAR at hippocampal dendritic spine in Alzheimer's disease. *PloS One*, *12*(8), p. e0182743.

[5] Goto Y., Niidome T., Akaike A., Kihara T. and Sugimoto H. (2006). Amyloid beta-peptide preconditioning reduces glutamate-induced neurotoxicity by promoting endocytosis of NMDA receptor. *Biochem. Biophys. Res. Commun.*, *351*(1), pp. 259–265.

[6] Hardingham G.E. and Bading H. (2010). Synaptic versus extrasynaptic NMDA receptor signalling: implications for neurodegenerative disorders. *Nature Rev. Neurosci.*, *11*(10), pp. 682–696.

[7] Sebollela A. *et al.* (2012). Amyloid-$\beta$ oligomers induce differential gene expression in adult human brain slices. *J. Biol. Chem.*, *287*(10), pp. 7436–7445.

[8] Rusakov D.A. and Kullmann D.M. (1998). Extrasynaptic glutamate diffusion in the hippocampus: ultrastructural constraints, uptake, and receptor activation. *J. Neurosci.*, *18*(9), pp. 3158–3170.

[9] Meldrum B.S. (2000). Glutamate as a neurotransmitter in the brain: review of physiology and pathology. *J. Nutrition*, *130*(4), p. 1007.

[10] Herman M.A. and Jahr C.E. (2007). Extracellular glutamate concentration in hippocampal slice. *J. Neurosci.*, *27*(36), pp. 9736–9741.

[11] Talantova M. *et al.* (2013). A$\beta$ induces astrocytic glutamate release, extrasynaptic NMDA receptor activation, and synaptic loss. *Proc. Natl. Acad. Sci. USA*, *110*(27), pp. E2518–E2527.

[12] Rusakov D.A. (2001). The role of perisynaptic glial sheaths in glutamate spillover and extracellular $Ca^{(2+)}$ depletion. *Biophys. J.*, *81*(4), pp. 1947–1959.

[13] Newpher T.M. and Ehlers M.D. (2008). Glutamate receptor dynamics in dendritic microdomains. *Neuron*, *58*(4), pp. 472–497.

[14] Sorra K.E. and Harris K.M. (2000). Overview on the structure, composition, function, development, and plasticity of hippocampal dendritic spines. *Hippocampus*, *10*(5), pp. 501–511.

[15] Holmes W.R. and Levy W.B. (1990). Insights into associative long-term potentiation from computational models of NMDA receptor-mediated calcium influx and intracellular calcium concentration changes. *J. Neurophysiology*, *63*(5), pp. 1148–1168.

[16] Sorra K.E. and Harris K.M. (2000). Overview on the structure, composition, function, development, and plasticity of hippocampal dendritic spines. *Hippocampus*, *10*(5), pp. 501–511.

[17] O'Donnell C., Nolan M.F. and van Rossum M.C.W. (2011). Dendritic spine dynamics regulate the long-term stability of synaptic plasticity. *J. Neurosci.*, *31*(45), pp. 16142–16156.

[18] Li S. *et al.* (2011). Soluble Abeta oligomers inhibit long-term potentiation through a mechanism involving excessive activation of extrasynaptic NR2B-containing NMDA receptors. *J. Neurosci.*, *31*(18), pp. 6627–6638.

[19] Chua J.J.E., Kindler S., Boyken J. and Jahn R. (2010). The architecture of an excitatory synapse. *J. Cell Sci.*, *123*(6), pp. 819–823.

[20] Cheng D. *et al.* (2006). Relative and absolute quantification of postsynaptic density proteome isolated from rat forebrain and cerebellum. *Mol. Cellular Proteomics*, *5*(6), pp. 1158–1170.

[21] Harris A.Z. and Pettit D.L. (2007). Extrasynaptic and synaptic NMDA receptors form stable and uniform pools in rat hippocampal slices. *J. Physiol.*, *584*(Pt 2), pp. 509–519.

[22] Tanaka J.-I. *et al.* (2005). Number and density of AMPA receptors in single synapses in immature cerebellum. *J. Neurosci.*, *25*(4), pp. 799–807.

[23] Erreger K., Dravid S.M., Banke T.G., Wyllie D.J. and Traynelis S.F. (2005). Subunit-specific gating controls rat NR1/NR2A and NR1/NR2B NMDA channel kinetics and synaptic signalling profiles. *J. Physiol.*, *563*(Pt 2), pp. 345–358.

[24] Jonas P., Major G. and Sakmann B. (1993). Quantal components of unitary EPSCs at the mossy fibre synapse on CA3 pyramidal cells of rat hippocampus. *J. Physiol.*, *472*, pp. 615–663.

[25] Petralia R.S. (2012). Distribution of extrasynaptic NMDA receptors on neurons. *Scientific World J.*, *2012*, Article 267120, p. 11.

[26] Takumi Y., Ramirez-Leon V., Laake P., Rinvik E. and Ottersen O.P. (1999). Different modes of expression of AMPA and NMDA receptors in hippocampal synapses. *Nat. Neurosci.*, *2*(7), pp. 618–624.

[27] Papouin T. *et al.* (2012). Synaptic and extrasynaptic NMDA receptors are gated by different endogenous coagonists. *Cell*, *150*(3), pp. 633–646.

[28] Bloodgood B.L. and Sabatini B.L. (2007). Nonlinear regulation of unitary synaptic signals by CaV(2.3) voltage-sensitive calcium channels located in dendritic spines. *Neuron*, *53*(2), pp. 249–260.

[29] Tempia F. *et al.* (1996). Fractional calcium current through neuronal AMPA-receptor channels with a low calcium permeability. *J. Neurosci.*, *16*(2), pp. 456–466.

[30] Koch C. (1998). *Biophysics of Computation: Information Processing in Single Neurons* (Oxford university press).

[31] Araya R., Jiang J., Eisenthal K.B. and Yuste R. (2006). The spine neck filters membrane potentials. *Proc. Natl. Acad. Sci. USA*, *103*(47), pp. 17961–17966.

[32] Sterratt D., Graham B., Gillies A. and Willshaw D. (2011). *Principles of Computational Modeling in Neuroscience* (Cambridge University Press).

[33] Jahr C.E. and Stevens C.F. (1990). Voltage dependence of NMDA-activated macroscopic conductances predicted by single-channel kinetics. *J. Neurosci.*, *10*(9), pp. 3178–3182.

[34] Mainen Z.F., Carnevale N.T., Zador A.M., Claiborne B.J. and Brown T.H. (1996). Electronic architecture of hippocampal CA1 pyramidal neurons based on three-dimensional reconstructions. *J. Neurophysiol.*, *76*(3), pp. 1904–1923.

[35] Schiegg A., Gerstner W., Ritz R. and van Hemmen J.L. (1995). Intracellular $Ca^{2+}$ stores can account for the time course of LTP induction: a model of $Ca^{2+}$ dynamics in dendritic spines. *J. Neurophysiol.*, *74*(3), pp. 1046–1055.

[36] Spruston N., Jonas P. and Sakmann B. (1995). Dendritic glutamate receptor channels in rat hippocampal CA3 and CA1 pyramidal neurons. *J. Physiol.*, *482*(Pt 2), pp. 325–352.

[37] Harris K.M., Jensen F.E. and Tsao B. (1992). Three-dimensional structure of dendritic spines and synapses in rat hippocampus (CA1) at postnatal day

15 and adult ages: implications for the maturation of synaptic physiology and long-term potentiation. *J. Neurosci.*, *12*(7), pp. 2685–2705.

[38] Ventura R. and Harris K.M. (1999). Three-dimensional relationships between hippocampal synapses and astrocytes. *J. Neurosci.*, *19*(16), pp. 6897–6906.

[39] Sabatini B.L., Oertner T.G. and Svoboda K. (2002). The life cycle of $Ca^{2+}$ ions in dendritic spines. *Neuron*, *33*(3), pp. 439–452.

[40] Blaustein M.P. and Lederer W.J. (1999). Sodium/calcium exchange: its physiological implications. *Physiolog. Rev.*, *79*(3), pp. 763–854.

[41] Carafoli E. (1992). The $Ca^{2+}$ pump of the plasma membrane. *J. Biol. Chem.*, *267*(4), pp. 2115–2118.

[42] Allbritton N.L., Meyer T. and Stryer L. (1992). Range of messenger action of calcium ion and inositol 1,4,5-trisphosphate. *Science (N.Y.)*, *258*(5089), pp. 1812–1815.

[43] Garaschuk O., Schneggenburger R., Schirra C., Tempia F. and Konnerth A. (1996). Fractional $Ca^{2+}$ currents through somatic and dendritic glutamate receptor channels of rat hippocampal CA1 pyramidal neurones. *J. Physiol.*, *491*(Pt 3), pp. 757–772.

[44] Kovalchuk Y., Eilers J., Lisman J. and Konnerth A. (2000). NMDA receptor-mediated subthreshold $Ca^{2+}$ signals in spines of hippocampal neurons. *J. Neurosci.*, *20*(5), pp. 1791–1799.

[45] Palmer L.M. and Stuart G.J. (2009). Membrane potential changes in dendritic spines during action potentials and synaptic input. *J. Neurosci.*, *29*(21), pp. 6897–6903.

[46] Masugi-Tokita M. *et al.* (2007). Number and density of AMPA receptors in individual synapses in the rat cerebellum as revealed by SDS-digested freeze-fracture replica labeling. *J. Neurosci.*, *27*(8), pp. 2135–2144.

[47] Palmer L.M. and Stuart G.J. (2009). Membrane potential changes in dendritic spines during action potentials and synaptic input. *J. Neurosci.*, *29*(21), pp. 6897–6903.

[48] Haario H., Saksman E. and Tamminen J. (2001). An adaptive metropolis algorithm. *Bernoulli*, pp. 223–242.

[49] Haario H., Laine M., Mira A. and Saksman E. (2006). DRAM: efficient adaptive MCMC. *Stat. Comput.*, *16*(4), pp. 339–354.

[50] Abramov E. *et al.* (2009). Amyloid-beta as a positive endogenous regulator of release probability at hippocampal synapses. *Nat. Neurosci.*, *12*(12), pp. 1567–1576.

[51] Matos M., Augusto E., Oliveira C.R. and Agostinho P. (2008). Amyloid-beta peptide decreases glutamate uptake in cultured astrocytes: involvement of oxidative stress and mitogen-activated protein kinase cascades. *Neurosci.*, *156*(4), pp. 898–910.

[52] Sheldon A.L. and Robinson M.B. (2007). The role of glutamate transporters in neurodegenerative diseases and potential opportunities for intervention. *Neurochem. Int.*, *51*(6–7), pp. 333–355.

[53] Mayer M.L., Vyklicky L. and Clements J. (1989). Regulation of NMDA receptor desensitization in mouse hippocampal neurons by glycine. *Nature*, *338*(6214), pp. 425–427.

[54] You H. *et al.* (2012). A$\beta$ neurotoxicity depends on interactions between copper ions, prion protein, and N-methyl-D-aspartate receptors. *Proc. Nat. Acad. Sci. USA*, *109*(5), pp. 1737–1742.
[55] Marino S., Hogue I.B., Ray C.J. and Kirschner D.E. (2008). A methodology for performing global uncertainty and sensitivity analysis in systems biology. *J. Theoret. Biol.*, *254*(1), pp. 178–196.
[56] Sobczyk A., Scheuss V. and Svoboda K. (2005). NMDA receptor subunit-dependent [$Ca^{2+}$] signaling in individual hippocampal dendritic spines. *J. Neurosci.*, *25*(26), pp. 6037–6046.
[57] Bordji K., Becerril-Ortega J., Nicole O. and Buisson A. (2010). Activation of extrasynaptic, but not synaptic, NMDA receptors modifies amyloid precursor protein expression pattern and increases amyloid-$\beta$ production. *J. Neurosci.*, *30*(47), pp. 15927–15942.
[58] Hardingham G.E. and Bading H. (2010). Synaptic versus extrasynaptic NMDA receptor signalling: implications for neurodegenerative disorders. *Nature Rev. Neurosci.*, *11*(10), pp. 682–696.
[59] Snyder E.M. *et al.* (2005). Regulation of NMDA receptor trafficking by amyloid-beta. *Nat. Neurosci*, *8*(8), pp. 1051–1058.
[60] Bayer K.U., De Koninck P., Leonard A.S., Hell J.W. and Schulman H. (2001). Interaction with the NMDA receptor locks CaMKII in an active conformation. *Nature*, *411*(6839), pp. 801–805.
[61] Fick A. (1855). Ueber Diffusion. *Ann. Phys.*, *170*(1), pp. 59–86.
[62] Rusakov D.A. (2001). The role of perisynaptic glial sheaths in glutamate spillover and extracellular $Ca^{2+}$ depletion. *Biophys. J.*, *81*(4), pp. 1947–1959.
[63] Guldberg C.M. and Waage P. (1864). Studies concerning affinity. *CM Forhandlinger: Videnskabs-Selskabet I Christiana*, *35*(1864), p. 1864.
[64] Michaelis L. and Menten M.L. (1913). Die kinetik der invertinwirkung. *Biochem. Z.*, *49*, pp. 333–369.
[65] Briggs G.E. and Haldane J.B.S. (1925). A note on the kinetics of enzyme action. *Biochem. J.*, *19*(2), pp. 338.
[66] Hill A.V. (1910). The possible effects of the aggregation of the molecules of haemoglobin on its dissociation curves. *J. Physiol. (Lond.)*, *40*, pp. 4–7.
[67] Weiss J.N. (1997). The Hill equation revisited: uses and misuses. *FASEB J*, *11*(11), pp. 835–841.
[68] Lytton J., Westlin M., Burk S.E., Shull G.E. and MacLennan D.H. (1992). Functional comparisons between isoforms of the sarcoplasmic or endoplasmic reticulum family of calcium pumps. *J. Biol. Chem.*, *267*(20), pp. 14483–14489.
[69] Tombola F., Pathak M.M. and Isacoff E.Y. (2006). How does voltage open an ion channel? *Annu. Rev. Cell Dev. Biol.*, *22*, pp. 23–52.
[70] Destexhe A., Mainen Z.F. and Sejnowski T.J. (1994). Synthesis of models for excitable membranes, synaptic transmission and neuromodulation using a common kinetic formalism. *J. Comput. Neurosci.*, *1*(3), pp. 195–230.
[71] Colquhoun D. and Hawkes A.G. (1982). On the stochastic properties of bursts of single ion channel openings and of clusters of bursts. *Philos. Trans. Roy. Soc. London B: Biol. Sci.*, *300*(1098), pp. 1–59.

[72] Qin F. and Li L. (2004). Model-based fitting of single-channel dwell-time distributions. *Biophys. J.*, *87*(3), pp. 1657–1671.

[73] Stevens C. (1978). Interactions between intrinsic membrane protein and electric field. An approach to studying nerve excitability. *Biophys. J.*, *22*(2), p. 295.

[74] Andrieu C., De Freitas N., Doucet A. and Jordan M.I. (2003). An introduction to MCMC for machine learning. *Machine Learning*, *50*(1–2), pp. 5–43.

[75] Metropolis N., Rosenbluth A.W., Rosenbluth M.N., Teller A.H. and Teller E. (1953). Equation of state calculations by fast computing machines. *J. Chem. Phys.*, *21*(6), pp. 1087–1092.

[76] Hastings W.K. (1970). Monte Carlo sampling methods using Markov chains and their applications. *Biometrika*, *57*(1), pp. 97–109.

[77] McKay M.D., Beckman R.J. and Conover W.J. (2000). A comparison of three methods for selecting values of input variables in the analysis of output from a computer code. *Technometrics*, *42*(1), pp. 55–61.

[78] Scheuss V., Yasuda R., Sobczyk A. and Svoboda K. (2006). Nonlinear [$Ca^{2+}$] signaling in dendrites and spines caused by activity-dependent depression of $Ca^{2+}$ extrusion. *J Neurosci.*, *26*(31), pp. 8183–8194.

[79] Harris M.E. *et al.* (1995). beta-Amyloid peptide-derived, oxygen-dependent free radicals inhibit glutamate uptake in cultured astrocytes: implications for Alzheimer's disease. *Neuroreport*, *6*(14), pp. 1875–1879.

[80] Harris M.E. *et al.* (1996). Amyloid *β* Peptide (25–35) Inhibits $Na^+$-Dependent Glutamate Uptake in Rat Hippocampal Astrocyte Cultures. *Journal of Neurochemistry*, *67*(1), pp. 277–286.

[81] Parpura-Gill A., Beitz D. and Uemura E. (1997). The inhibitory effects of *β*- amyloid on glutamate and glucose uptakes by cultured astrocytes. *Brain Research*, *754*(1–2), pp. 65–71.

[82] Matos M., Augusto E., Oliveira C. and Agostinho P. (2008). Amyloid-beta peptide decreases glutamate uptake in cultured astrocytes: involvement of oxidative stress and mitogen-activated protein kinase cascades. *Neuroscience*, *156*(4), pp. 898–910.

[83] Fernández-Tomé P., Brera B., Arévalo M. a.-A. and de Ceballos M.a.L. (2004). *β*-Amyloid 25-35 inhibits glutamate uptake in cultured neurons and astrocytes: modulation of uptake as a survival mechanism. *Neurobiology of Disease*, *15*(3), pp. 580–589.

[84] McKay M.D., Beckman R.J. and Conover W.J. (2000). A comparison of three methods for selecting values of input variables in the analysis of output from a computer code. *Technometrics*, *42*(1), pp. 55–61.

# Chapter 8

# Computational Modeling of Fluxes in ER

## 8.1. Introduction

The dendritic smooth endoplasmic reticulum (ER) extends into the spine head through the spine neck and in some spine heads, forms a specialized organelle called the spine apparatus, which comprises a stack of membrane disks [1, 2]. About 20% of the pyramidal dendritic spine contains the ER and these ER-containing spines tend to have much larger volumes than other spines [1, 3]. Experimental studies suggest that the spine apparatus may be involved in the synaptic signaling through modifying $Ca^{2+}$ influx from extracellular space in response to stimulation [4, 5]. The ER releases $Ca^{2+}$ to the cytosol through receptor-mediated $Ca^{2+}$-induced $Ca^{2+}$ release (CICR) and membrane $Ca^{2+}$ leakage, and uptakes cytosol $Ca^{2+}$ by SERCA pumps. The opposed $Ca^{2+}$ fluxes lead to a dual role of the ER that functions as a $Ca^{2+}$ source and functions as a $Ca^{2+}$ sink [6].

In research on Alzheimer's disease, the alterations in ER function have been observed in AD and are suggested to contribute to AD pathophysiology (reviewed in Section 7.5 in Chapter 7). In the dendritic spine of the pyramidal neuron in the hippocampus, these alterations may partly account for the disturbance in synaptic $Ca^{2+}$ signaling in AD by abnormal responses to $Ca^{2+}$ influx via NMDAR [7, 8] (Fig. 8.1). The dysregulation of ER $Ca^{2+}$ handling is suggested to cause defects in the structure of the spine head or synaptic function which, ultimately, synapse loss, neuronal death and cognitive deficits seen in AD [9, 10]. For instance, studies of cultured hippocampal

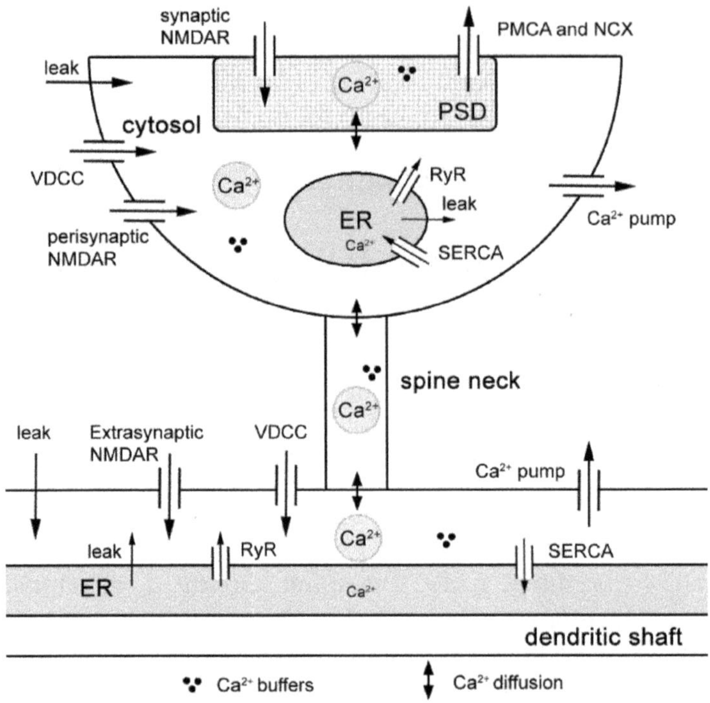


Fig. 8.1. $Ca^{2+}$ signaling in an ER-containing spine. $Ca^{2+}$ transient in the dendritic spine head is a production of $Ca^{2+}$ fluxes from the external and internal $Ca^{2+}$ sources via membrane $Ca^{2+}$ channels (external: NMDAR and VDCC; internal: RyR in this case), passive membrane leaking, binding of intracellular $Ca^{2+}$ buffers and extrusion by membrane $Ca^{2+}$ pumps (external: PMCA and NCX; internal: SERCA).

neurons have shown that the increase in $Ca^{2+}$ responses by RyR causes mitochondrial fragmentation and neuronal death [11]. Besides, the Bezprozvanny research group [12, 13] proposes that alterations in the contribution of the ER to $Ca^{2+}$ signaling in ageing and AD relate to a shift in the balance from CaMKII-mediated "LTP-like" synaptic signaling to CaN-mediated "LTD-like" synaptic signaling. The latter signal weakens and destabilizes the synapse and, consequently, leads to memory deficits.

It is important to understand the role of dysregulated ER in $Ca^{2+}$ signaling in the dendritic spine in AD pathology. In particular, how different alterations affect the NMDAR-mediated $Ca^{2+}$ signaling and how these effects may contribute to synapse dysfunction. In this chapter, we select three major alterations of the ER reported in AD: (1) ER $Ca^{2+}$ overload; (2) upregulation of RyR expression; and (3) reduction in SERCA pumps. Therefore, we investigate these alterations by plugging the ER submodel into the model described in Chapter 7. Using the extended model, we examine the interactions among the different $Ca^{2+}$ sources between the healthy condition and the AD condition. By analyzing the computational experimental results, it provides insights into the contribution of external and internal $Ca^{2+}$ sources in synaptic transmission and how dysregulation of the ER function affects the postsynaptic $Ca^{2+}$ response.

In this chapter, we first describe the ER submodel in Section 8.2. In Section 8.3, we give the details of the model integration, including parameter calibration and estimation, and present the model performance under healthy condition. In Section 8.4, we investigate the effects of the ER $Ca^{2+}$ overload and dysregulations in RyR and SERCA in AD conditions using the extended model and, in Section 8.5, we give a brief discussion and the conclusions.

## 8.2. Model Development

We extend the $Ca^{2+}$ model from Chapter 7 with an intracellular ER component, as described in Fig. 8.2. The ER consists of three parts: (1) $Ca^{2+}$ release to the cytosol via membrane

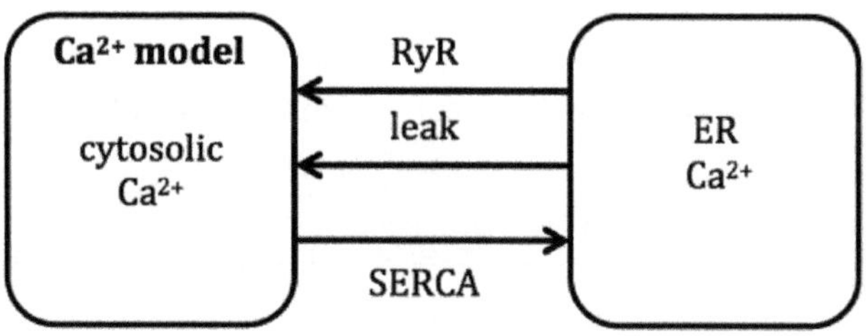


Fig. 8.2. The conceptual framework of the extended model by including ER into the $Ca^{2+}$ model in Chapter 7.

receptors; (2) membrane leaking; and (3) $Ca^{2+}$ uptake by SERCA pumps. In hippocampal pyramidal CA1 neurons, RyRs are expressed throughout the axons, dendritic shaft and spines, whereas $IP_3$Rs are primarily present in dendritic shafts and cell bodies, but absent in dendritic spines [14–16]. Therefore, we set the location of RyRs in the ER in the compartments of the spine head and dendritic shaft. We did not include $IP_3$Rs in the dendritic shaft compartment in this model. This is because the presynaptic stimulation cannot induce high enough increase in the $IP_3$ concentration to induce $IP_3$R opening in the dendritic shaft.

We place the ER in the spine head compartment and in the adjacent dendritic shaft compartment. For reasons of simplicity, we assume there is no store in the spine neck compartment. To take the ER into consideration, we modified Eq. (7.12) by adding a new term, $J_{ER,j}$, to govern the $Ca^{2+}$ dynamics in the spine head and dendritic shaft:

$$\frac{d[\mathrm{Ca}^{2+}]_j}{dt} = J_{I_{\mathrm{Ca}},j} + J_{\mathrm{mem},j} + J_{\mathrm{diffusion},j} - J_{\mathrm{buffer},j} + J_{\mathrm{ER},j}, \tag{8.1}$$

where $j = 2$ or $4$ indicates the compartment index ($j = 2$, cytosol; 4, dendritic shaft). $J_{\mathrm{ER},j}$ denotes the net ER $Ca^{2+}$ flux to the cytosol, and is the balance between the outward $Ca^{2+}$ flux via RyRs ($J_{\mathrm{RyR}}$) and membrane leakage ($J_{leak\mathrm{ER}}$), and the inward $Ca^{2+}$ flux via SERCA pumps ($J_{\mathrm{SERCA}}$):

$$J_{\mathrm{ER,j}} = J_{\mathrm{RyR,j}} - J_{\mathrm{SERCA,j}} + J_{leak\mathrm{ER},j}, \quad j = 2, 4. \tag{8.2}$$

The $Ca^{2+}$ dynamics of ERs are governed as follows:

$$\frac{d[\mathrm{Ca}^{2+}]_{\mathrm{ER},j}}{dt} = \rho J_{\mathrm{ER},j}, \tag{8.3}$$

where $\rho$ is the ratio of ER volume to the cytosol volume. Parameters used in this section are summarized in Table 8.1.

Table 8.1. Parameters values used in this chapter under healthy conditions.

| Description | Symbol | Value | Ref. |
|---|---|---|---|
| Ratio of ER volume to the cytosol volume | $\rho$ | 0.032 | [3] |
| RyR $Ca^{2+}$ release rate | $V_{\mathrm{RyR},2}$ | 0.005 $\mathrm{ms}^{-1}$ | [17] |
| RyR kinetic constants | $k_{af}$ | 1.5 $\mu\mathrm{M}^{-4}\mathrm{ms}^{-1}$ | [17] |
| | $k_{ab}$ | 0.0288 $\mathrm{ms}^{-1}$ | [17] |
| | $k_{bf}$ | 1.5 $\mu\mathrm{M}^{-3}\mathrm{ms}^{-1}$ | [17] |
| | $k_{bb}$ | 0.3859 $\mathrm{ms}^{-1}$ | [17] |
| | $k_{cf}$ | 0.00175 $\mathrm{ms}^{-1}$ | [17] |
| | $k_{cb}$ | 0.0001 $\mathrm{ms}^{-1}$ | [17] |
| Maximum rate of SERCA | $v_{\mathrm{SERCA}}$ | 0.9003 $\mu\mathrm{Mms}^{-1}$ | MCMC estimated |
| Dissociation constant for SERCA | $K_{\mathrm{SERCA}}$ | 0.27 $\mu$M | [18] |
| Leak rate constant of ER | $v_{leak,\mathrm{ER}}$ | 0.0004 $\mathrm{ms}^{-1}$ | MCMC estimated |

### 8.2.1. *The RyR gating*

We simulate the RyR mediated CICR as:

$$J_{\mathrm{RyR,i}} = \mathrm{V}_{\mathrm{RyR},j} Po_{\mathrm{RyR},j}([\mathrm{Ca}^{2+}]_{\mathrm{ER}} - [\mathrm{Ca}^{2+}]_j), \tag{8.4}$$

where $\mathrm{V}_{\mathrm{RyR}}$ is the rate constant of RyRs and is proportional to the number of RyRs. The opening of RyR by $Ca^{2+}$ is simulated using the four-state model developed by Keizer and Levine [17] (Fig. 8.3).

$Po_{\mathrm{RyR},j}$ is the fraction of RyR that is opened by cytosolic $Ca^{2+}$:

$$Po_{\mathrm{RyR},j} = w_j \frac{1+\left(\frac{[\mathrm{Ca}^{2+}]_i}{\mathrm{K}_b}\right)^3}{1+\left(\frac{\mathrm{K}_a}{[\mathrm{Ca}^{2+}]_j}\right)^4+\left(\frac{[\mathrm{Ca}^{2+}]_j}{\mathrm{K}_b}\right)^3}, \tag{8.5}$$

where $w_j$ denotes the fraction of channels not in state $\mathrm{C}_2$, and the three dissociation constants, $\mathrm{K}_i$, are $\mathrm{K}_a{}^4 = \frac{k_{ab}}{k_{af}}$, $\mathrm{K}_b{}^3 = \frac{k_{bb}}{k_{bf}}$, and $\mathrm{K}_c = \frac{k_{cb}}{k_{cf}}$. The change in $w_j$ with time is calculated as

$$\frac{dw_j}{dt} = \frac{1}{\tau_w\left([\mathrm{Ca}^{2+}]_{\mathrm{j}}\right)}\left(w_\infty\left([\mathrm{Ca}^{2+}]_j\right) - w_j\right), \tag{8.6}$$

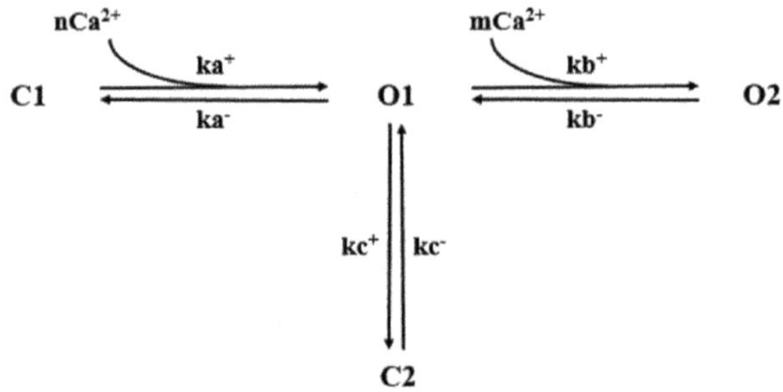


Fig. 8.3. Schematic diagram of the model of RyR developed by Keizer and Levine [17]. This model contains two closed states, C1 and C2, and two open states, O1 and O2. The two open states are assumed to have the same single channel conductance. Therefore, the fraction of RyRs in the open state is the summation of the fraction of RyRs that are in state O1 and O2. The transition from C1 to O1 and O1 to O2 are $Ca^{2+}$ dependent. The transitions between C1 and O1 and between O1 and O2 are assumed to be fast (in milliseconds), and reach equilibrium rapidly. In contrast, the transition between C2 and O1 is assumed to be slow (in seconds).

where

$$w_\infty\left([\mathrm{Ca}^{2+}]_j\right) = \frac{1 + \frac{K_a^4}{[\mathrm{Ca}^{2+}]_j^4} + \frac{[\mathrm{Ca}^{2+}]_j^4}{K_b^3}}{1 + \frac{1}{K_c} + \frac{K_a^4}{[\mathrm{Ca}^{2+}]_j^4} + \frac{[\mathrm{Ca}^{2+}]_j^3}{K_b^3}} \quad \text{and}$$

$$\tau_w\left([\mathrm{Ca}^{2+}]_\mathrm{j}\right) = \frac{w_\infty\left(\left[\mathrm{Ca}^{2+}\right]_j\right)}{k_{cb}}.$$

### 8.2.2. *The SERCA pump*

The activity of the SERCA pump is modeled by a Hill equation with a Hill coefficient of 2 [18]:

$$J_{\mathrm{SERCA,j}} = v_{\mathrm{SERCA}} \frac{\left[\mathrm{Ca}^{2+}\right]_j^2}{K_{\mathrm{SERCA}}^2 + \left[\mathrm{Ca}^{2+}\right]_i^2}, \tag{8.7}$$

where $v_{\mathrm{SERCA}}$ is the maximum rate of $Ca^{2+}$ uptake by ER via the SERCA pump, and $K_{\mathrm{SERCA}}$ is the affinity of the SERCA pump for $Ca^{2+}$ ions (see Appendix C.4 in Chapter 7 for the explanation).

The value $v_{\mathrm{SERCA}}$ is proportional to the availability of SERCA and is estimated in the next section.

### 8.2.3. $Ca^{2+}$ *passive leak*

The flux of the passive $\mathrm{Ca}^{2+}$ leak from the ER is described as

$$J_{leak,j}=v_{leak,\mathrm{ER}}([\mathrm{Ca}^{2+}]_{\mathrm{ER},j}-[\mathrm{Ca}^{2+}]_j), \tag{8.8}$$

where $v_{leak,\mathrm{ER}}$ is the leak rate constant of ER.

Therefore, we substitute Eqs. (8.4), (8.7) and (8.9) into Eq. (8.2) and obtain

$$\begin{aligned} J_{\mathrm{ER},j} &= \mathrm{V}_{\mathrm{RyR}} Po_{\mathrm{RyR},i}([\mathrm{Ca}^{2+}]_{\mathrm{ER}} - [\mathrm{Ca}^{2+}]_j) \\ &\quad - v_{\mathrm{SERCA}} \frac{\left[\mathrm{Ca}^{2+}\right]_j^2}{K_{\mathrm{SERCA}}^2 + \left[\mathrm{Ca}^{2+}\right]_j^2} \\ &\quad + v_{leak,\mathrm{ER}}([\mathrm{Ca}^{2+}]_{\mathrm{ER}.j} - [\mathrm{Ca}^{2+}]_j), \end{aligned} \tag{8.9}$$

where $j = 2$ or 4. At the resting state ($J_{\mathrm{ER},j} = 0$), the $\mathrm{Ca}^{2+}$ leaking flux together with RyR-mediated $\mathrm{Ca}^{2+}$ release balances the $\mathrm{Ca}^{2+}$ influx by the SERCA pumps, according to which we can calculate the corresponding value of $v_{leak,\mathrm{ER}}$.

## 8.3. Model Parameter Estimation and Calibration

Keizer and Levine [17] used different values for the relevant parameters for the small and large ER, respectively. The $\rho$ they used for the small ER, 0.02, is close to the ratio observed experimentally (3.2±0.2% [3]). Assuming the basal $\mathrm{Ca}^{2+}$ concentration of cytosol ($\left[\mathrm{Ca}^{2+}\right]_{\mathrm{cyto0}}$) is 0.07 $\mu$M, then the basal $\mathrm{Ca}^{2+}$ concentration of the ER ($\left[\mathrm{Ca}^{2+}\right]_{\mathrm{ER0}}$) is between 56.2 and 66.2 $\mu$M for different sized ERs, and is calculated based on these constraints as

$$\left[\mathrm{Ca}^{2+}\right]_{\mathrm{ER0}} = \frac{1}{\rho}\left(\left[\mathrm{Ca}^{2+}\right]_{\mathrm{cyto0}} - \left[\mathrm{Ca}^{2+}\right]_{\mathrm{total}}\right),$$

where $\left[\mathrm{Ca}^{2+}\right]_{\mathrm{total}}$ is the total $\mathrm{Ca}^{2+}$ concentration in the dendrites. We choose $\left[\mathrm{Ca}u^{2+}\right]_{\mathrm{ER0}}$ to be 60 $\mu$M under control conditions and $\rho$ to be 0.032, which gives $\left[\mathrm{Ca}^{2+}\right]_{\mathrm{total}}$ as 1.99 $\mu$M. This is slightly higher than the value for the small ER used in the model of Keizer and Levine [17], 1.2 $\mu$M, since the value for $\rho$ in our model is slightly higher than that in their model.

We assume that the densities of RyR and the SERCA pump and the surface area to volume ratios of the ER are the same in the spine head and the dendritic shaft. Therefore, the rate constants of RyRs, $V_{\text{RyR}}$, in Eq. (8.4) shaft are the same in the spine head and dendritic shaft, which is assumed to be 0.005 ms$^{-1}$ [17]. The maximum rate of $Ca^{2+}$ uptake by SERCA, $v_{\text{SERCA}}$, can be calculated by

$$v_{\text{SERCA,j}} = V_{\max} Ps_{\text{SERCA}} \frac{A_j}{V_j},$$

where $V_{\max}$ is the maximum turnover rate of $Ca^{2+}$ by SERCA, $Ps_{\text{SERCA}}$ is its surface density and $\frac{A_j}{V}$ is the ratio of surface area to the volume of compartment $j$. Based on our assumption, the value of $v_{\text{SERCA}}$ in the spine head and the dendritic shaft are identical.

### 8.3.1. *Estimate parameter values using MCMC*

In Chapter 7, we estimate ten parameters using MCMC from the experimental data. The estimation is based on the assumption of the absence of the ER; therefore, we need to re-estimate their values by considering the presence of the ER.

Sabatini, Oertner and Svoboda [19] observed that in response to a single bAP, about 30% of the $Ca^{2+}$ ions are extruded from the cytosol via SERCA in the spine, while the rest are extruded via PMCA and NCX. We add this as an additional target in the experimental data listed in Table 7.6 in Chapter 7 to estimate the value for $v_{\text{SERCA}}$ as well as the other ten parameters (Table 8.2). The range of $v_{\text{SERCA}}$ is from 0.1 to 1 $\mu$M ms$^{-1}$ [17]. The procedure for estimation is the same as described in Section 7.3.4 in Chapter 7. The means of MCMC samples and the parameter set with the lowest MAPE are listed in the last two columns in Table 8.3. The simulation results show that the model with the parameter set with the lowest MAPE performs much better than the one using the mean of MCMC samples (Table 8.2). Therefore, we choose parameters with the lowest MAPE for our model under control conditions.

Table 8.2. Values of target experimental data, means of MCMC samples and the parameter set with the lowest mean absolute percentage error (MAPE).

| Target experimental data | Target value | Mean of MCMC samples | Set with the lowest MAPE |
|---|---|---|---|
| $\triangle\,[\mathrm{Ca}]_{bAP}$ in spine ($\mu$M) | 1.7 ($\pm$0.6) | 1.257 | 1.647 |
| $\triangle\,[\mathrm{Ca}]_{bAP}$ in dendrite ($\mu$M) | 1.5 ($\pm$0.5) | 1.277 | 1.606 |
| $\tau_{\mathrm{decay}}$ in spine (ms) | 12 ($\pm$4) | 14.731 | 12.486 |
| $\tau_{\mathrm{decay}}$ in dendrite (ms) | 15 ($\pm$5) | 17.803 | 14.862 |
| $\frac{[Ca]_{PMCA}}{[Ca]_{NCX}}$ in spine | 1.070 | 2.464 | 1.200 |
| $\frac{[Ca]_{PMCA}}{[Ca]_{NCX}}$ in dendrite | 0.825 | 1.425 | 0.826 |
| $\frac{[\mathrm{Ca}^{2+}]_{\mathrm{SERCA}}}{[\mathrm{Ca}^{2+}]_{\mathrm{SERCA}}+[\mathrm{Ca}^{2+}]_{PMCA}+[\mathrm{Ca}^{2+}]_{NCX}}$ | 0.3 | 0.168 | 0.307 |
| MAPE | | 0.377 | 0.040 |

### 8.3.2. *Model performance under control conditions*

In this section, we present the simulation results of the ER-included model under control conditions, using parameter values from Table 8.1. We evaluate the model performance in response to different types of presynaptic stimulation from aspects of the $\mathrm{Ca}^{2+}$ dynamics in the spine head and its parent dendritic shaft, and the contributions of the different $\mathrm{Ca}^{2+}$ sources to the $\mathrm{Ca}^{2+}$ dynamics.

#### 8.3.2.1. *A single bAP stimulation*

Sabatini, Oertner and Svoboda [19] applied the cyclopiazonic acid (CPA) to block the SERCA pumps to study the relative contribution of SERCA in $\mathrm{Ca}^{2+}$ transient in the spine head induced by a single bAP. The CPA is a specific inhibitor of SERCA pumps, and it will deplete the $\mathrm{Ca}^{2+}$ content of the ER. We mimic the block of SERCA pumps by CPA by letting vSERCA be zero and running the model long enough to let all variables reach their steady levels. We then use these steady levels as the initial values of each variable. The model is simulated under the conditions of the absence (immobile buffer only) and presence of mobile buffers ($CaM$ and CaD), respectively. After the application of CPA, the peaks of the $\mathrm{Ca}^{2+}$ transients evoked by a single bAP in the spine head and shaft are not

Table 8.3. Parameter ranges for MCMC and the values estimated by MCMC.

| Parameter | Range for MCMC | Mean of MCMC samples | Set with the lowest MAPE |
|---|---|---|---|
| VDCC density in spine head | 2.23–4.66 $\mu m^{-2}$ | 3.630 $\mu m^{-2}$ | 3.169 $\mu m^{-2}$ |
| VDCC density in shaft | 7.29–14.57 $\mu m^{-2}$ | 11.661 $\mu m^{-2}$ | 10.188 $\mu m^{-2}$ |
| PMCA density in spine head | 200–2000 $\mu m^{-2}$ | 555.014 $\mu m^{-2}$ | 283.878 $\mu m^{-2}$ |
| PMCA density in shaft | 200–2000 $\mu m^{-2}$ | 979.994 $\mu m^{-2}$ | 490.837 $\mu m^{-2}$ |
| NCX density in spine head | 10–700 $\mu m^{-2}$ | 0.350 $\mu m^{-2}$ | 0.410 $\mu m^{-2}$ |
| NCX density in shaft | 100–1000 $\mu m^{-2}$ | 508.219 $\mu m^{-2}$ | 433.938 $\mu m^{-2}$ |
| Immobile buffer concentration in spine head | 50–200 $\mu$M | 115.779 $\mu$M | 83.871 $\mu$M |
| Immobile buffer concentration in shaft | 50–200 $\mu$M | 130.911 $\mu$M | 102.409 $\mu$M |
| Forward rate constant of immobile buffer | 0.05–1 $\mu M^{-1} ms^{-1}$ | 0.267 $\mu M^{-1} ms^{-1}$ | 0.227 $\mu M^{-1} ms^{-1}$ |
| Backward rate constant of immobile buffer | 0.05–1 $ms^{-1}$ | 0.520 $ms^{-1}$ | 0.509 $ms^{-1}$ |
| Maximum rate of SERCA | 0.1–1 $\mu Mms^{-1}$ | 0.684 $\mu Mms^{-1}$ | 0.903 $\mu Mms^{-1}$ |

affected under either conditions (Fig. 8.4(a)) and the decay time constants to resting values are increased. These results are consistent with the experimental observations in Sabatini, Oertner and Svoboda [19].

Figure 8.4(b) shows that the model performs well in predicting the contribution of different $Ca^{2+}$ sources to the $Ca^{2+}$ transients in response to a single bAP in comparison to the experimental observations [19]. The increase of cytosolic $Ca^{2+}$ is mainly via VDCC in the spine head and shaft. There is only a minor $Ca^{2+}$ release from the ER by RyRs to the cytosolic $Ca^{2+}$ transients (Fig. 8.4(b) middle), which is very similar to the experimental observation that

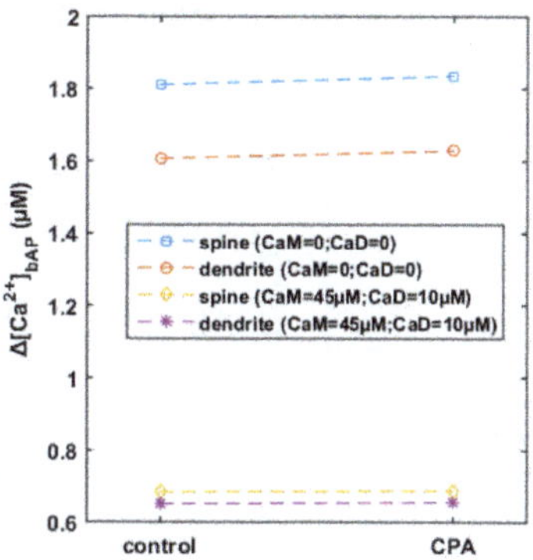


(a)

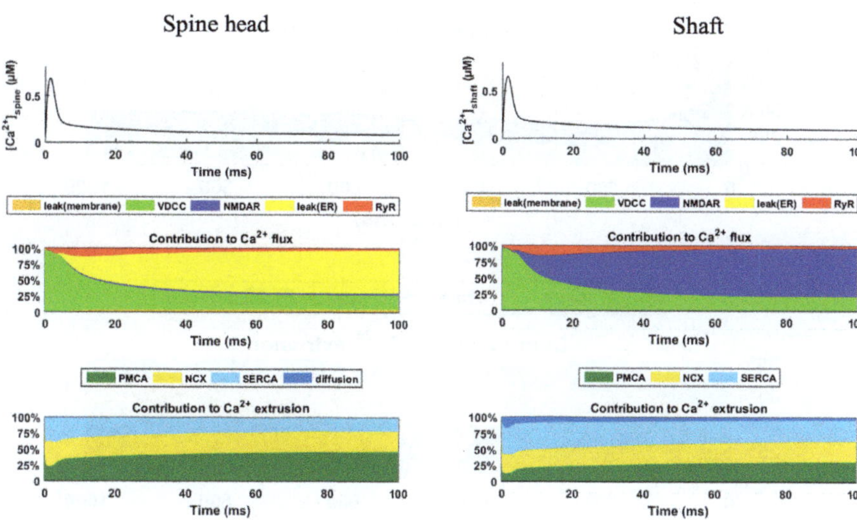


(b)

Fig. 8.4. $Ca^{2+}$ response in the spine head and the adjacent dendritic shaft evoked by a single bAP. (a) Effects of blocking of $Ca^{2+}$ uptake into the ER via SERCA pumps in response to a single bAP. (b) $Ca^{2+}$ transient (top), fractional contribution of $Ca^{2+}$ influx from different sources (middle) and $Ca^{2+}$ extrusion by different mechanisms (bottom) to the cytosolic $Ca^{2+}$ transients in the spine head and the adjacent dendritic shaft.

CICR does not contribute to the $Ca^{2+}$ elevation stimulated by bAP [5, 19]. Moreover, about 30% and 35% of cytosolic $Ca^{2+}$ extrusion is taken up by the ER during the transient in the spine head and shaft, respectively (Fig. 8.4(b) bottom).

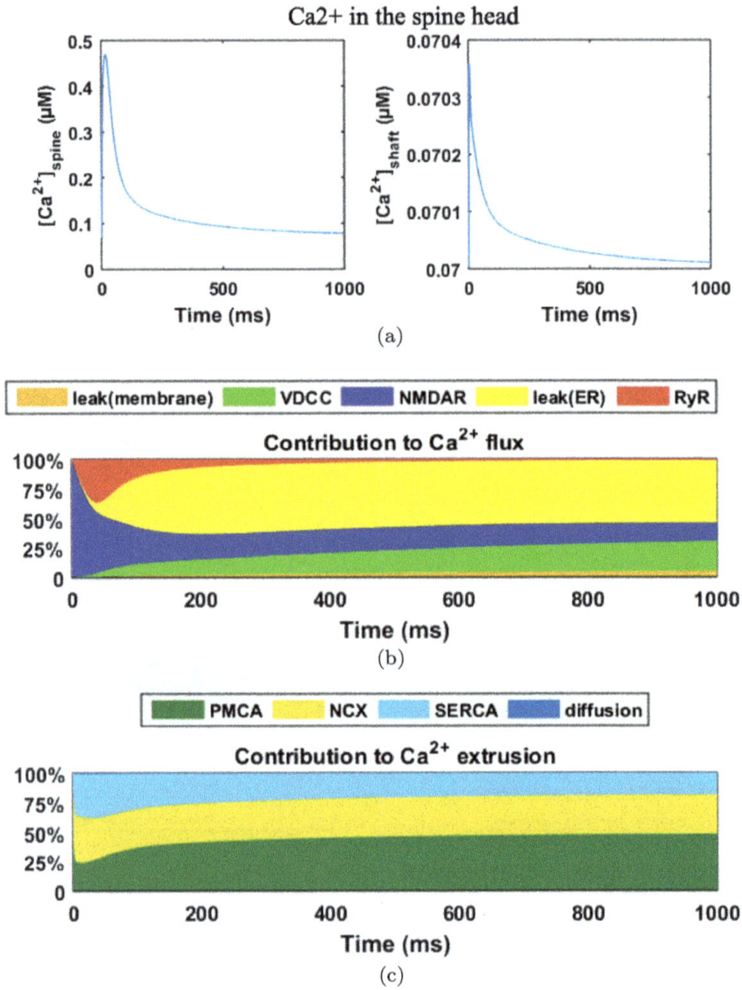


Fig. 8.5. $Ca^{2+}$ response in the spine head and the adjacent dendritic shaft evoked by a single presynaptic stimulation. (a) $Ca^{2+}$ concentration in the spine head and the adjacent dendritic shaft. Contributions of $Ca^{2+}$ influx from different sources (b) and $Ca^{2+}$ extrusion by different mechanisms (c) to the cytosolic $Ca^{2+}$ transients in the spine head.

### 8.3.2.2. *1 s of presynaptic stimulation*

CICR by ER can be triggered by $Ca^{2+}$ influx via NMDAR in response to the presynaptic stimulation [5]. We use the model to simulate the $Ca^{2+}$ dynamics in a dendritic spine in response to 1 s of presynaptic stimulation at 1 Hz (1 pulse), 10 Hz (10 pulses) and 100 Hz (100 pulses), respectively, under healthy conditions. At a 1 Hz stimulation, the peak $Ca^{2+}$ transient is about 0.48 $\mu$M in the spine head (Fig. 8.5(a) left) at the presence of mobile buffers, which is located in the range estimated under experimental conditions [20]. The $Ca^{2+}$ elevation is mainly achieved by the $Ca^{2+}$ influx via NMDARs, and RyRs and ER membrane leakage show higher contributions to the $Ca^{2+}$ elevation in response to a single presynaptic stimulation (Fig. 8.5(b)) than in response to a single bAP stimulation (Fig. 8.4(b)). VDCCs do not contribute to the $Ca^{2+}$ elevation because the small membrane depolarization is not sufficient to open them. About one-third of cytosolic $Ca^{2+}$ ions are removed by SERCA pumps during the $Ca^{2+}$ transient, while the rest is by the membrane pumps (PMCA and NCX) (Fig. 8.5(c)). The $Ca^{2+}$ response in the dendritic shaft is relatively less noticeable (Fig. 8.5(a) right), showing that the $Ca^{2+}$ response to a single presynaptic stimulation is highly restricted to the spine head [21]. Therefore, in following sections, we only focus on the $Ca^{2+}$ dynamics in the spine head.

At 10 and 100 Hz stimulation, the maximum $Ca^{2+}$ levels are about 0.83 and 3.7 $\mu$M, respectively (Figs. 8.6(a) top and (b) top). At 100 Hz stimulation, after reaching a peak, the $Ca^{2+}$ concentration decreases and stays on a plateau at a lower level (about 1.1 $\mu$M) until the end of stimulation, which is because of the large desensitization of synaptic NMDAR by high-frequency stimulation. RyRs show a larger contribution to the cytosolic $Ca^{2+}$ transients at the decay phase at a 10 Hz stimulation than at a 100 Hz one (Figs. 8.6(a) middle and (b) middle). The contribution of $Ca^{2+}$ extrusion by different mechanisms to the cytosolic $Ca^{2+}$ transients is very similar between 10 and 100 Hz stimulation (Figs. 8.6(a) bottom and (b) bottom).

## 8.4. Computational Experiments

We investigate three selected effects of ER alteration on $Ca^{2+}$ signaling in the dendritic spine individually and collectively by simulating the model under the following five conditions: (1) ER $Ca^{2+}$ overload (Section 8.4.1); (2) upregulation in RyR response

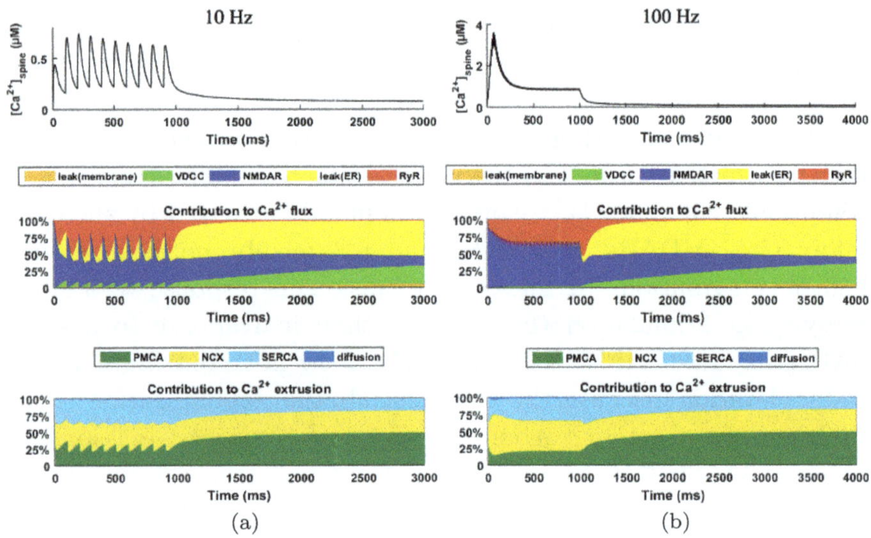


Fig. 8.6. $Ca^{2+}$ response in the spine head and the adjacent dendritic shaft evoked by 1 s of stimulation at 10 Hz (a) and 100 Hz (b), respectively. $Ca^{2+}$ concentration in the spine head (top). The fractional contribution of $Ca^{2+}$ influx from different sources (middle) and $Ca^{2+}$ extrusion by different mechanisms (bottom) to the cytosolic $Ca^{2+}$ transients in the spine head.

(Section 8.4.2); (3) combined effects of (1), (2) (Section 8.4.3); (4) SERCA pumps (Section 8.4); and (5) combined effects of (1), (4) (Section 8.4.5). Simulations are carried in response to a 1 s presynaptic stimulation at 1 Hz (1 pulse), 10 Hz (10 pulses) and 100 Hz (100 pulses), respectively. The results are compared with those of the control produced by the parameters in Table 8.1 and the ones used in Chapter 7.

The outputs we chose to collect after simulation are as follows:

(1) $Ca^{2+}$ elevation in the spine head ($[Ca^{2+}]_{cyto}$). This includes $[Ca^{2+}]_{cyto}$ peaks in response to each stimulation pulse.
(2) $Ca^{2+}$ responses of RyR ($J_{RyR}$; positive outward flux), the SERCA pump ($J_{SERCA}$; negative, inward flux), the ER membrane leaking ($J_{leakER}$; positive outward flux), and the net $Ca^{2+}$ flux from the ER to the cytosol ($J_{ER}$) (Eq. (8.9)). This indicates the role the ER plays in the cytosolic $Ca^{2+}$ dynamics. A positive

$J_{\mathrm{ER}}$ indicates that the ER mainly acts as a $\mathrm{Ca}^{2+}$ source and a negative $J_{\mathrm{ER}}$ indicates it acts as a $\mathrm{Ca}^{2+}$ sink.

(3) The fraction of RyRs ($Po_{\mathrm{RyR}}$) opened by cytosolic $\mathrm{Ca}^{2+}$ and the total open time. The total open time of RyR is calculated from the area under the $Po_{\mathrm{RyR}}$ curve against time.

(4) The ratio of total $\mathrm{Ca}^{2+}$ ions entered via RyR to those via the NMDAR ($[\mathrm{Ca}^{2+}]_{\mathrm{RyR}} : [\mathrm{Ca}^{2+}]_{NMDAR}$ ratio). This indicates the contribution between the internal and external $\mathrm{Ca}^{2+}$ sources to cytosolic $\mathrm{Ca}^{2+}$ elevation. The total $\mathrm{Ca}^{2+}$ ion is calculated from the area under the $\mathrm{Ca}^{2+}$ flux curve against time.

### 8.4.1. *ER $Ca^{2+}$ overload*

An increase in the resting $\mathrm{Ca}^{2+}$ level of the ER has been observed in AD neurons as well as in ageing neurons [22]. Several studies have linked the ER $\mathrm{Ca}^{2+}$ overload to mutations in PS (see Section 6.3.4.2 in Chapter 6). PS is suggested to serve as a $\mathrm{Ca}^{2+}$ leak channel on the ER membrane, which balances the activity of the SERCA pumps and RyRs to maintain $\mathrm{Ca}^{2+}$ homeostasis of the ER at the resting level [23]. Therefore, PS mutation will diminish its leak function, leading to a $\mathrm{Ca}^{2+}$ overload in the ER and an alteration in $\mathrm{Ca}^{2+}$ homeostasis [23–26]. We mimic the alteration in ER $\mathrm{Ca}^{2+}$ by varying the resting $\mathrm{Ca}^{2+}$ of the ER ($[\mathrm{Ca}^{2+}]_{\mathrm{ER0}}$) from 30 $\mu$M to 600 $\mu$M ($[\mathrm{Ca}^{2+}]_{\mathrm{ER0}}$ as the healthy condition is assumed to be 150 $\mu$M), and keep other parameters at standard values.

The ER $\mathrm{Ca}^{2+}$ leak rate constant, $v_{leak,\mathrm{ER}}$, is calculated by rearranging Eq. (8.9):

$$v_{leak\mathrm{ER}} = \mathrm{V}_{\mathrm{RyR}} Po_{\mathrm{RyR}} - v_{\mathrm{SERCA}} \times \frac{[\mathrm{Ca}^{2+}]^2_{\mathrm{cyto}}}{K^2_{\mathrm{SERCA}} + [\mathrm{Ca}^{2+}]^2_{\mathrm{cyto}}} \frac{1}{[\mathrm{Ca}^{2+}]_{\mathrm{ER}} - [\mathrm{Ca}^{2+}]_{\mathrm{cyto}}}. \quad (8.10)$$

For fixed values of RyR and the SERCA pump-related parameter and resting cytosolic $\mathrm{Ca}^{2+}$ concentration ($[\mathrm{Ca}^{2+}]_{\mathrm{cyto}}$), $v_{leak\mathrm{ER}}$ is negatively dependent on the concentration gradient between the cytosol and ER under resting conditions (Fig. 8.7).

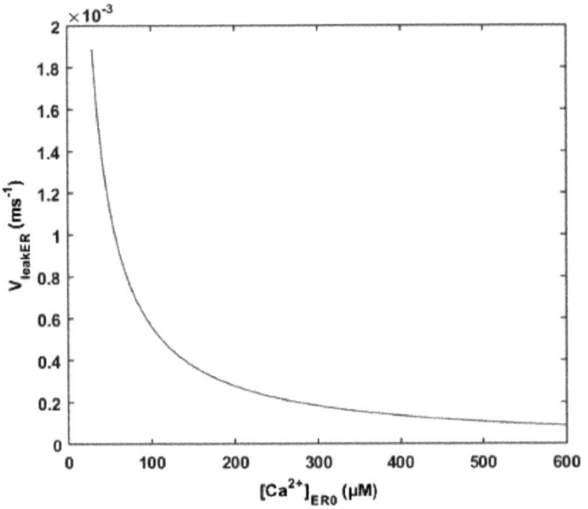


Fig. 8.7. The membrane $Ca^{2+}$ leak rate constant of the ER ($v_{leak\mathrm{ER}}$) at various levels of resting $Ca^{2+}$ levels of the ER ($\left[Ca^{2+}\right]_{\mathrm{ER0}}$). The value of $v_{leak\mathrm{ER}}$ is calculated using Eq. (8.10). $V_{\mathrm{RyR}} = 0.005\ \mathrm{ms}^{-1}$. $v_{\mathrm{SERCA}} = 0.9\ \mu\mathrm{M\ ms}^{-1}$. $K_{\mathrm{SERCA}} = 0.27\ \mu\mathrm{M}$. $\left[Ca^{2+}\right]_{\mathrm{cyto}} = 0.07\ \mu\mathrm{M}$. $Po_{\mathrm{RyR}}$ is calculated from Eqs. (8.5) and (8.6) at steady state.

The peak of $\left[Ca^{2+}\right]_{\mathrm{cyto}}$ elevation in the postsynaptic spine head in response to a 1 Hz presynaptic stimulation increases with the rise in $\left[Ca^{2+}\right]_{\mathrm{ER0}}$ (Fig. 8.8(a)), from 0.32 $\mu$M at standard $\left[Ca^{2+}\right]_{\mathrm{ER0}}$ to 0.42 $\mu$M when $\left[Ca^{2+}\right]_{\mathrm{ER0}}$ increases to 600 $\mu$M. At the standard and lower level of $\left[Ca^{2+}\right]_{\mathrm{ER0}}$, the net ER $Ca^{2+}$ flux is negative/inward during the rising of $\left[Ca^{2+}\right]_{\mathrm{cyto}}$ and becomes positive during the $Ca^{2+}$ decay stage. In contrast, at high levels of $\left[Ca^{2+}\right]_{\mathrm{ER0}}$ (300 and 600 $\mu$M), a positive peak appears during the rising stage of $\left[Ca^{2+}\right]_{\mathrm{cyto}}$ (Fig. 8.8(b)). This is mainly contributed to by the increasing level of RyR response (Figs. 8.8(b) and (c)). The peak RyR response increases linearly with increases in $\left[Ca^{2+}\right]_{\mathrm{ER0}}$ while the peak activity of the SERCA pump is nearly unaffected across the range of $\left[Ca^{2+}\right]_{\mathrm{ER0}}$ tested (Fig. 8.8(c)). This suggests that under a single presynaptic stimulation, when the ER is overfilled with $Ca^{2+}$ at the resting level, it shifts its role from a $Ca^{2+}$ sink, which

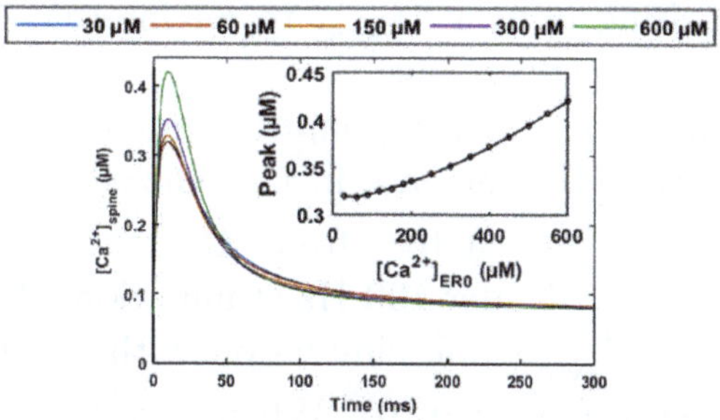


(a)

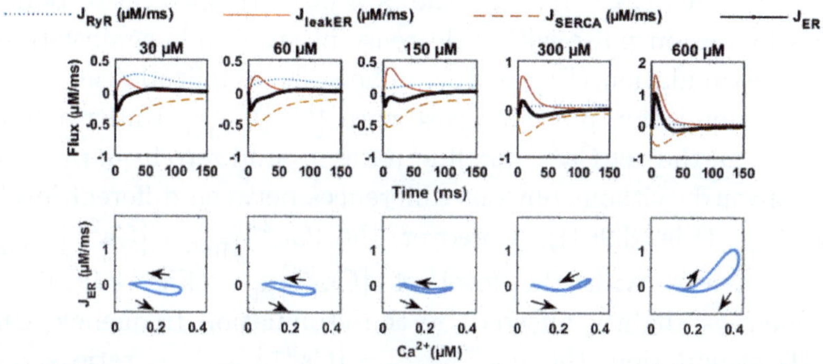


(b)

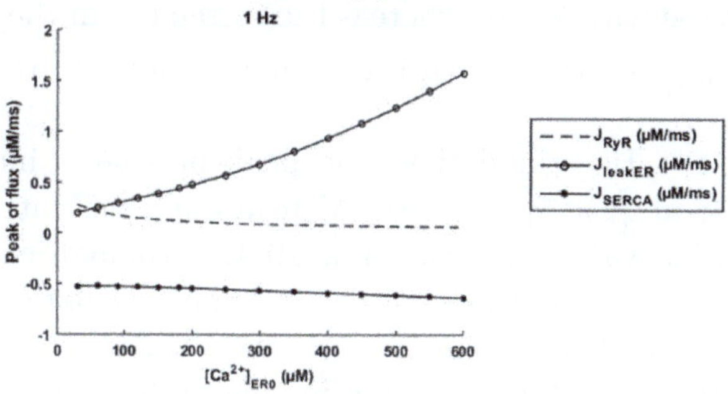


(c)

Fig. 8.8. Effects of rising $[Ca^{2+}]_{ER0}$ on postsynaptic response by a single presynaptic stimulation pulse (1 Hz). (a) The $[Ca^{2+}]_{cyto}$ dynamics in the spine head and the relationship between $[Ca^{2+}]_{ER0}$ and the $[Ca^{2+}]_{cyto}$ peaks (inset). The corresponding $Ca^{2+}$ responses of RyR ($J_{RyR}$), the SERCA pump ($J_{SERCA}$), the ER membrane leaking ($J_{leakER}$), and the net $Ca^{2+}$ flux from the ER to the cytosol ($J_{ER}$) are shown in (b). Arrows in the bottom panel indicate the direction of time flow. The peaks of $J_{RyR}$, $J_{SERCA}$ and $J_{leakER}$ at different levels of $[Ca^{2+}]_{ER0}$ are in (c).

attenuates the rising of $[Ca^{2+}]_{cyto}$, to an internal $Ca^{2+}$ source in addition to the extracellular one.

When we apply 1 s of presynaptic stimulations at 10 Hz and 100 Hz, respectively, the increase in normalized peak of $[Ca^{2+}]_{cyto}$ with $[Ca^{2+}]_{ER0}$ is attenuated from the second pulse onward (Figs. 8.9(a) and (b)). Under 100 Hz stimulation, there is no difference in peak $[Ca^{2+}]_{cyto}$ upon simulation with different $[Ca^{2+}]_{ER0}$ after the 10th pulse (Fig. 8.9(b)). When $[Ca^{2+}]_{ER0}$is 150 $\mu$M and higher, the net ER $Ca^{2+}$ flux shows a positive spike in response to each stimulation pulse with a decrease in amplitude against time at a 10 Hz stimulation (Fig. 8.9(c)). The spike amplitudes in response to each stimulation pulse increase with $[Ca^{2+}]_{ER0}$. Under a 100 Hz stimulation, the net $Ca^{2+}$ flux fluctuates around zero from the second pulse onward, without obvious differences between different levels of $[Ca^{2+}]_{ER0}$ (Fig. 8.9(d)). Moreover, the $[Ca^{2+}]_{RyR} : [Ca^{2+}]_{NMDAR}$ ratio increases with the level of $[Ca^{2+}]_{ER0}$ (Fig. 8.9(e)). The increment is slightly affected by the stimulation frequency. Under a 1 Hz stimulation, the $[Ca^{2+}]_{RyR} : [Ca^{2+}]_{NMDAR}$ ratio shows a linear increase with $V_{RyR}$, whereas the increment in the $[Ca^{2+}]_{RyR} :$ $[Ca^{2+}]_{NMDAR}$ ratio by $V_{RyR}$ is attenuated under 10 Hz and 100 Hz stimulations.

Under a 1 Hz stimulation, the peak of $Po_{RyR}$ increases from around 0.38 at $[Ca^{2+}]_{ER0}$= 30 $\mu$M to around 0.67 at $[Ca^{2+}]_{ER0}$= 300 $\mu$M (Fig. 8.10(a) inset). At a 10 Hz stimulation, the $Po_{RyR}$ reaches a peak in response to the second stimulation pulse, and afterwards, the increase in $Po_{RyR}$ with $[Ca^{2+}]_{ER0}$ is attenuated (Fig. 8.10(b)). At a 100 Hz stimulation, more than 90% of RyRs open in response to the second pulse and gradually decrease to around 30% through the stimulation (Fig. 8.10(c)). There is no obvious difference in $Po_{RyR}$ among the different values of $[Ca^{2+}]_{ER0}$. The total open time of RyR increases with $[Ca^{2+}]_{ER0}$ at 1 Hz and 10 Hz stimulations, whereas at a 100 Hz stimulation, it decreases and rises again between 451.2 and 453.5 ms (Fig. 8.10(d)).

The reasons that cause these differences are the fast activation and slow inactivation of RyRs. Once a presynaptic stimulation pulse

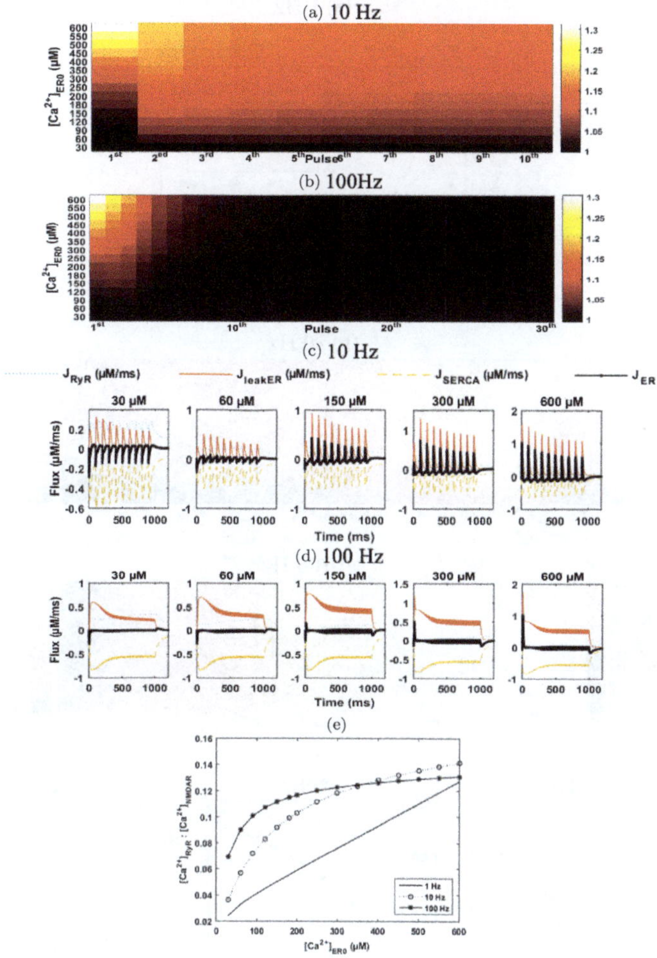


Fig. 8.9. Effects of rising $[Ca^{2+}]_{ER0}$ on postsynaptic response under 1 s presynaptic stimulation at 10 Hz and 100 Hz, respectively. The $[Ca^{2+}]_{cyto}$ peaks in response to different stimulation pulses at a 10 Hz (a) and a 100 Hz (b) stimulation, respectively, are normalized to the one with the lowest $[Ca^{2+}]_{ER0}$ at each pulse. The color density represents the values of $[Ca^{2+}]_{cyto}$ peaks in the spine head. Under the 100 Hz stimulation, only the first 30 pulses are shown in the figure since there is no difference afterwards. The corresponding $Ca^{2+}$ responses of RyR ($J_{RyR}$), the SERCA pump ($J_{SERCA}$), the ER membrane leaking ($J_{leakER}$), and the net $Ca^{2+}$ flux from the ER to the cytosol ($J_{ER}$) are shown in (c) and (d), respectively. (e) Effects of rising $[Ca^{2+}]_{ER0}$ on $[Ca^{2+}]_{RyR}$:$[Ca^{2+}]_{NMDAR}$ ratio.

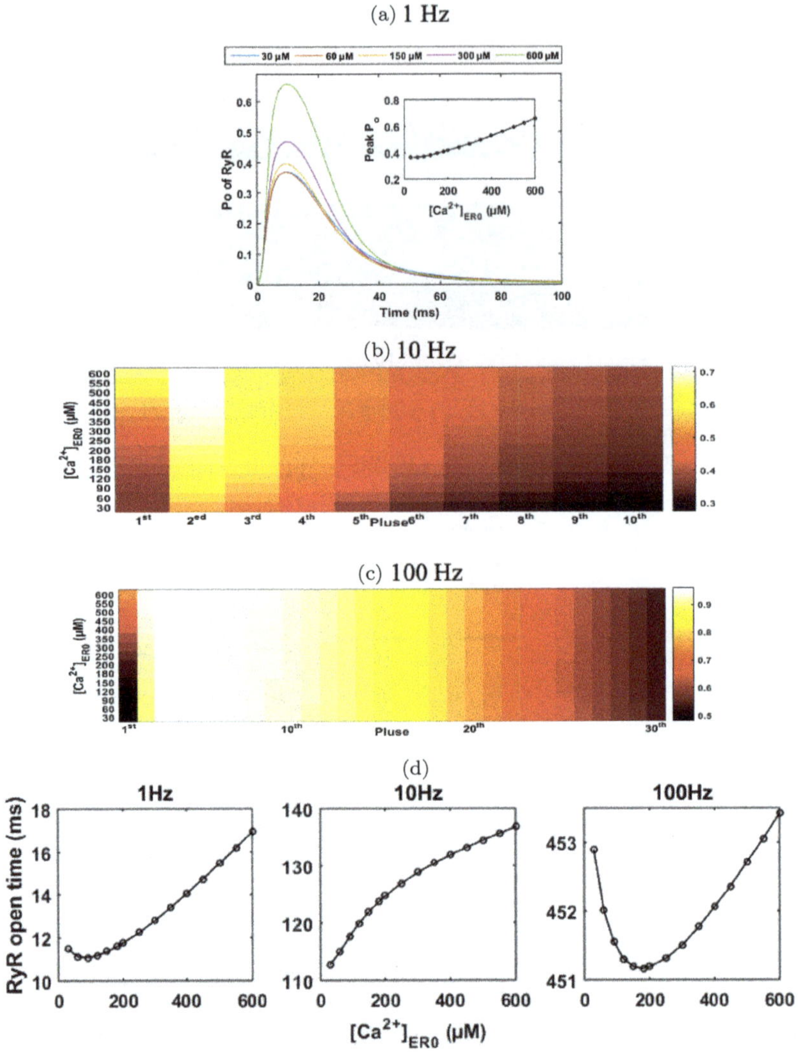


Fig. 8.10. Effects of rising $[Ca^{2+}]_{ER0}$ on RyR opening. (a) The $Po_{RyR}$ with time in the spine head and the relationship between $[Ca^{2+}]_{ER0}$ and the $Po_{RyR}$ peaks (inset) at a 1 Hz stimulation. The heatmaps show the maximums of $Po_{RyR}$ in response to different stimulation pulses at a 10 Hz (b) and a 100 Hz (c) stimulation, respectively. The color density represents the values of $Po_{RyR}$ peaks in the spine head. Under the 100 Hz stimulation, only the first 30 pulses are shown in the figure since there is no difference afterwards. (d) The corresponding total opening times of RyR at stimulations with different frequencies.

arrives at a resting synapse, it leads to a rapid $Ca^{2+}$ elevation and immediate RyR response. When the time duration between the two stimulation pulses shortens under the higher frequency stimulation, this allows a fraction of RyR that is activated by the previous pulse to still be open when the next pulse arrives. This explains why higher $Po_{\mathrm{RyR}}$ peaks are obtained at a 100 Hz stimulation (Fig. 8.10(c)) than at a 10 Hz stimulation (Fig. 8.10(b)).

### 8.4.2. *Enhanced $Ca^{2+}$ release via RyRs*

For a fixed maximum rate of a single receptor, the maximum rate of total RyRs is in proportion to the number of RyRs. In this section, we mimic the upregulation in RyR expression by increasing $V_{\mathrm{RyR}}$ from its standard level of 0.005 $\mathrm{ms}^{-1}$ to 0.3 $\mathrm{ms}^{-1}$. We also include lower levels of $V_{\mathrm{RyR}}$ (below 0.005 $\mathrm{ms}^{-1}$) in the computational experiments.

Under a 1 Hz stimulation, the peak of $[Ca^{2+}]_{\mathrm{cyto}}$ increases from 0.3 $\mu$M, when $V_{\mathrm{RyR}}$ is 0.001 $\mathrm{ms}^{-1}$, to around 0.37 $\mu$M, when $V_{\mathrm{RyR}}$ is 0.03 $\mathrm{ms}^{-1}$ (Fig. 8.11(a)). Further increases in $V_{\mathrm{RyR}}$ do not lead to a higher $[Ca^{2+}]_{\mathrm{cyto}}$ elevation. ER contributes mainly as a $Ca^{2+}$ sink at standard and lower levels of $V_{\mathrm{RyR}}$, while, with an increase in $V_{\mathrm{RyR}}$, it is more like an internal $Ca^{2+}$ source during stimulation (Fig. 8.11(b)). The peak $Ca^{2+}$ flux via RyR, increases dramatically with the value of $V_{\mathrm{RyR}}$ and fluctuates at high levels of $V_{\mathrm{RyR}}$, whereas the peak activity of the SERCA pumps is not affected (Fig. 8.11(c)).

Under a 10 Hz stimulation, from the second pulse onward, both the normalized $[Ca^{2+}]_{\mathrm{cyto}}$ peak (Fig. 8.12(a)) and the $Po_{\mathrm{RyR}}$ peak (Fig. 8.13(b)) show a maximum value when $V_{\mathrm{RyR}}$ is between 0.05 and 0.03 $\mathrm{ms}^{-1}$. Positive spikes in the net $Ca^{2+}$ flux from ER, $J_{\mathrm{RyR}}$, appear during stimulation (Fig. 8.12(c)). At 100 Hz, from the second pulse onward, the degree of increase in $[Ca^{2+}]_{\mathrm{cyto}}$ peaks with $V_{\mathrm{RyR}}$ being attenuated over time (Fig. 8.12(b)). There is no difference in $[Ca^{2+}]_{\mathrm{cyto}}$ dynamics across the testing range of $V_{\mathrm{RyR}}$ values after the 15th pulse. $J_{\mathrm{ER}}$ fluctuates around zero (Fig. 8.12(d)) and shows a balance between release and uptake of $Ca^{2+}$ by ER. The $Po_{\mathrm{RyR}}$ does not show differences in response to a 100 Hz stimulation across the tested ranges of $V_{\mathrm{RyR}}$ from the second pulse onward

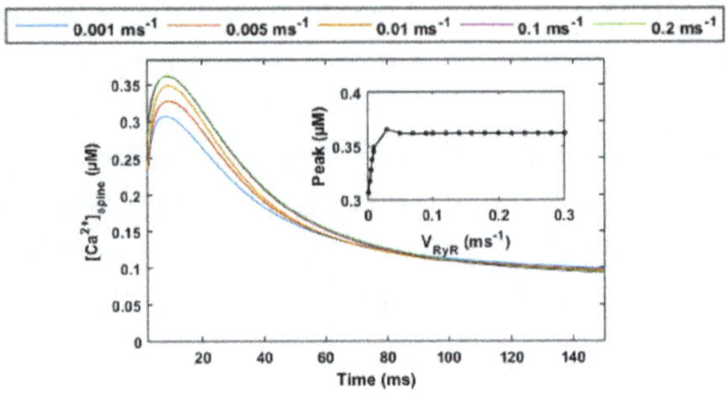


(a)

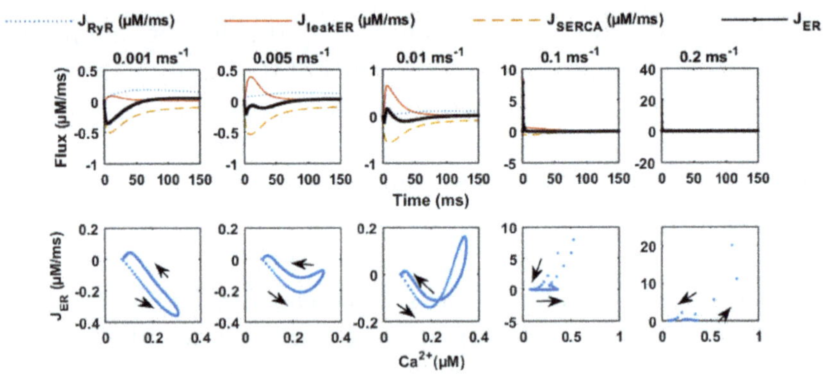


(b)

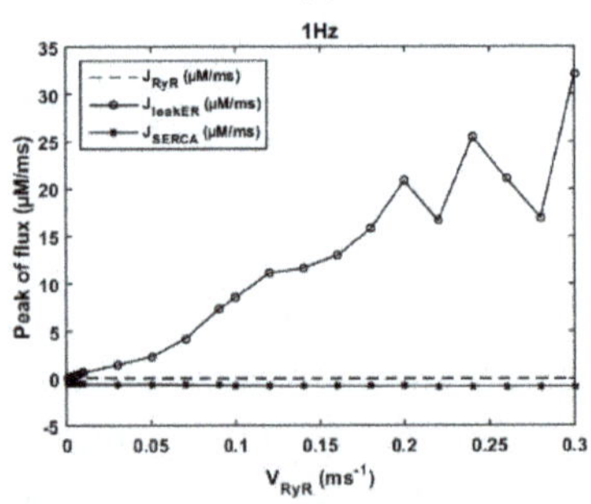


(c)

Fig. 8.11. Effects of $V_{\mathrm{RyR}}$ on postsynaptic response by a single presynaptic stimulation pulse (1 Hz). (a) The $[\mathrm{Ca}^{2+}]_{\mathrm{cyto}}$ dynamics in the spine head and the relationship between $V_{\mathrm{RyR}}$ and the $[\mathrm{Ca}^{2+}]_{\mathrm{cyto}}$ peaks (inset). The corresponding $\mathrm{Ca}^{2+}$ responses of RyR ($J_{\mathrm{RyR}}$), the SERCA pump ($J_{\mathrm{SERCA}}$), the ER membrane leaking ($J_{leak\mathrm{ER}}$), and the net $\mathrm{Ca}^{2+}$ flux from the ER to the cytosol ($J_{\mathrm{ER}}$) are shown in (b). Arrows in the bottom panel indicate the direction of time flow. The peaks of $J_{\mathrm{RyR}}$, $J_{\mathrm{SERCA}}$ and $J_{leak\mathrm{ER}}$ at different levels of $V_{RyR}$ are in (c).

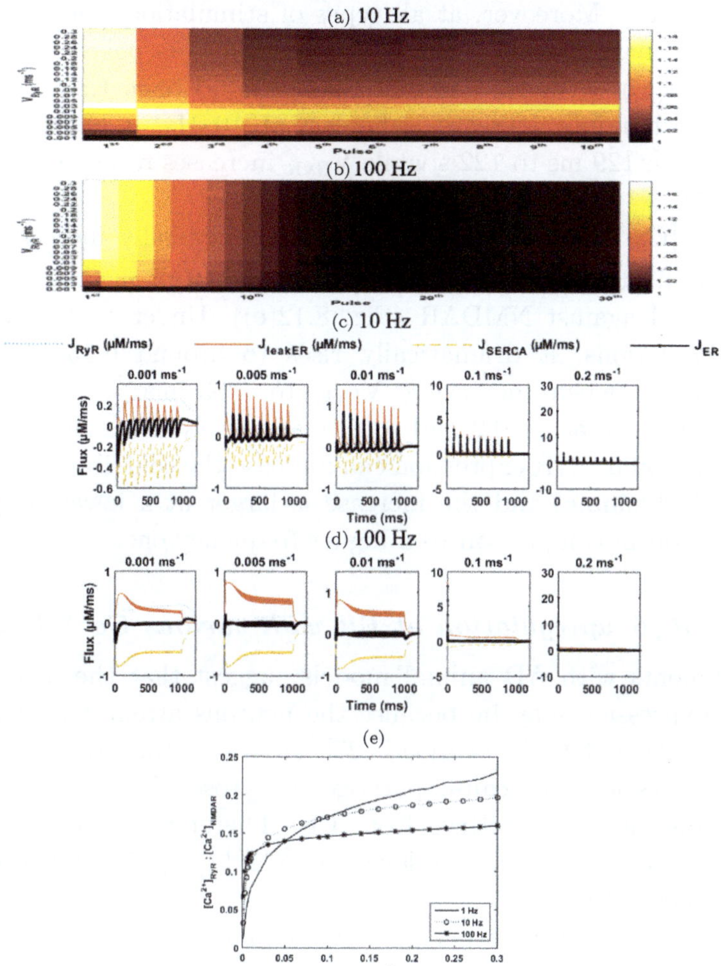


Fig. 8.12. Effects of $V_{\mathrm{RyR}}$ on postsynaptic response under 1 s presynaptic stimulation at 10 Hz and 100 Hz, respectively. The $\left[\mathrm{Ca}^{2+}\right]_{\mathrm{cyto}}$ peaks in response to different stimulation pulses at a 10 Hz (a) and a 100 Hz (b) stimulation, respectively, are normalized to the one with the lowest $V_{\mathrm{RyR}}$ at each pulse. The color density represents the values of $\left[\mathrm{Ca}^{2+}\right]_{\mathrm{cyto}}$ peaks in the spine head. Under the 100 Hz stimulation, only the first 30 pulses are shown in the figure since there is no difference afterwards. The corresponding $\mathrm{Ca}^{2+}$ responses of RyR ($J_{\mathrm{RyR}}$), the SERCA pump ($J_{\mathrm{SERCA}}$), the ER membrane leaking ($J_{leak\mathrm{ER}}$), and the net $\mathrm{Ca}^{2+}$ flux from the ER to the cytosol ($J_{\mathrm{ER}}$) are shown in (c) and (d), respectively. (e) Effects of rising $V_{\mathrm{RyR}}$ on $\left[\mathrm{Ca}^{2+}\right]_{\mathrm{RyR}} : \left[\mathrm{Ca}^{2+}\right]_{NMDAR}$ ratio.

(Fig. 8.13(c)). Moreover, at all types of stimulation, the open time increases sharply with $V_{RyR}$ and when $V_{RyR}$ is below 0.01 $ms^{-1}$ (Fig. 8.13(c)). At 1 Hz and 100 Hz, the total open time is almost stable with minor increases whereas, at 10 Hz, it decreases gradually from 129 ms to 122 s while $V_{RyR}$ increases from 0.01 $ms^{-1}$ to 0.3 $ms^{-1}$.

The $[Ca^{2+}]_{RyR}$ : $[Ca^{2+}]_{NMDAR}$ ratio generally increases with $V_{RyR}$, indicating a increasing contribution of RyR in $Ca^{2+}$ entering the cytosol against NMDAR (Fig. 8.12(e)). Under 10 Hz and 100 Hz stimulations, it dramatically rises to around 0.13 and 0.14, respectively, when increasing $V_{RyR}$ to 0.03 $ms^{-1}$, and then it gradually approaches 0.2 and 0.16, respectively. This suggests that upregulation in RyR expression can increase the contribution of RyR to $Ca^{2+}$ dynamics and the increase is larger at a lower frequency stimulation in comparison to a higher frequency one.

### 8.4.3. *RyR upregulation at ER with various $Ca^{2+}$ load*

Experiments with AD animal models suggest that the increase in RyRs expression may be because the neurons attempt to compensate for the ER $Ca^{2+}$ overload [27]. However, the upregulation of RyR expression can cause pathogenic cytosol $Ca^{2+}$ elevation by increasing the $Ca^{2+}$ release from ER. Therefore, we examine the system responses at various levels of $[Ca^{2+}]_{ER0}$ (30–600 $\mu$M) and $V_{RyR}$ (0.001–0.15 $ms^{-1}$).

At a 1 Hz stimulation or the stimulation by the first pulse of multi-pulse stimulation, the $[Ca^{2+}]_{cyto}$ peak generally increases with higher levels of $[Ca^{2+}]_{ER0}$ and $V_{RyR}$ (Fig. 8.14(a)). At a 10 Hz stimulation, the $[Ca^{2+}]_{cyto}$ peak increases with $V_{RyR}$ and reaches the first maximum level when $V_{RyR}$ is around 0.02 $ms^{-1}$ (Fig. 8.14(b)). It then decreases with further increases in $V_{RyR}$; then when $[Ca^{2+}]_{ER0}$ is above 200 $\mu$M, it increases again with $V_{RyR}$. The $[Ca^{2+}]_{cyto}$ peak–$V_{RyR}$ relationship is transformed from a single-peak to a double-peak, with the increase of the $[Ca^{2+}]_{ER0}$. Under a 100 Hz stimulation, the $[Ca^{2+}]_{cyto}$ peaks show similar patterns in response to different simulation pulses from the second pulse onward

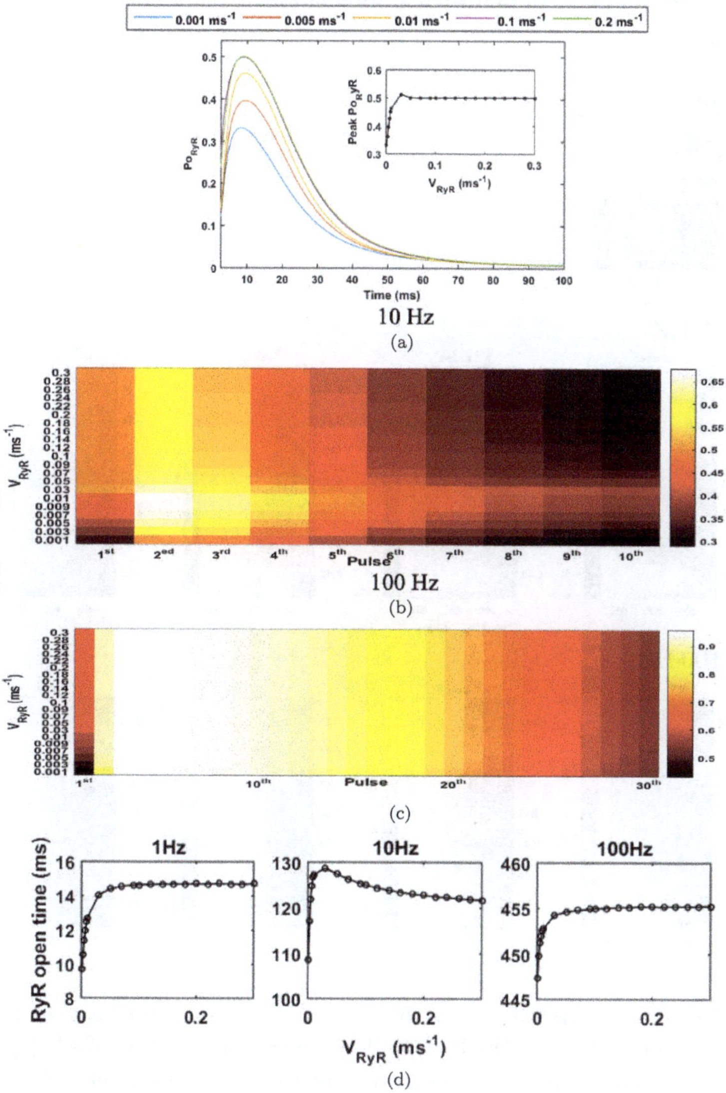


Fig. 8.13. Effects of $V_{\mathrm{RyR}}$ on RyR opening. (a) The $Po_{\mathrm{RyR}}$ with time in the spine head and the relationship between $V_{\mathrm{RyR}}$ and the $Po_{\mathrm{RyR}}$ peaks (inset) at a 1 Hz stimulation. The heatmaps show the maximums of $Po_{\mathrm{RyR}}$ in response to different stimulation pulses at a 10 Hz (b) and a 100 Hz (c) stimulation, respectively. The color density represents the values of $Po_{\mathrm{RyR}}$ peaks in the spine head. Under the 100 Hz stimulation, only the first 30 pulses are shown in the figure since there is no difference afterwards. (d) The corresponding total opening times of RyR at stimulations with different frequencies.

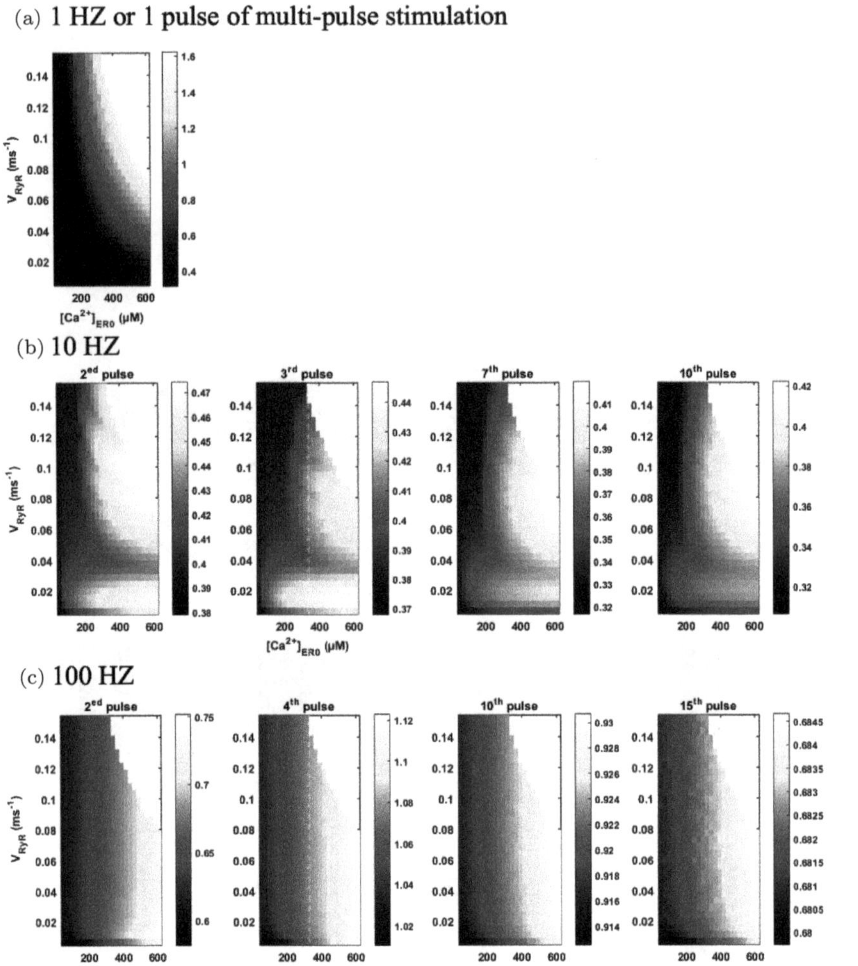


Fig. 8.14. $Ca^{2+}$ response at various level of $[Ca^{2+}]_{ER0}$ and $V_{RyR}$ at (a) 1 Hz, (b) 10 Hz and (c) 100 Hz stimulations. Figures (b) and (c) show $[Ca^{2+}]_{cyto}$ peaks at selected stimulation pulse. The gray scale represents the values of $[Ca^{2+}]_{cyto}$ peaks in the spine head.

(Fig. 8.14(c)). At each degree of $[Ca^{2+}]_{ER0}$, the $[Ca^{2+}]_{cyto}$ peak increases with $V_{RyR}$ when $V_{RyR}$ is below 0.02 $ms^{-1}$, and is not affected by $V_{RyR}$ when $V_{RyR}$ is greater. In contrast, as the degree of $[Ca^{2+}]_{ER0}$ becomes higher, the $[Ca^{2+}]_{cyto}$ peaks increase across the tested ranges of $V_{RyR}$.

The $[Ca^{2+}]_{RyR} : [Ca^{2+}]_{NMDAR}$ ratio increases with $[Ca^{2+}]_{ER0}$ and $V_{RyR}$ at all tested stimulation frequencies (Fig. 8.15(a)). As the frequency is higher, the $[Ca^{2+}]_{RyR} : [Ca^{2+}]_{NMDAR}$ ratio is easier to obtain a higher level than at a lower degree of $[Ca^{2+}]_{ER0}$ and $V_{RyR}$. This suggests that ER, as the internal $Ca^{2+}$ source, makes a high contribution to the cytosolic response under conditions of RyR expression and ER $Ca^{2+}$ overload. The results of the total open time of RyR under presynaptic stimulation (Fig. 8.15(b)) show similar pattern to ones of the $[Ca^{2+}]_{cyto}$ peaks in Fig. 8.14. Under 10 Hz stimulation, a peak value appears when $V_{RyR}$ is around 0.02 $ms^{-1}$. At 100 Hz, RyR opens for a longer time as the increase in the $[Ca^{2+}]_{ER0}$ and $V_{RyR}$, where when $V_{RyR}$ is above 0.03 $ms^{-1}$, it does not affect the open time with a fixed $[Ca^{2+}]_{ER0}$.

### 8.4.4. *SERCA pump reduction*

We manipulate the maximum rate of SERCA pump, $v_{SERCA}$, to see the effect of $Ca^{2+}$ uptake by the ER on $Ca^{2+}$ response. The $[Ca^{2+}]_{cyto}$ peak decreases from 0.37 $\mu$M to 0.3 $\mu$M when the $v_{SERCA}$ increases from 0.01 to 1.5 $\mu$M/ms (Fig. 8.16(a)) at 1 Hz stimulation. The net ER $Ca^{2+}$ flux is negative but has a lower amplitude as the value of $v_{SERCA}$ increases (Fig. 8.16(b)). The $Ca^{2+}$ fluxes via SERCA pump as well as via RyR and membrane leaking are enhanced by larger $v_{SERCA}$ (Fig. 8.16(c)). Under a 10 Hz stimulation, the normalized $[Ca^{2+}]_{cyto}$ peak slightly increases with $v_{SERCA}$ from the second pulse onward (Fig. 8.17(a)), whereas with a 100 Hz stimulation, a reduced degree of reduction in the $[Ca^{2+}]_{cyto}$ peak with $v_{SERCA}$ from the second pulse onward, and after the fourth pulse, there is no obvious difference in $[Ca^{2+}]_{cyto}$ peak with $v_{SERCA}$ (Fig. 8.17(b)). The net ER $Ca^{2+}$ flux is negative with decreasing amplitude when $v_{SERCA}$ is below 0.9 under a 10 Hz stimulation (Fig. 8.17(c)). At a high level of $v_{SERCA}$ (1.5 $\mu$M/ms), positive spikes in the net ER $Ca^{2+}$ flux appear from the second stimulation pulse onward. This is because the RyR response to $Ca^{2+}$ is greatly enhanced by a larger $v_{SERCA}$ in comparison with the $Ca^{2+}$ flux via SERCA pumps and membrane leaking (Fig. 8.17(e)). In contrast, with a 100 Hz

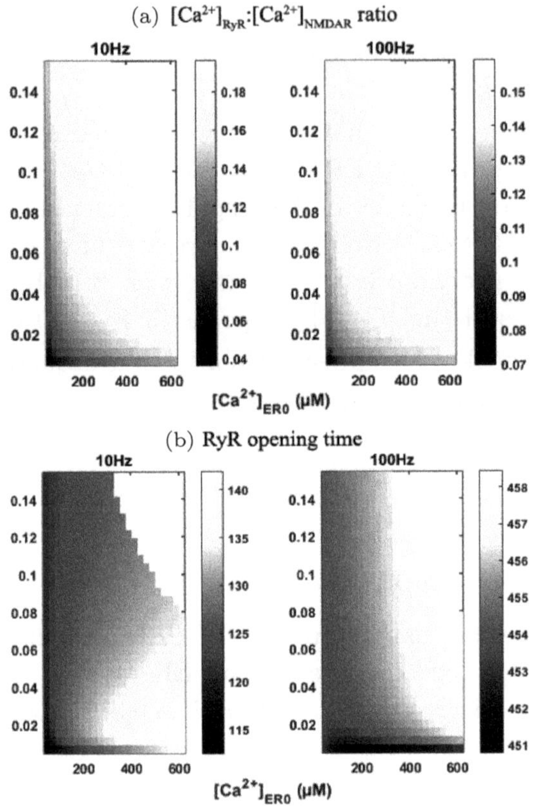


Fig. 8.15. (a) $[Ca^{2+}]_{RyR}:[Ca^{2+}]_{NMDAR}$ ratio and (b) RyR opening time at various level of $[Ca^{2+}]_{ER0}$ and $V_{RyR}$. The gray scale represents the values of the $[Ca^{2+}]_{RyR}$:$[Ca^{2+}]_{NMDAR}$ ratio and the RyR opening time in the spine head, respectively.

stimulation, the peak $Ca^{2+}$ flux via SERCA pumps stays at around $-1.5\mu$M/ms across the testing range of $v_{SERCA}$ values (Fig. 8.17(f)), which is around $-1$ $\mu$M/ms at the 1 and 10 Hz stimulations. This is because of the high $Ca^{2+}$ influx via NMDAR under a 100 Hz stimulation. which leads to a high inward net $Ca^{2+}$ flux from the cytosol to the ER. The peak of the net flux decreases with $v_{SERCA}$ (Fig. 8.17(d)). Moreover, the $[Ca^{2+}]_{RyR}$:$[Ca^{2+}]_{NMDAR}$ ratio shows a linear increase with $v_{SERCA}$ at all frequencies of stimulation, with a larger slope at the higher frequency (Fig. 8.18).

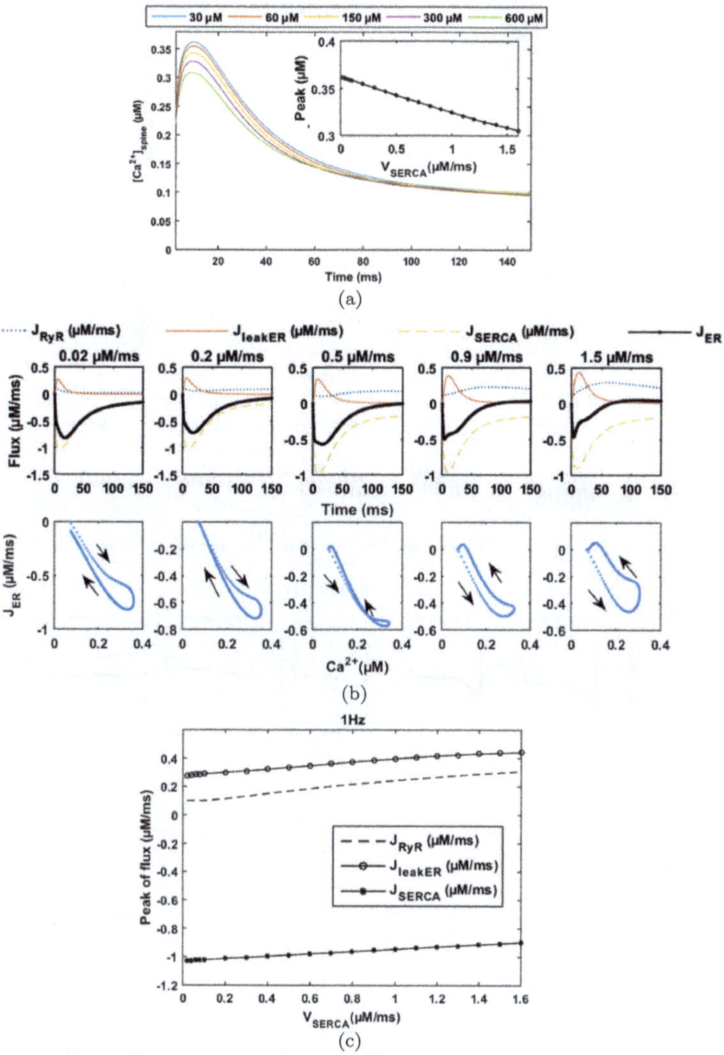


Fig. 8.16. Effects of $v_{SERCA}$ on postsynaptic response by a single presynaptic stimulation pulse (1 Hz). (a) The $\left[\mathrm{Ca}^{2+}\right]_{\mathrm{cyto}}$ dynamics in the spine head and the relationship between $v_{\mathrm{SERCA}}$ and the $\left[\mathrm{Ca}^{2+}\right]_{\mathrm{cyto}}$ peaks (inset). The corresponding $\mathrm{Ca}^{2+}$ responses of RyR ($J_{\mathrm{RyR}}$) and the SERCA pump ($J_{\mathrm{SERCA}}$), ER membrane leaking ($J_{leak\mathrm{ER}}$), and the net $\mathrm{Ca}^{2+}$ flux from the ER to the cytosol ($J_{\mathrm{ER}}$) are shown in (b). Arrows in the bottom panel indicate the direction of time flow. The peaks of $J_{\mathrm{RyR}}$, $J_{\mathrm{SERCA}}$ and $J_{leak\mathrm{ER}}$ at different levels of $v_{\mathrm{SERCA}}$ are in (c).

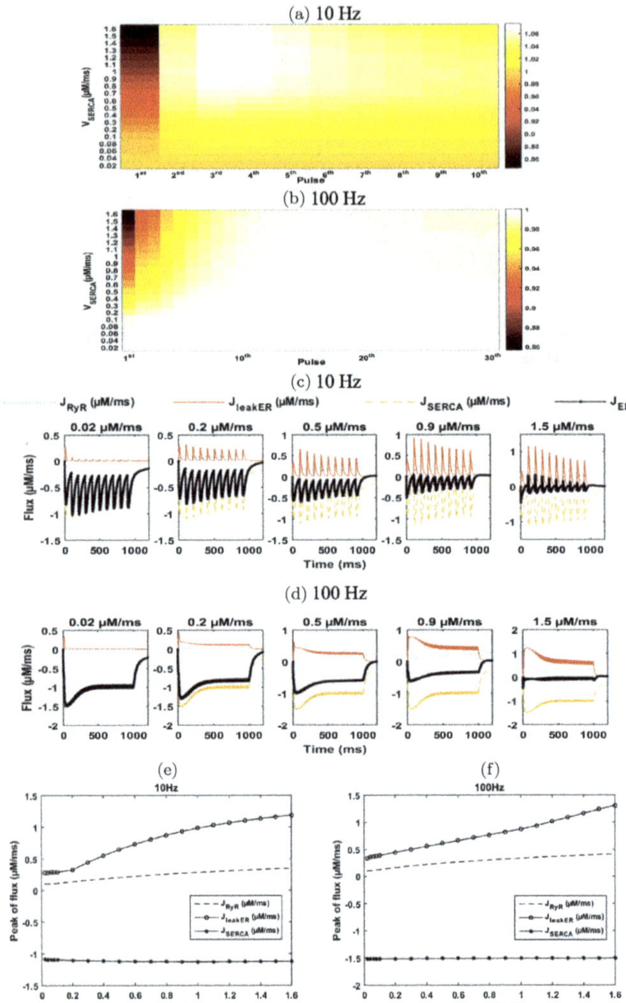


Fig. 8.17. Effects of $v_{\mathrm{SERCA}}$ on postsynaptic response under 1 s presynaptic stimulation at 10 Hz and 100 Hz, respectively. The $[\mathrm{Ca}^{2+}]_{\mathrm{cyto}}$ peaks in response to different stimulation pulses at a 10 Hz (a) and a 100 Hz (b) stimulation, respectively, are normalized to the one with the lowest $v_{\mathrm{SERCA}}$ at each pulse. The color density represents the values of $[\mathrm{Ca}^{2+}]_{\mathrm{cyto}}$ peaks in the spine head. Under the 100 Hz stimulation, only the first 30 pulses are shown in the figure since there is no difference afterwards. The corresponding $\mathrm{Ca}^{2+}$ responses of RyR ($J_{\mathrm{RyR}}$), the SERCA pump ($J_{\mathrm{SERCA}}$), the ER membrane leaking ($J_{leak\mathrm{ER}}$), and the net $\mathrm{Ca}^{2+}$ flux from the ER to the cytosol ($J_{\mathrm{ER}}$) are shown in (c) and (d), respectively. The peaks of $J_{\mathrm{RyR}}$, $J_{\mathrm{SERCA}}$ and $J_{leak\mathrm{ER}}$ at different levels of $v_{\mathrm{SERCA}}$ are in (e) and (f).

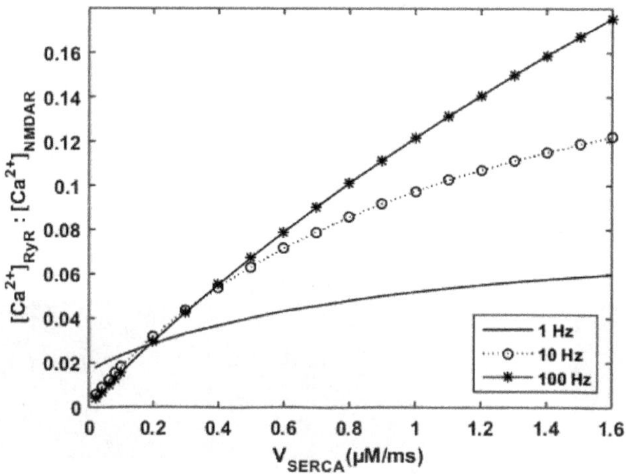


Fig. 8.18. Effects of $v_{\mathrm{SERCA}}$ on $[\mathrm{Ca}^{2+}]_{\mathrm{RyR}}$:$[\mathrm{Ca}^{2+}]_{\mathrm{NMDAR}}$ ratio at 1 Hz, 10 Hz and 100 Hz stimulations, respectively.

The fraction of RyR opened by the 1 Hz stimulation shows a decrease in the peak, from 0.5 to 0.34, when $v_{\mathrm{SERCA}}$ increases from 0.01 to 1.5 $\mu$M/ms (Fig. 8.19(a)). At 100 Hz, the maximum $Po_{\mathrm{RyR}}$ increases to a peak of around 95% from the second stimulation pulse to the eighth one, and decreases gradually afterwards. There is no difference in the maximum of $Po_{\mathrm{RyR}}$ from different $Po_{\mathrm{RyR}}$ (Fig. 8.19(c)). However, under the 10 Hz stimulation, the pattern of maximum $Po_{\mathrm{RyR}}$ switches to increasing with $v_{\mathrm{SERCA}}$ from the second pulse onward (Fig. 8.19(b)). The total open time of RyR decreases by 4.4 ms and 5.3 ms when increasing $v_{\mathrm{SERCA}}$ from 0.01 to 1.5 $\mu$M/ms at 1 and 100 Hz stimulations, respectively (Fig. 8.19(d)). Under 10 Hz stimulation, a bell-shaped relationship between the RyR open time and $v_{\mathrm{SERCA}}$, with the peak value, appears when $v_{\mathrm{SERCA}}$ is around 0.6 $\mu$M/ms.

### 8.4.5. *Alteration in SERCA pumps at ER with various $Ca^{2+}$ load*

Green, Demuro, Akbari, Hitt, Smith, Parker and LaFerla [28] reported that PS interacts with SERCA pump to modulate its

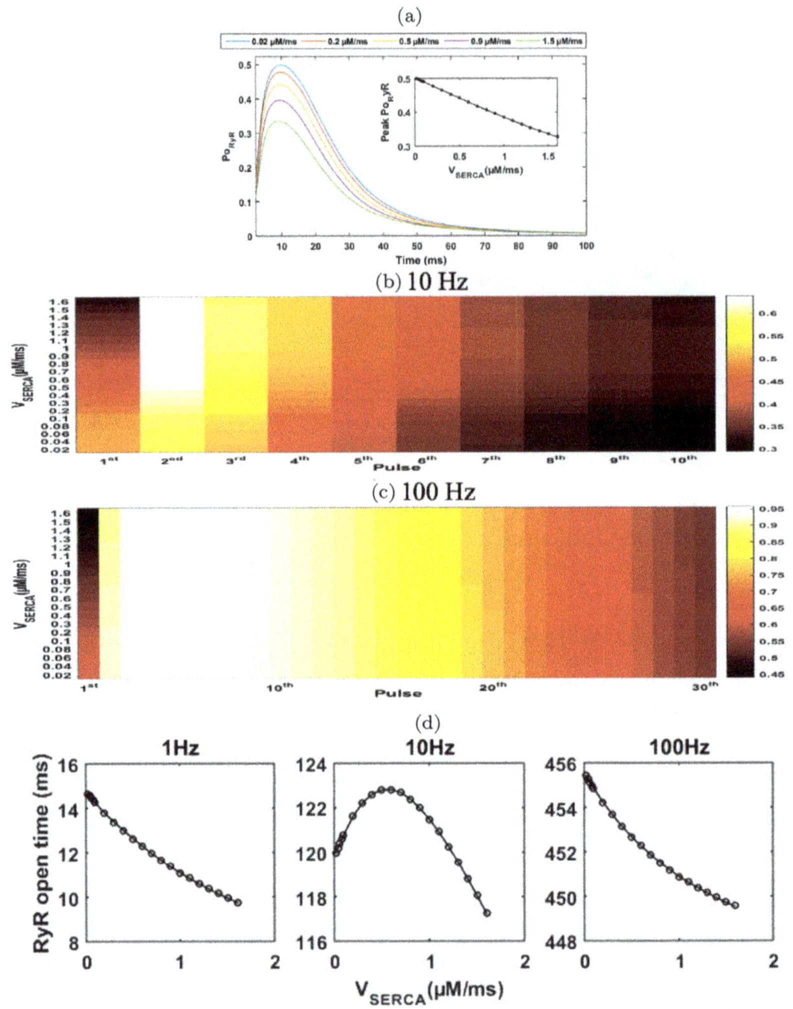


Fig. 8.19. Effects of $v_{\mathrm{SERCA}}$ on RyR opening. (a) The $Po_{\mathrm{RyR}}$ with time in the spine head and the relationship between $v_{\mathrm{SERCA}}$ and the $Po_{\mathrm{RyR}}$ peaks (inset) at a 1 Hz stimulation. The heatmaps show the maximums of $Po_{\mathrm{RyR}}$ in response to different stimulation pulses at a 10 Hz (b) and a 100 Hz (c) stimulation, respectively. The color density represents the values of $Po_{\mathrm{RyR}}$ peaks in the spine head. Under the 100 Hz stimulation, only the first 30 pulses are shown in the figure since there is no difference afterwards. (d) The corresponding total opening times of RyR at stimulations with different frequencies.

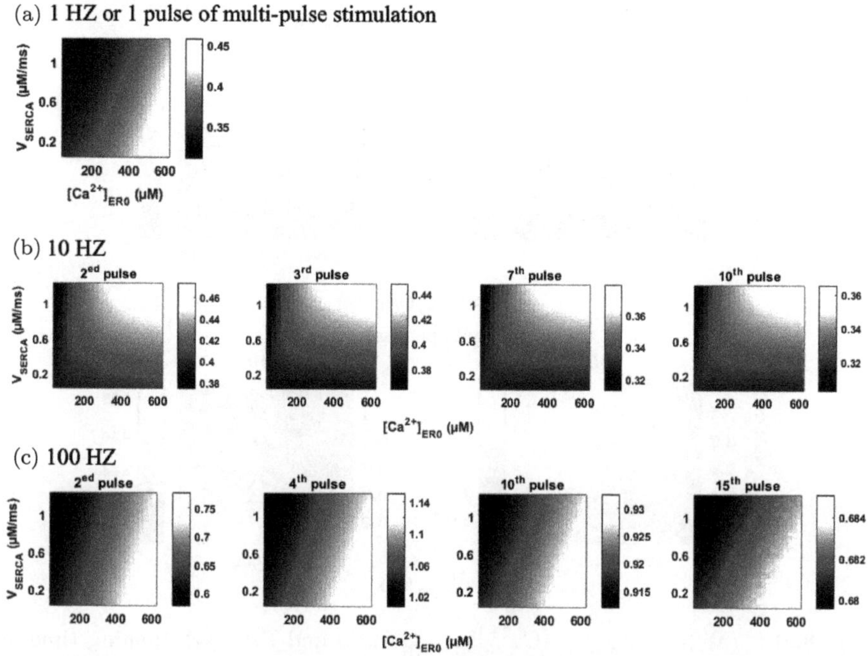


Fig. 8.20. $Ca^{2+}$ response at various level of $[Ca^{2+}]_{ER0}$ and $v_{SERCA}$ at (a) 1 Hz, (b) 10 Hz and (c) 100 Hz stimulations. Figures (b) and (c) show $[Ca^{2+}]_{cyto}$ peaks at selected stimulation pulses. The gray scale represents the values of $[Ca^{2+}]_{cyto}$ peaks in the spine head.

function and lack of PS impairs the SERCA pumps' function in *Xenopus laevis* oocytes. However, it is not clear how the PS mutation in AD will affect SERCA pumping. However, Brunello, Zampese, Florean, Pozzan, Pizzo and Fasolato [29] produced opposite results that both wild-type and mutant PS2 could reduce SERCA activity. Therefore, we examine the system responses at various levels of $[Ca^{2+}]_{ER0}$ (30–600 $\mu$M) and $v_{SERCA}$ (0.01–1.2 $\mu$M/ms), to investigate the possible contribution of SERCA alteration in the $Ca^{2+}$ elevation in the spine head.

At a 1 Hz stimulation or the stimulation by the first pulse of multi-pulse stimulation, the $[Ca^{2+}]_{cyto}$ peak generally increases with higher levels of $[Ca^{2+}]_{ER0}$ and lower levels of $v_{SERCA}$ (Fig. 8.20(a)). The results show similar pattern in response to the second pulse

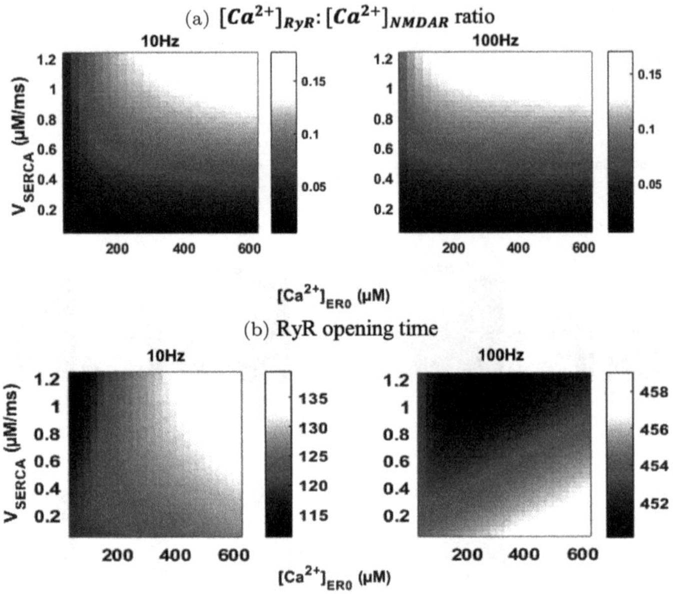


Fig. 8.21. (a) $[\mathrm{Ca}^{2+}]_{\mathrm{RyR}}$ : $[\mathrm{Ca}^{2+}]_{\mathrm{NMDAR}}$ ratio and (b) RyR opening time at various levels of $[\mathrm{Ca}^{2+}]_{\mathrm{ER0}}$ and $v_{\mathrm{SERCA}}$. The gray scale represents the values of the $[\mathrm{Ca}^{2+}]_{\mathrm{RyR}}$:$[\mathrm{Ca}^{2+}]_{\mathrm{NMDAR}}$ ratio and the RyR opening time in the spine head, respectively.

onward at 100 Hz stimulation, (Fig. 8.20(c)). In contrast, at a 10 Hz stimulation, the $[\mathrm{Ca}^{2+}]_{\mathrm{cyto}}$ peak from the second pulse onward increases with the decrease in $v_{\mathrm{SERCA}}$ when $[\mathrm{Ca}^{2+}]_{\mathrm{ER0}}$ is below 150 $\mu$M and shift to decrease with the decrease in $v_{\mathrm{SERCA}}$ at higher levels of $[\mathrm{Ca}^{2+}]_{\mathrm{ER0}}$ (Fig. 8.20(b)).

At both 10 and 100 Hz stimulations, the $[\mathrm{Ca}^{2+}]_{\mathrm{RyR}}$: $[\mathrm{Ca}^{2+}]_{NMDAR}$ ratio increases with $v_{\mathrm{SERCA}}$, but only increases with $[\mathrm{Ca}^{2+}]_{\mathrm{ER0}}$ at lower levels of $[\mathrm{Ca}^{2+}]_{\mathrm{ER0}}$ (Fig. 8.21(a)). This suggests that the expression level of SERCA pumps promotes the role as an internal $\mathrm{Ca}^{2+}$ source by ER at various resting $\mathrm{Ca}^{2+}$ level of ER. Figure 8.21(b) shows that the results of the RyR opening time have a similar pattern to ones of the $[\mathrm{Ca}^{2+}]_{\mathrm{cyto}}$ peaks in Fig. 8.20, in response to different simulation pulses. At 10 Hz stimulation, lower levels of $v_{\mathrm{SERCA}}$ lead to higher levels of $[\mathrm{Ca}^{2+}]_{\mathrm{RyR}}$:$[\mathrm{Ca}^{2+}]_{NMDAR}$ when $[\mathrm{Ca}^{2+}]_{\mathrm{ER0}}$ is below 150 $\mu$M, while, at higher levels of $[Ca^{2+}]_{ER0}$, higher levels of $v_{SERCA}$ lead to higher levels of $[Ca^{2+}]_{RyR}$:$[Ca^{2+}]_{NMDAR}$. At a 100 Hz stimulation, the RyR opening time generally increases with $[Ca^{2+}]_{ER0}$ and decreases with $v_{SERCA}$.

## 8.5. Discussion and Conclusions

In this chapter, we use an extended model from the one in Chapter 7 to investigate the effects of the alteration of several of the main factors in the ER in the postsynaptic response. We look at the relationship of $Ca^{2+}$ transient levels, net ER $Ca^{2+}$ fluxes, RyR opening and its contribution to cytosolic $Ca^{2+}$ elevation and the relationship between them and selected alterations of ER function, as reported in AD.

The simulation predicts that the ER $Ca^{2+}$ overload and upregulation in RyR expression will enhance its role as an internal $Ca^{2+}$ source to amplify the external $Ca^{2+}$ influx. This is on account of the longer open times of RyR or the reduction in $Ca^{2+}$ uptake by SERCA pumps. The increase in RyR response suggests a reduced CICR threshold, which means RyR shows higher sensitivity to the NMDAR-mediated $Ca^{2+}$ signal under these conditions. This alteration can promote the processing of amyloid precursor protein (APP) to facilitate the production of A$\beta$ peptides [30]. In addition, experimental studies suggest that upregulation of RyR expression may be a compensatory mechanism to the $Ca^{2+}$ overload in the ER in the early stages of AD [31]. Our simulation predicts that at high levels of ER $Ca^{2+}$, low levels of upregulation in RyR expression will increase all outputs. However, when the upregulation in RyR expression is higher, it shows less of an effect on all outputs. This suggests that, although an increase in RyR expression may be beneficial to ease the stress of the ER from $Ca^{2+}$ overloading, it may, however, make the cell more vulnerable by amplifying the cytosolic $Ca^{2+}$ level.

A stimulation frequency-dependent effect in the outputs has been observed when applying multiple stimulation pulses at various stimulation frequencies. The overall enhanced effect on the postsynaptic $Ca^{2+}$ response is higher under low-frequency stimulation (10 Hz) in comparison with high-frequency stimulation (100 Hz). For example, manipulating the selected factors shows no obvious effects on the peak of $Ca^{2+}$ elevation and maximum fraction of RyR opened in response to presynaptic stimulation at 100 Hz from the second pulse onward; whereas a clear pattern with the level of selected factors can be observed under 10 Hz stimulation. This implies that the alteration is more likely to amplify the NMDAR-mediated $Ca^{2+}$ signal with a lower frequency and lower amplitude, consistent with the experimental observations [32]. This will affect the basal synaptic transmission or may bring the basal synaptic transmission to a long-term depression.

Less study has been done on the alteration of SERCA pumps than of RyRs and ER $Ca^{2+}$ load in the ER. Green, Demuro, Akbari, Hitt, Smith, Parker and LaFerla [28] and Brunello, Zampese, Florean, Pozzan, Pizzo and Fasolato [29] suggest opposites effects on SERCA pumps by PS and PS mutation in AD. Here, we predict opposite effects by the alteration of SERCA pumps in the $Ca^{2+}$ signaling in the spine head, which shows an interesting frequency-dependence. Higher levels of SERCA pumps can increase the $Ca^{2+}$ uptake by the ER. At the same time, it leads to a faster refilling of $Ca^{2+}$ by the ER and thus, promotes the $Ca^{2+}$ release from the ER at the same time. At high-frequency stimulation, an increase in the level of SERCA pumps causes a shorter opening for RyR; however, it speeds up the replenishment of the ER with $Ca^{2+}$, leading to a higher $Ca^{2+}$ release via RyR from the ER. Even though the increase in SERCA will lead to an exceeding amount of $Ca^{2+}$ ions removed from the cytosol to the ER, this overcomes the effect of increasing $Ca^{2+}$ release by RyR and accounts for the decreasing in the elevation of cytosol $Ca^{2+}$ transients. At low-frequency stimulation, because the $Ca^{2+}$ influx via NMDAR is lower than at high frequency, the $Ca^{2+}$ uptake by ER enhanced by the upregulation in the SERCA pumps is not high enough to compromise the increasing $Ca^{2+}$ release from ER. Therefore, the cytosolic $Ca^{2+}$ signal is amplified by the upregulation in SERCA pumps.

The simulation results suggest that inhibit SERCA pumps may rescue the spine head from dysregulation in $Ca^{2+}$ signaling by ER

$Ca^{2+}$ overload. At low-level stimulation, inhibition of SERCA pumps will reduce the $Ca^{2+}$ from ER by attenuating the refilling of $Ca^{2+}$ by ER. At high stimulation, although inhibiting SERCA pumps will increase the $Ca^{2+}$ release from ER, long-term inhibition will decrease the $Ca^{2+}$ load of ER. Besides, Green, Demuro, Akbari, Hitt, Smith, Parker and LaFerla [28] reported that decrease the SERCA function can reduce the production of A$\beta$, which may be a consequence of reduced ER $Ca^{2+}$ signaling.

## References

[1] Holbro N., Grunditz Å. and Oertner T.G. (2009). Differential distribution of endoplasmic reticulum controls metabotropic signaling and plasticity at hippocampal synapses. *Proc. Natl. Acade. Sci. USA*, *106*(35), pp. 15055–15060.

[2] Spacek J. and Harris K.M. (1997). Three-dimensional organization of smooth endoplasmic reticulum in hippocampal CA1 dendrites and dendritic spines of the immature and mature rat. *J. Neurosci.*, *17*(1), pp. 190–203.

[3] Cooney J.R., Hurlburt J.L., Selig D.K., Harris K.M. and Fiala J.C. (2002). Endosomal compartments serve multiple hippocampal dendritic spines from a widespread rather than a local store of recycling membrane. *J. Neurosci.*, *22*(6), pp. 2215–2224.

[4] Matsuzaki M., Honkura N., Ellis-Davies G.C. and Kasai H. (2004). Structural basis of long-term potentiation in single dendritic spines. *Nature*, *429*(6993), pp. 761–766.

[5] Emptage N., Bliss T.V. and Fine A. (1999). Single synaptic events evoke NMDA receptor–mediated release of calcium from internal stores in hippocampal dendritic spines. *Neuron*, *22*(1), pp. 115–124.

[6] Friel D.D. and Tsien R.W. (1992). A caffeine- and ryanodine-sensitive $Ca^{2+}$ store in bullfrog sympathetic neurones modulates effects of $Ca^{2+}$ entry on $[Ca^{2+}]$i. *J. Physiology*, *450*, pp. 217–246.

[7] Goussakov I., Miller M.B. and Stutzmann G.E. (2010). NMDA-mediated $Ca^{2+}$ influx drives aberrant ryanodine receptor activation in dendrites of young Alzheimer's disease mice. *J. Neurosci.*, *30*(36), pp. 12128–12137.

[8] Supnet C. and Bezprozvanny I. (2010). The dysregulation of intracellular calcium in Alzheimer disease. *Cell Calcium*, *47*(2), pp. 183–189.

[9] Popugaeva E. and Bezprozvanny I. (2013). Role of endoplasmic reticulum $Ca^{2+}$ signaling in the pathogenesis of Alzheimer disease. *Frontiers Mol. Neurosci.*, *6*, p. 29.

[10] Liang J., Kulasiri D. and Samarasinghe S. (2015). $Ca^{2+}$ dysregulation in the endoplasmic reticulum related to Alzheimer's disease: A review on experimental progress and computational modeling. *Biosystems*, *134*, pp. 1–15.

[11] Chan S.L., Mayne M., Holden C.P., Geiger J.D. and Mattson M.P. (2000). Presenilin-1 mutations increase levels of ryanodine receptors and calcium release in PC12 cells and cortical neurons. *J. Biolo. Chem.*, *275*(24), pp. 18195–18200.

[12] Bezprozvanny I. and Hiesinger P.R. (2013). The synaptic maintenance problem: membrane recycling, $Ca^{2+}$ homeostasis and late onset degeneration. *Mole. Neurodegeneration*, *8*(1), pp. 1–14.

[13] Popugaeva E., Supnet C. and Bezprozvanny I. (2012). Presenilins, deranged calcium homeostasis, synaptic loss and dysfunction in Alzheimer's disease. *Messenger*, *1*(1) pp. 53–62.

[14] Sabatini B.L., Maravall M. and Svoboda K. (2001). Ca(2+) signaling in dendritic spines. *Curr. Opinion Neurobiol.*, *11*(3), pp. 349–356.

[15] Sharp A.H. *et al.* (1993). Differential immunohistochemical localization of inositol 1,4,5-trisphosphate- and ryanodine-sensitive Ca2+ release channels in rat brain. *J. Neurosci.*, *13*(7), pp. 3051–3063.

[16] Hertle D.N. and Yeckel M.F. (2007). Distribution of inositol-1, 4,5-trisphosphate receptor isotypes and ryanodine receptor isotypes during maturation of the rat hippocampus. *Neuroscience*, *150*(3), pp. 625–638.

[17] Keizer J. and Levine L. (1996). Ryanodine receptor adaptation and $Ca^{2+}$(-) induced $Ca^{2+}$ release-dependent $Ca^{2+}$ oscillations. *Biophys. J.*, *71*(6), pp. 3477–3487.

[18] Lytton J., Westlin M., Burk S.E., Shull G.E. and MacLennan D.H. (1992). Functional comparisons between isoforms of the sarcoplasmic or endoplasmic reticulum family of calcium pumps. *J. Biol. Chem.*, *267*(20), pp. 14483–14489.

[19] Sabatini B.L., Oertner T.G. and Svoboda K. (2002). The life cycle of $Ca^{2+}$ Ions in dendritic spines. *Neuron*, *33*(3), pp. 439–452.

[20] Sterratt D., Graham B., Gillies A. and Willshaw D. (2011). *Principles of Computational Modelling in Neuroscience* (Cambridge University Press).

[21] Nevian T. and Sakmann B. (2004). Single spine $Ca^{2+}$ signals evoked by coincident EPSPs and backpropagating action potentials in spiny stellate cells of layer 4 in the juvenile rat somatosensory barrel cortex. *J. Neurosci.*, *24*(7), pp. 1689–1699.

[22] Popugaeva E. and Bezprozvanny I. (2013). Role of endoplasmic reticulum $Ca^{2+}$ signaling in the pathogenesis of Alzheimer disease. *Front Mol. Neurosci.*, *6*(29).

[23] Tu H.P. *et al.* (2006). Presenilins form ER $Ca^{2+}$ leak channels, a function disrupted by familial Alzheimer's disease-linked mutations. *Cell*, *126*(5), pp. 981–993.

[24] Nelson O. *et al.* (2011). Mutagenesis mapping of the presenilin 1 calcium leak conductance pore. *J. Biol. Chem.*, *286*(25), pp. 22339–22347.

[25] Nelson O. *et al.* (2007). Familial Alzheimer disease-linked mutations specifically disrupt $Ca^{2+}$ leak function of presenilin 1. *J. Clin. Invest.*, *117*(5), pp. 1230–1239.

[26] Zhang H., Sun S., Herreman A., De Strooper B. and Bezprozvanny I. (2010). Role of presenilins in neuronal calcium homeostasis. *J. Neurosci.*, *30*(25), pp. 8566–8580.

[27] Zhang H., Sun S., Herreman A., De Strooper B. and Bezprozvanny I. (2010). Role of presenilins in neuronal calcium homeostasis. *J. Neurosci.*, *30*(25), pp. 8566–8580.

[28] Green K. N. *et al.* (2008). SERCA pump activity is physiologically regulated by presenilin and regulates amyloid $\beta$ production. *J. Cell Biol.*, *181*(7), pp. 1107–1116.

[29] Brunello L. *et al.* (2009). Presenilin-2 dampens intracellular $Ca^{2+}$ stores by increasing $Ca^{2+}$ leakage and reducing $Ca^{2+}$ uptake. *Journal of Cellular and Molecular Medicine*, *13*(9b), pp. 3358–3369.

[30] Querfurth H.W., Jiang J., Geiger J.D. and Selkoe D.J. (1997). Caffeine stimulates amyloid $\beta$-peptide release from $\beta$-amyloid precursor protein-transfected HEK293 cells. *J. Neurochem.*, *69*(4), pp. 1580–1591.

[31] Del Prete D., Checler F. and Chami M. (2014). Ryanodine receptors: physiological function and deregulation in Alzheimer disease. *Mol. Neurodegeneration*, *9*, p. 21.

[32] Chakroborty S., Goussakov I., Miller M.B. and Stutzmann G.E. (2009). Deviant ryanodine receptor-mediated calcium release resets synaptic homeostasis in presymptomatic 3xTg-AD mice. *J. Neurosci.*, *29*(30), pp. 9458–9470.

# Chapter 9

# Dynamics of CaMKII under AD Conditions

## 9.1. Introduction

NMDAR-dependent LTP (NMDAR-LTP) is a widely studied form of LTP which can be found in the glutamatergic synapse in the hippocampus [1–3]. Recent studies that suggest the roles of NMDAR in synaptic plasticity are determined by their subunit composition and location of NMDARs [5, 6]. There have been contradictory experimental observations made about which subunit composition predominantly mediates the NMDAR-LTP. For instance, preferential blocking of NR2A-NMDARs inhibits LTP in slices of the hippocampal CA1 area [9] and the perirhinal cortex of rats [10]. Blocking NR2B by ifenprodil abolishes LTD but does not affect LTP, suggesting that NR2B-NMDAR is only required for LTD induction [9]. However, overexpression of NR2B in the transgenic mouse forebrain shows enhanced LTP in response to stimulation at 10–100 Hz [12]. Knock-down of the NR2B subunit in the hippocampus of young rats suppresses LTP formation [14]. These conflicting results may arise from the differences in the types and concentrations of the antagonists, the experimental conditions, the developmental stages and the location of brain regions of the experimental materials.

One possible reason underlying the different contributions of NR2A- and NR2B-NMDAR to the synaptic plasticity may be because of the different kinetic properties of NR2A-NMDAR and

NR2B-NMDAR [6, 15]. Moreover, the different preferences by other key molecular players in synaptic plasticity may also determine their different contributions [5]. Therefore, it is important to investigate the different roles of NMDAR collectively. In this chapter, we explore the effects of disturbances on NMDAR availability under AD conditions on the $Ca^{2+}$ signaling and downstream CaMKII activation, a key event in NMDAR-LTP induction and maintenance [16].

NMDAR-LTP in the hippocampus is induced by the NMDAR-mediated $Ca^{2+}$ influx in the dendritic spine. The transient $Ca^{2+}$ signal is then transferred into a long-lasting enhancement of synaptic strength, which is believed to contribute to memory formation [16]. The induction of LTP as a long-lasting increase in synaptic strength requires CaMKII activation; however, the mechanism of its maintenance is under debate [5, 17, 18]. The autophosphorylation of CaMKII, and its translation into the PSD in the CA1 region of hippocampal slices, have been observed after NMDAR-mediated $Ca^{2+}$ transients [19–21]. CaMKII is a holoenzyme consisting of 12 subunits that are autoinhibited at resting $Ca^{2+}$ levels. The $Ca^{2+}$ ions entering the cytosol will bind to calmodulin (CaM) to form the fully-bound $Ca_4CaM$ complex, which can activate CaMKII by binding to its CaM footprint. This leads to the exposure of the threonine 286 (Thr286) site of CaMKII and the subunit can then be autophosphorylated by active neighboring subunits [22]. The autophosphorylated CaMKII is CaM-independent and shows persistent activity long after the $Ca^{2+}$ transients [23, 24].

The translocation of CaMKII into the PSD is also mediated by $Ca_4CaM$. This process is reversible, while autophosphorylation of CaMKII at Thr286 can persist its presence in the PSD for a longer time. CaMKII in the PSD can phosphorylate synaptic AMPARs, which can enhance their conductance and increase their numbers in the PSD by mediating their trafficking [22]. This will, consequently, lead to an enhanced AMPAR-mediated transmission during LTP [25]. Once the CaMKII is translocated into the PSD, it can also bind to synaptic NR2B at the T-site [26]. The CaMKII-NMDAR complex anchors AMPAR by creating new anchoring sites for additional synaptic AMPAR [27, 28]. We do not consider the binding of CaMKII to NR2A-NMDAR because of its lower affinity in comparison to NR2B-NMDAR [5].

The role of synaptic NR2B in forming CaMKII-NMDAR complex in the PSD is independent from its role as a $Ca^{2+}$ channel [29]. Given the dual role of NR2B-NMDAR in NMDAR-LTP formation and the differential disturbances in NR2A- and NR2B-NMDARs in AD, it raises questions: how do different disturbances of NMDAR reported in AD affect CaMKII ST? Which disturbance predominantly suppresses NMDAR-LTP formation? What pathological consequences in AD may these effects link to? To answer these questions, we incorporate a CaMKII state transition (ST) model into our $Ca^{2+}$ model (Fig. 9.1). In this chapter, we use the term $Ca^{2+}$ model to refer to the model we present in Chapters 7 and 8. The simulation results show that both NR2A- and NR2B-NMDAR are necessary for CaMKII ST and it requires cooperation between them. The model provides insights into the potential contribution of the different roles of NMDAR in the formation of NMDAR-mediated-LTP. The

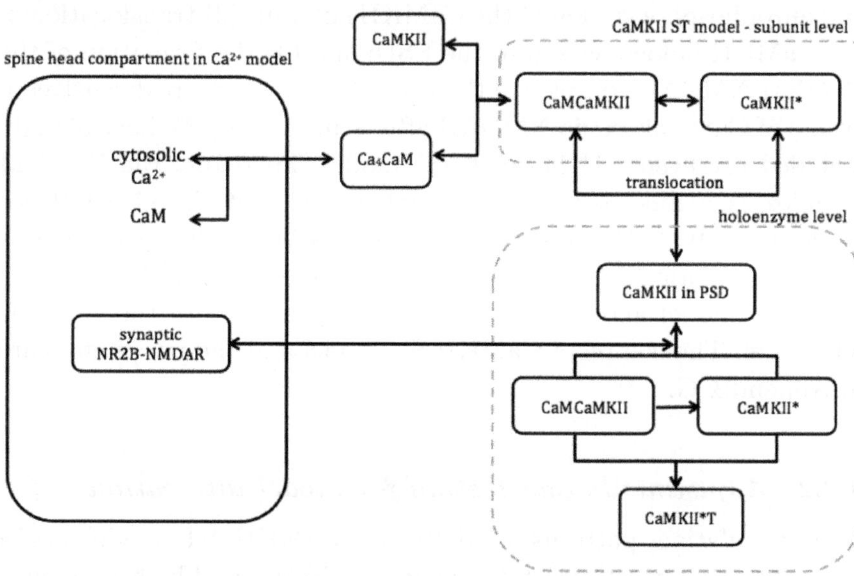


Fig. 9.1. Schematics of the model integration. Schematic diagram of the CaMKII state transition (ST) model [7].

simulation results also suggest that the disturbance in synaptic NMDAR in AD could be a potential cause of other critical changes in the pathology of the disease, which are related to the learning and memory deficit in AD patients.

In this chapter, we briefly introduce the CaMKII ST model by He, Kulasiri and Samarasinghe [7] and the integration of it to our $Ca^{2+}$ model in Section 9.2. In Section 9.3, we analyze the disturbance of the synaptic NMDAR numbers in the CaMKII-NMDAR complex formation and, in Section 9.4, we give a brief discussion and summary of the results.

## 9.2. Model Integration

### 9.2.1. *CaMKII state transition (ST) model*

The CaMKII ST model developed by He, Kulasiri and Samarasinghe [7] simulates the formation of the CaMKII-NMDAR complex in the PSD in response to presynaptic stimulation. It consists of a series of key events: (1) formation of the $Ca_4CaM$ complex; (2) activation and autophosphorylation of the CaMKII subunit; (3) translocation of the CaMKII holoenzyme into the PSD; and (4) the formation of the CaMKII-NMDAR complex in the PSD. These events are downstream events of $Ca^{2+}$ influx via NMDAR after stimulation. We have already included the event (1) in our $Ca^{2+}$ model. Therefore, the $Ca_4CaM$ complex is a link between our $Ca^{2+}$ model and the CaMKII ST model (Fig. 9.1). Besides, the other link between the two model is the postsynaptic NR2B-NMDAR, which serves as a $Ca^{2+}$ channel in the $Ca^{2+}$ model and as a CaMKII binding partner in the CaMKII ST model. The complete CaMKII ST model and the parameters are in Appendix G.

### 9.2.2. *Adjustments and testing for model integration*

The stimulation patterns used in the CaMKII ST model are a tetanus of 100 pulses at low-frequency (1, 10 Hz) and high-frequency (100 Hz), respectively. The $Ca^{2+}$ dynamics in the dendritic spine

($[\mathrm{Ca}^{2+}]_{\mathrm{cyto}}$) in response to a stimulation tetanus is governed by

$$[\mathrm{Ca}^{2+}]_{\mathrm{cyto}} = [\mathrm{Ca}^{2+}]_{\mathrm{rest}} + A\sum_{i=1}^{n} \exp\left(-\frac{i}{f\tau}\right), \tag{9.1}$$

where $[\mathrm{Ca}^{2+}]_{\mathrm{rest}}$ is the cytosolic $\mathrm{Ca}^{2+}$ concentration at the resting level, $A$ is the amplitude of $\mathrm{Ca}^{2+}$ concentration induced by one stimulation pulse, $\tau$ is the decay time constant, $f$ is the stimulation frequency and $n$ is the total number of the stimulation pulses [30]. The intracellular $\mathrm{Ca}^{2+}$ dynamics in response to a 100-pulse tetanus at 100 Hz is in Fig. 9.2(a). The $\mathrm{Ca}^{2+}$ level increases with time during the stimulation until it reaches a maximum level of about 20 $\mu$M, and after the last pulse, it decreases exponentially back to the resting level ($\tau = 200$ ms). We apply a high-frequency presynaptic stimulation (HFS; 100 pulses at 100 Hz) and a high-frequency presynaptic stimulation paired with postsynaptic stimulation (pairing HFS; 100 pulses at 100 Hz) to our $\mathrm{Ca}^{2+}$ model, respectively, and the results of the $\mathrm{Ca}^{2+}$ dynamics are shown in Fig. 9.2(b). We use a paired pre/postsynaptic stimulation protocol because experimental evidence showed that postsynaptic membrane depolarization triggered by presynaptic stimulation alone was not sufficient to induce LTP [31, 32]. Paired stimulation at both the presynaptic and postsynaptic neurons is used to create pairing of the EPSP and the backpropagation of action potentials (bAP), which leads to a large depolarization and $\mathrm{Ca}^{2+}$ elevation by *NMDAR* in the postsynaptic spine head [33]. When simulating the pairing HFS protocol, a 2 ms-delayed bAP is introduced after each presynaptic stimulus pulse (the formula for bAP was introduced in Chapter 3).

The result produced by Eq. 9.1 did not show any desensitization of NMDARs (Fig. 9.2(a)). This may be because of assumptions that NMDARs are not the major $\mathrm{Ca}^{2+}$ channels, or that NMDARs recover from desensitization completely between two pulses. Both of these assumptions conflict with the setting of our model that NMDARs are the major $\mathrm{Ca}^{2+}$ channels in the spine head and

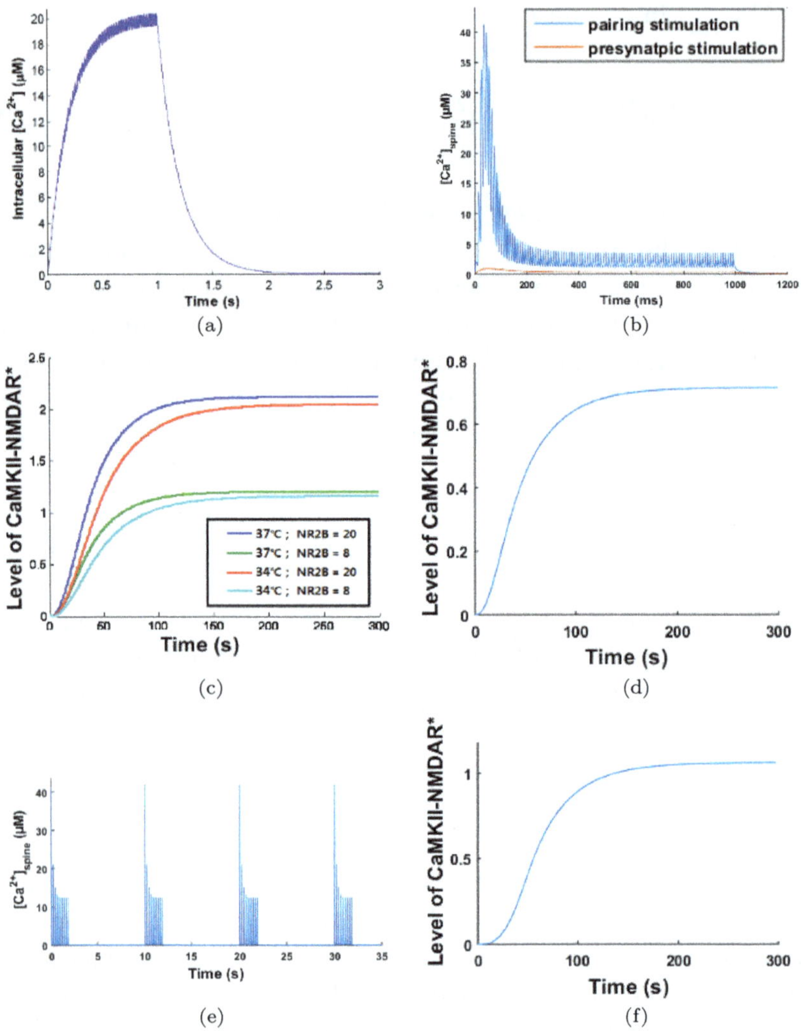


Fig. 9.2. $Ca^{2+}$ elevation and CaMKII-NMDAR complex formation in the spine head in response to presynaptic stimulation. $Ca^{2+}$ dynamics in the spine head in response to $Ca^{2+}$ by Eq. 9.1, (a) HFS, (b) pairing HFS and (e) four trains of TBS, respectively, and the corresponding CaMKII-NMDAR complex production (c, d and f).

the experimental observation that both NR2A-NMDAR and NR2B-NMDAR will be desensitized under high-frequency stimulation [15]. The $Ca^{2+}$ response of $Ca^{2+}$ model to HFS and pairing HFS reaches a peak level, decreases, then stays on a plateau at a lower level until the end of stimulation, which reflects the large desensitization of synaptic NMDAR by high-frequency stimulation (Fig. 9.2(b)). Under HFS, the elevation in $[Ca^{2+}]_{cyto}$ is much less than in the original $Ca^{2+}$ response. In contrast, under the pairing HFS, although the peak level of $[Ca^{2+}]_{cyto}$ can be above 40 $\mu$M and is higher than the maximum $[Ca^{2+}]_{cyto}$ for the original $Ca^{2+}$ response (around 20 $\mu$M), the plateau level (around 5 $\mu$M) is much lower than in the original response during the same stimulation time period.

We next compare the levels of the CaMKII-NMDAR complex in the original and our $Ca^{2+}$ results. Before simulation, we have made several adjustments before connecting the $Ca^{2+}$ model to the CaMKII ST model. In the CaMKII ST model, all proteins are in units of particle numbers (#). The time-dependent changes in the concentration of these proteins are calculated in particle numbers and all concentration-based rate constants are in $\#^{-1}s^{-1}$. Therefore, we convert the $Ca_4CaM$ concentration (in $\mu$M) from $Ca^{2+}$ model into particle numbers (in #) before calculating the translocation of CaMKII. The conversion is according to the following formula:

$$particle\ number = concentration \times N_A \times Vol,$$

where $N_A$ is the Avogadro constant ($6.022140857 \times 10^{23}$ $mol^{-1}$) and $Vol$ is the volume of the spine head (0.1 fL).

The rate constants used in the CaMKII ST model are based on 37°C, therefore, we adjusted them to 34°C using a $Q_{10}$ of 2.15 [4] to be consistent with the conditions in the previous chapters. The details of temperature correction on reaction rate constants are in Section 3.2.1. The simulation results of the CaMKII ST model show a minor difference in the level of CaMKII-NMDAR complex formation at 34°C and 37°C (Fig. 9.2 (c)).

Furthermore, we adjust the NR2B number in the CaMKII ST model from 20 to 8, which is the standard value based on our previous assumptions in Chapter 7. This decreases the level of the CaMKII-NMDAR complex by CaMKII ST model at $t = 300$ s by 1 (Fig. 9.2 (c)). In contrast, the level of the CaMKII-NMDAR complex by our $Ca^{2+}$ model in response to pairing HFS is about 0.7 (Fig. 9.2(d)) lower than in response to the original $Ca^{2+}$ input

(Fig. 9.2(c)). The difference arises because a larger fraction of NMDARs are desensitized during pairing HFS and this leads to fewer $Ca^{2+}$ ions entering the spine than in the original input in CaMKII ST. There is no CaMKII-NMDAR formation in response to presynaptic HFS alone, because of the insufficient amount of $Ca^{2+}$ ions entering the cytosol.

Moreover, we also use theta-burst stimulation (TBS) as an optional stimulation protocol in this chapter (see Chapter 2 for the detail of TBS). In a train of TBS, pulses are grouped into several bursts, and the time duration (200 ms) between two bursts allows desensitized NMDARs to partially recover. TBS is considered to be a more physiologically relevant stimulus, which is close to the frequency of the endogenous hippocampal rhythm that triggers LTP [34, 35]. One train of TBS consists of 10 stimulus bursts at 5 Hz (200 ms separation between bursts) and each burst consists of four pulses at 100 Hz. A 2 ms-delayed bAP is introduced into the model after each presynaptic stimulus pulse. Four trains of TBS are delivered at 0.1 Hz (10 s separation between trains), which is used to induce LTP experimentally [35]. The changes in $Ca^{2+}$ level in the spine head by four trains of TBS and the corresponding production of CaMKII-NMDAR complex are shown in Figs. 9.2(e) and 9.2(f), respectively. The level of the CaMKII-NMDAR complex shows a good agreement to that produced by the original $Ca^{2+}$ input (Fig. 9.2(c)).

## 9.3. Computational Experiments

In this section, we mimic disturbances in the availability of synaptic NMDAR in AD to investigate the consequential effects on CaMKII ST and study the contribution of NMDARs with different subunit compositions in CaMKII ST. For each experiment, we run the model under 1 s of pairing HFS (100 pulses at 100 Hz) and 4 trains of pairing TBS (4 TBS), respectively. In both stimulation protocols, a 2 ms-delayed bAP is introduced into the model after each presynaptic stimulus pulse. All simulations are run for 300 s.

After each simulation, we collect the $Ca^{2+}$ elevation in the spine head ($[Ca^{2+}]_{spine}$) and four key outputs from the downstream events as following:

(1) Number of $Ca_4CaM$ complexes. This is a determining factor for CaMKII activation and, subsequently, autophosphorylation.
(2) Number of autophosphorylated CaMKII subunits. Because once a CaMKII is autophosphorylated, its activity is independent from CaM and it shows persistent activity after the removal of stimulation. This is critical for the induction and maintenance of LTP, which requires a much longer time course in comparison with the stimulation.
(3) Number of CaMKII in PSD. Because CaMKII needs to be translocated into the PSD to affect its target, AMPAR, and bind to NR2B-NMDAR, once a CaMKII enters the PSD, its autophosphorylation is suggested to be irreversible [36]. The level of translocated CaMKII can be an indicator of the potential ability to form CaMKII-NMDAR complexes.
(4) Number of CaMKII-NMDAR complex in the PSD at time $t = 300$ s. The persistent localization of CaMKII in the PSD is suggested to be a critical factor for LTP induction and maintenance, and NR2B may play a key role in the synaptic CaMKII maintenance and recruitment [26]. Therefore, the level of CaMKII-NMDAR complex numbers is an important output that can provide insights into the induction and maintenance of NMDAR-LTP.

### 9.3.1. *NR2A/NR2B-NMDAR*

The numbers of NR2A-NMDAR and NR2B-NMDAR in the PSD under healthy conditions are assumed to be 12 and 8, respectively, which are consistent with the conditions in the previous chapters. We decrease the number of one of them from the standard value to zero to mimic the increasing degree of reduction in the availability of a particular NMDAR type in the PSD.

Reduction in the numbers of NR2A-NMDAR in PSD greatly reduces the maximum amplitude of $Ca^{2+}$ elevation during the first 10 stimulation pulses of pairing HFS, and this effect diminishes as more pulses arrive (Fig. 9.3(a)). The decrease in the $Ca^{2+}$ elevation level during first few stimulation pulses reduces all four other outputs (Fig. 9.3(b)). The amplitudes of $Ca^{2+}$ elevation in response to

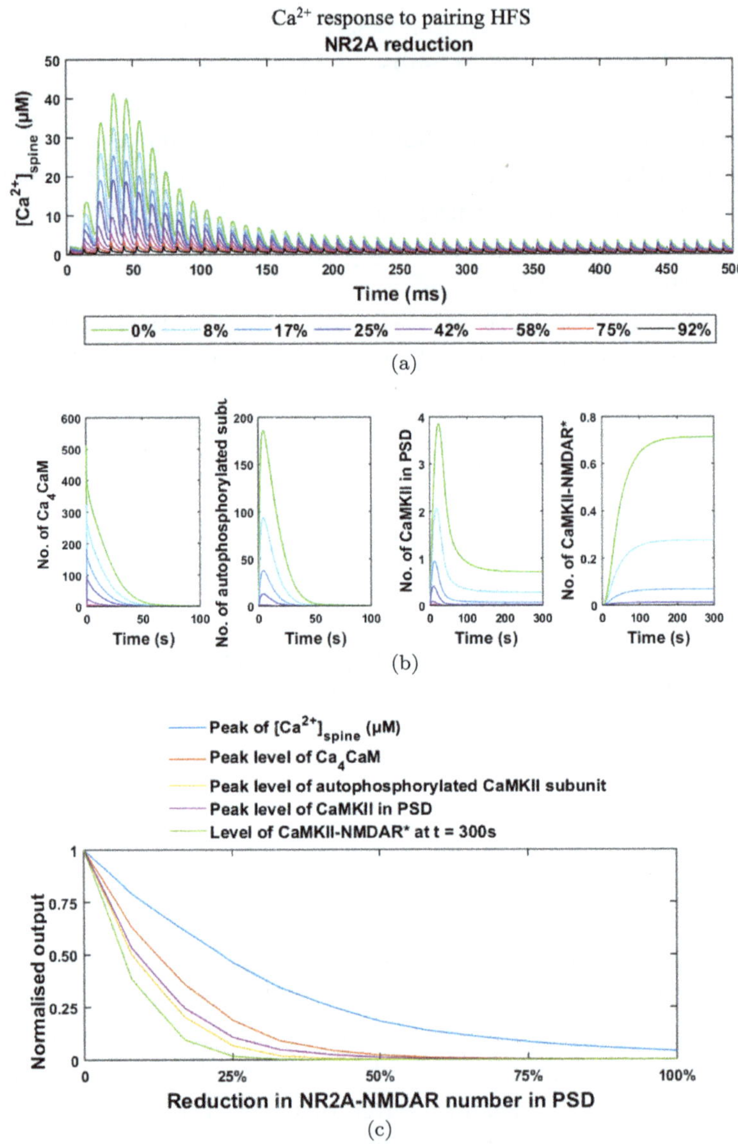


Fig. 9.3. Effects of reduction in NR2A-NMDAR numbers in response to pairing HFS. (a) $Ca^{2+}$ responses in spine head and (b) four outputs from downstream events with a different level of reduction in NR2A-NMDAR. (c) The relationship between the reduction level in NR2A-NMDAR numbers and selected typical values of the outputs. The results are normalized to those at the healthy condition (NR2A-NMDAR = 12; 0% reduction).

four TBS show similar decreases with the degree of reduction in the NR2A-NMDAR number (Fig. 9.4(a)). During both stimulation protocols, a 50% reduction (6 NR2A-NMDAR left in the PSD) can block all downstream events (Figs. 9.3(c) and 9.4(c)). Especially, the formation of CaMKII-NMDAR is mostly sensitive to NR2A-NMDAR reduction, a 25% reduction (9 NR2A-NMDAR left in the PSD) can lead to no production of CaMKII-NMDAR. Even an 8% reduction (11 NR2A-NMDAR left) can reduce the level of CaMKII-NMDAR formation by 60% (pairing HFS; Figs. 9.3(b) and 9.3(c)) and 75% (four TBS; Figs. 9.4(b) and 9.4(c)), respectively.

In contrast, a reduction in NR2B-NMDAR numbers affects all outputs more lightly than the reduction in NR2A-NMDAR numbers. Even when reducing the NR2B-NMDAR from 8 to 1 (88% reduction), the amplitude of $Ca^{2+}$ elevation only decreases by about 35% (from 42 $\mu$M to 27.3 $\mu$M) (Figs. 9.5(a) and 9.6(a)). This decrease is not able to fully block most downstream events, except for the formation of the CaMKII-NMDAR complex (Figs. 9.5(c) and 9.6(c)). The level of $Ca_4CaM$ formation, CaMKII autophosphorylation and translocation are reduced by about 65%, 78 % and 75%, respectively (Figs. 9.5(b) 9.5(c) and Figs. 9.6(b) and 9.6(c)). The formation of CaMKII-NMDAR complex is largely reduced by the reduction in NR2B-NMDAR numbers, where a 50% reduction leads to a 70% reduction in the final level CaMKII-NMDAR complex at $t = 300$ s.

### 9.3.2. *The ratio of NR2B:NR2A*

The ratio of NR2A- to NR2B-NMDAR (NR2A/NR2B ratio) is precisely regulated by the production, trafficking and degradation of NMDARs, and the different ratios are linked to the preferences in the induction of different types of synaptic plasticity [37]. Therefore, we investigate the effects of a disturbance on the NR2A/NR2B ratio in the CaMKII-NMDAR complex formation by keeping the total number of NMDARs in the PSD constant (20 NMDARs) and varying the NR2A/NR2B ratio from 1:19 to 19:1.

The amplitude of $[Ca^{2+}]_{spine}$ by stimulation increases with the NR2A/NR2B ratio (Figs. 9.7(a) and 9.8(b)). When the NR2A/NR2B

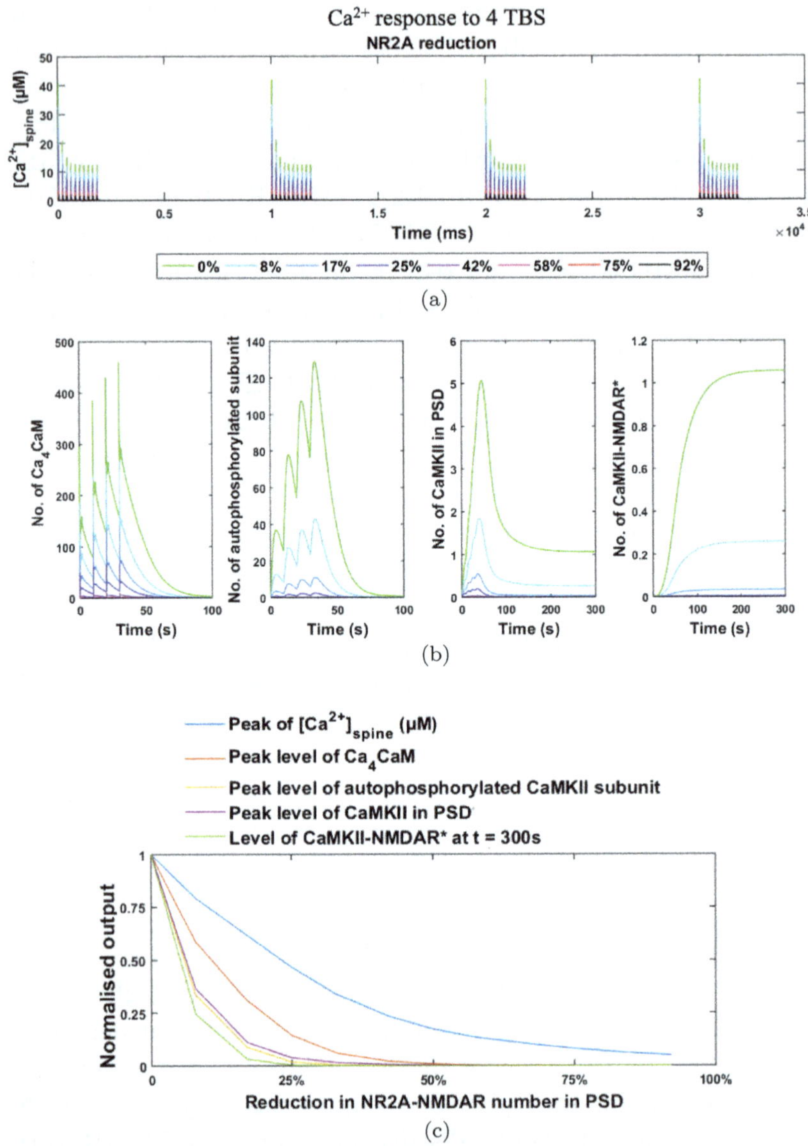


Fig. 9.4. Effects of reduction in NR2A-NMDAR numbers in response to four TBS. (a) $Ca^{2+}$ responses in spine head and (b) four outputs from downstream events with a different level of reduction in NR2A-NMDAR. (c) The relationship between the reduction level in NR2A-NMDAR numbers and selected typical values of the outputs. The results are normalized to those at the healthy condition (NR2A-NMDAR = 12; 0% reduction).

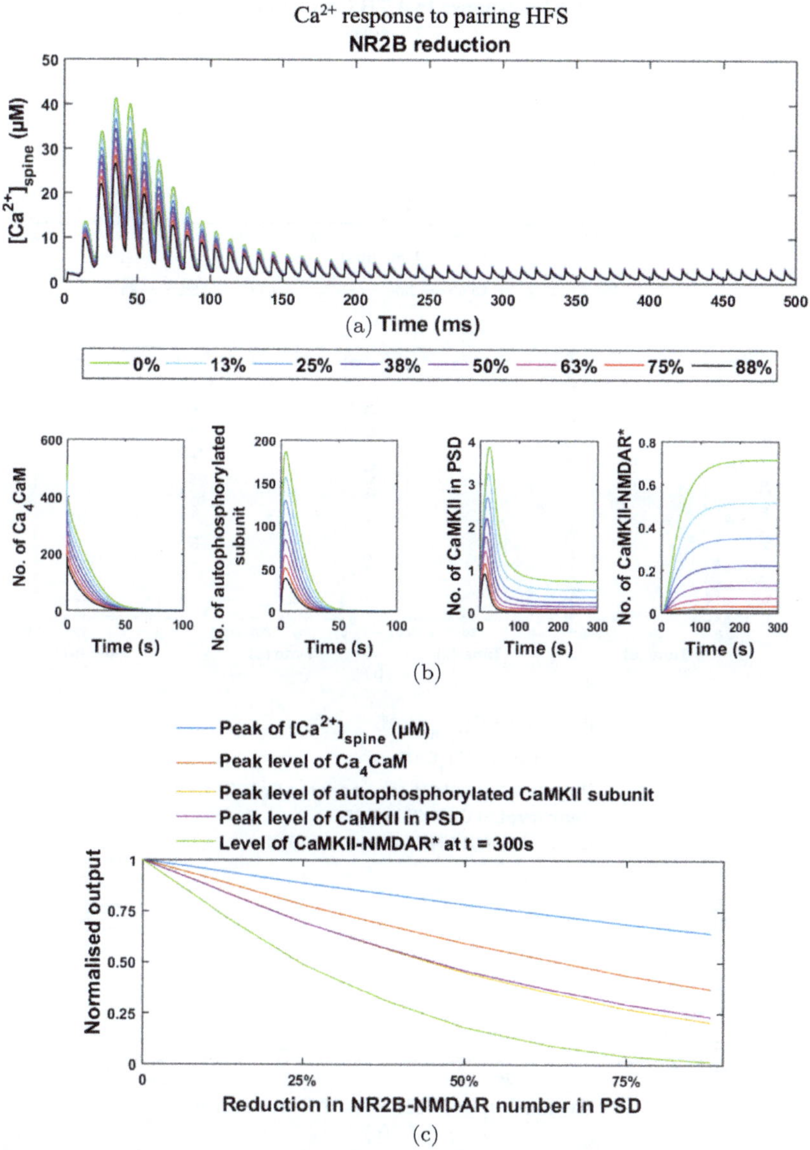


Fig. 9.5. Effects of reduction in NR2B-NMDAR numbers in response to pairing HFS. (a) $Ca^{2+}$ responses in spine head and (b) four outputs from downstream events with a different level of reduction in NR2B-NMDAR. (c) The relationship between the reduction level in NR2B-NMDAR numbers and selected typical values of the outputs. The results are normalized to those at the healthy condition (NR2B-NMDAR = 8; 0% reduction).

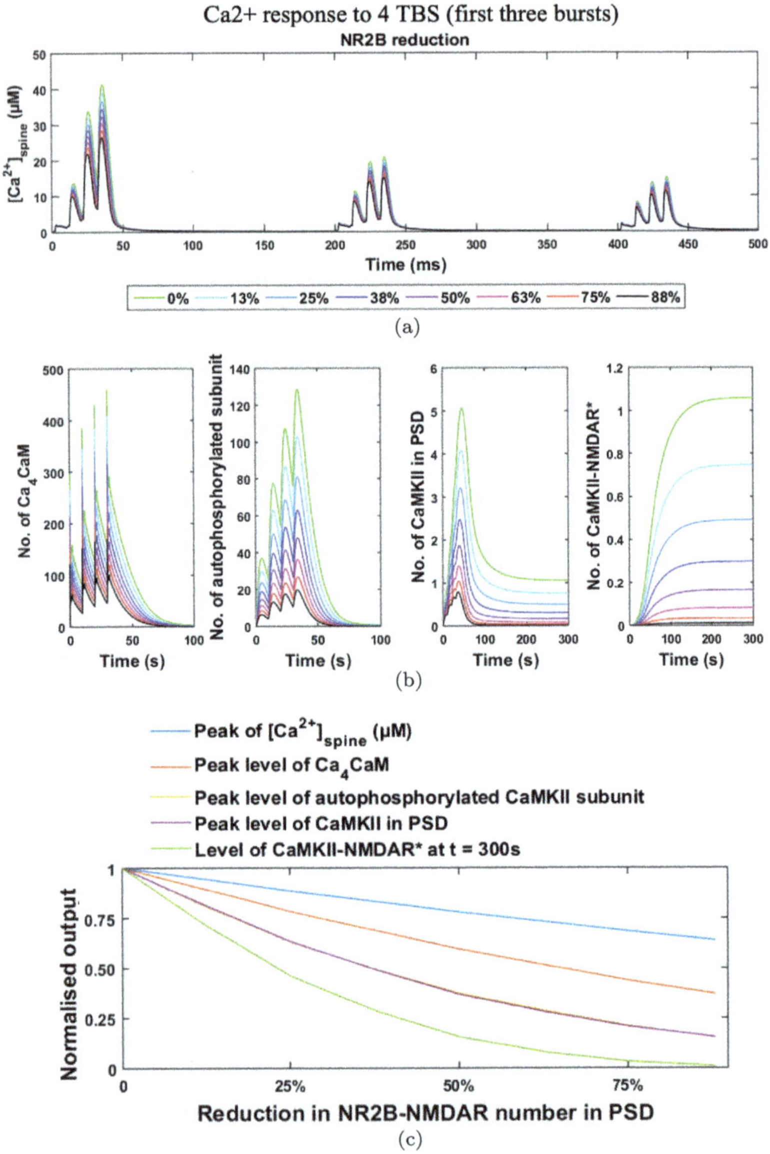


Fig. 9.6. Effects of reduction in NR2B-NMDAR numbers in response to 4 TBS. (a) $Ca^{2+}$ responses in spine head (only first three bursts are shown here to display the difference clearly). (b) Four outputs from downstream events with a different level of reduction in NR2B-NMDAR. (c) The relationship between the reduction level in NR2B-NMDAR numbers and selected typical values of the outputs. The results are normalized to those at the healthy condition (NR2B-NMDAR = 8; 0% reduction).

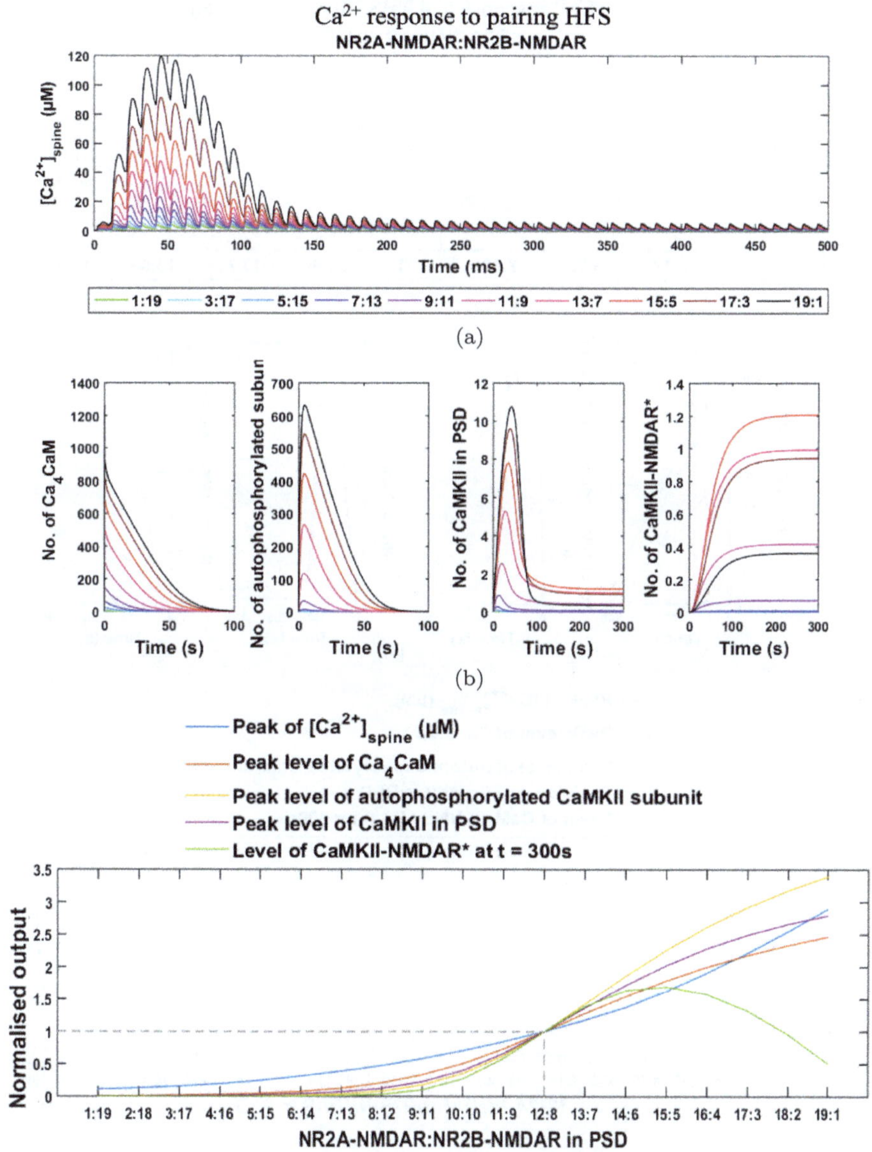


Fig. 9.7. Effects of variation in the NR2A/NR2B ratio in response to pairing HFS. (a) $Ca^{2+}$ responses in spine head and (b) four outputs from downstream events with different NR2A/NR2B ratio. (c) The relationship between the NR2A/NR2B ratio and selected typical values of the outputs. The results are normalized to those at the healthy condition (NR2A-NMDAR:NR2B-NMDAR = 12:8).

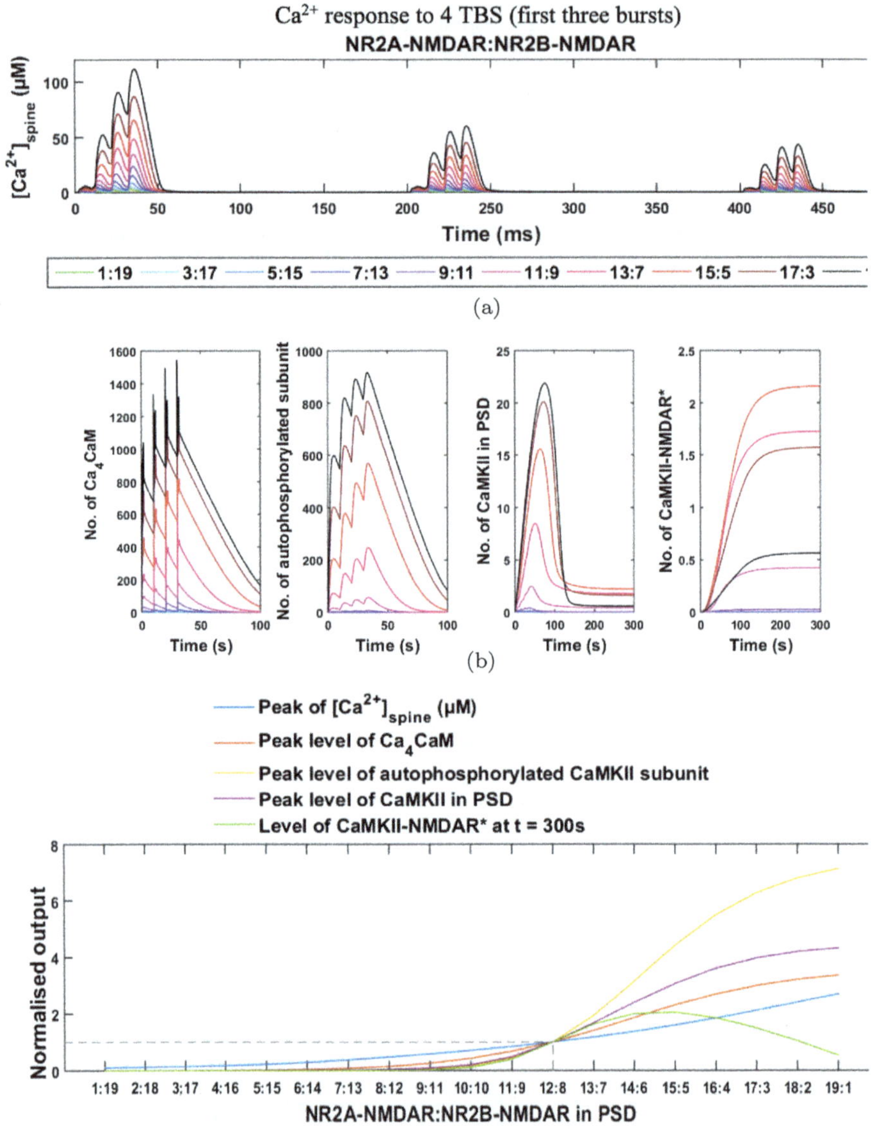


Fig. 9.8. Effects of variation in the NR2A/NR2B ratio in response to 4 TBS. (a) $Ca^{2+}$ responses in spine head (only first three bursts are shown here to display the difference clearly). (b) Four outputs from downstream events with different NR2A/NR2B ratio. (c) The relationship between the NR2A/NR2B ratio and selected typical values of the outputs. The results are normalized to those at the healthy condition (NR2A-NMDAR:NR2B-NMDAR = 12:8).

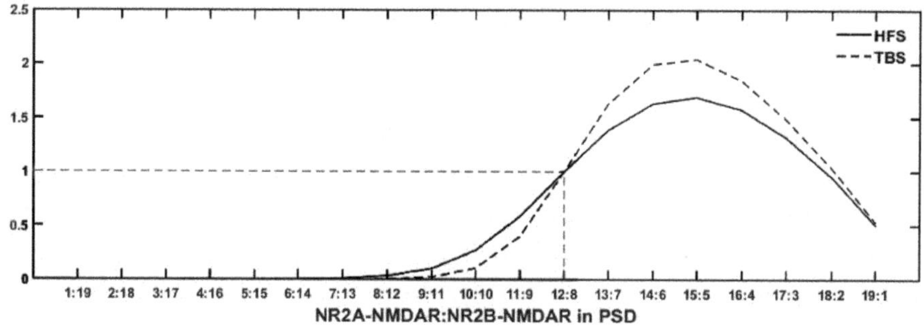


Fig. 9.9. Effect of variation in the NR2A/NR2B ratio on the level of CaMKII-NMDAR complex production.

ratio below 6:14, the peak $[Ca^{2+}]_{spine}$ in responses to each stimulation pulse are the same, and are less than with the standard NR2A/NR2B ratio (12:8). Productions of the other four outputs are blocked when the NR2A/NR2B ratio bellows 6:14 (Figs. 9.7(b) and 9.7(c) and Figs. 9.8(b) and 9.8(c)). With the increase in the NR2A/NR2B ratio peak levels of all outputs, but not the production of CaMKII-NMDAR complex, rise with different slopes in response to a pairing HFS (Fig. 9.7). The results from the four TBS simulation show similar trends, but higher increases in comparison with the ones from the pairing HFS (Fig. 9.8(c)). In contrast, the normalized final level of CaMKII-NMDAR after stimulation increases to a maximum level of 1.69 (pairing HFS) and 2.04 (pairing TBS), respectively, at the NR2A/NR2B ratio of 15:5, and decreases afterwards (Fig. 9.9). This suggests the existence of an optimal NR2A/NR2B ratio in the generation of the CaMKII-NMDAR complex.

## 9.4. Discussion and Summary

In this chapter, we extend our $Ca^{2+}$ model presented in the previous chapter with a CaMKII TS model, to investigate the roles of NMDARs in $Ca^{2+}$ transients in the dendritic spine head and the subsequent CaMKII TS in response to repeated paired pre/postsynaptic stimulation. We simulate the model under the conditions of the dysregulated level of NMDARs to study their role as $Ca^{2+}$ channels (NR2A- and NR2B-NMDAR) and as a scaffold in the PSD to anchor CaMKII. We also predict their contributions to a series of selected downstream events leading to CaMKII-NMDAR complex formation. The simulation results of the levels of the selected outputs are useful to gain insights into the different contributions of NR2A- and NR2B-NMDAR in the induction of LTP and the ability for maintenance later on. It also helps us to better understand the potential outcomes of the disturbances in NMDAR observed in AD and to explain the possible linkages between those disturbances and the pathology of the disease.

Simulation of the reduction in NR2A- and NR2B-NMDAR show negative effects on the activation, autophosphorylation and translocation of CaMKII and the formation of CaMKII-NMDAR complexes at different degrees. This suggests that both types of NMDAR are necessary for LTP formation, while contributing to its induction and maintenance in different ways. Specifically, the role of NR2A-NMDARs is to allow sufficient $Ca^{2+}$ influx to trigger the downstream $Ca^{2+}$–CaM interaction, which determines the activation of CaMKII. NR2B-NMDAR contributes less as a $Ca^{2+}$ channel than NR2A-NMDAR and itself alone is insufficient to trigger the activation of CaMKII. However, it is required to function as a scaffold in PSD that anchors CaMKII by forming a CaMKII-NMDAR complex. This is consistent with the experimental findings and hypothesis that the opening of NR2B-NMDAR may not be necessary for LTP induction [29, 38]. Simulation with different NR2A/NR2B ratios suggests that although a high NR2A/NR2B ratio facilities activation, autophosphorylation and translocation of CaMKII, it still requires a certain level of NR2B-NMDAR to anchor CaMKII in the PSD. Overall, these results provide a clearer picture that NMDAR-LTP may require cooperation between the NR2A- and NR2B-NMDAR.

The internalization of synaptic NMDAR seen in AD [39, 40] could underly some of the critical alterations in the pathology of the disease. For instance, in AD transgenic mice, it has been observed that A$\beta$ can alter the CaMKII distribution and reduce the synaptic CaMKII level, which may be the mechanism by which A$\beta$ induces the loss of synaptic AMPARs [41]. It is unclear how the distribution of

CaMKII is altered by A$\beta$. Here, the simulation results predict that a reduction in synaptic NMDARs by A$\beta$ can reduce the CaMKII-NMDAR formation in the PSD by affecting the upstream $Ca^{2+}$ response to stimulation or the scaffold function of NR2B-NMDAR. Consequently, this may contribute to the deficits of LTP and loss of synapse in AD [42, 43].

## Appendix G. Complete CaMKII ST Model and Parameters

### G.1. Reaction Rates

There are 7 ($R_1 - R_7$) and 12 reaction rates ($R_8 - R_{19}$) of the reactions governing the subunit state transitions and holoenzyme state transitions of CaMKII, respectively. All reaction rates are given in a unit of # $s^{-1}$.

#### G.1.1. *Subunit state transitions CaMKII*

$Ca^{2+}$ attaches to CaM to form CaCaM,

$$R_1 = k_{6f}\mathrm{Ca}^{2+} \cdot \mathrm{CaM} - k_{6b}\mathrm{CaCaM}. \tag{G.1}$$

$Ca^{2+}$ attaches to CaCaM to form $Ca_2CaM$,

$$R_2 = k_{7f}\mathrm{Ca}^{2+} \cdot \mathrm{CaCaM} - k_{7b}\mathrm{Ca_2CaM}. \tag{G.2}$$

$Ca^{2+}$ attaches to $Ca_2CaM$ to form $Ca_3CaM$,

$$R_3 = k_{8f}\mathrm{Ca}^{2+} \cdot \mathrm{Ca_2CaM} - k_{8b}Ca_3\mathrm{CaM}. \tag{G.3}$$

$Ca^{2+}$ attaches to $Ca_3CaM$ to form $Ca^{2+}$/CaM complex,

$$R_4 = k_{8f}\mathrm{Ca}^{2+} \cdot \mathrm{Ca_3CaM} - k_{9b}\mathrm{Ca_4CaM}. \tag{G.4}$$

$Ca^{2+}$/CaM complex binds to iCaMKII to form CaMKIICaM,

$$R_5 = k_{1f}\mathrm{Ca_4CaM} \cdot i\mathrm{CaMKII} - k_{1b}\mathrm{CaMKIICaM}. \tag{G.5}$$

Autophosphorylation of CaMKIICaM to form CaMKII*[4],

$$R_6 = \frac{K_{cat1}(CaMKIICaM.P.ATP)}{K_{m1} + ATP}, \tag{G.6}$$

where $P = 1 - (\text{iCaMKII}/\text{CaMKII}_{\text{Total}})^2$.

Dephosphorylation of CaMKII* into CaMKIICaM by PP1 [4],

$$R_7 = \frac{K_{cat2}(CaMKII^* \cdot PP1)}{K_{m2} + CaMKII^*}. \tag{G.7}$$

### G.1.2. *Holoenzyme state transitions of CaMKII*

Translocation of CaMKII into the PSD

$$R_8 = k_{2f}EoPSD\frac{\Gamma_2 + \Gamma_3}{K_{m3} + \Gamma_2 + \Gamma_3} - k_{2b}EiPSD\frac{K_{m4}}{K_{m4} + \Gamma_3}, \tag{G.8}$$

where $\Gamma_2$ and $\Gamma_3$ are the fraction of CaMKIICaM to the total number of unbound subunits and the fraction of CaMKII* to the total number of unbound subunits, respectively. ($\Gamma_1$ is the fraction of iCaMKII to the total number of unbound subunits).

$$\Gamma_1 = \frac{iCaMKII}{CaMKII_{Total} - (CaMKIIS + CaMKII^*S + CaMKII^*T},$$

$$\Gamma_2 = \frac{CaMKIICaM}{CaMKII_{Total} - (CaMKIIS + CaMKII^*S + CaMKII^*T},$$

$$\Gamma_3 = \frac{CaMKII^*}{CaMKII_{Total} - (CaMKIIS + CaMKII^*S + CaMKII^*T},$$

and

$$\Gamma_1 + \Gamma_2 + \Gamma_3 = 1.$$

Binding between CaMKII and NMDAR to form CaMKII-NMDAR complex:

$$R_9 = k_{3f}\Omega_1 EiPSD.NMDAR - k_{3ba}CaMKIIS, \tag{G.9}$$

$$R_{10} = k_{3f}\Omega_2 EiPSD.NMDAR - k_{3bb}CaMKII^*S, \tag{G.10}$$

where $\Omega_1$ and $\Omega_2$ are the modifier for average formation rate of CaMKIIS with respect to 91 possible conformations of CaMKII and the modifier for average formation rate of CaMKII*S with respect to 91 possible conformations of CaMKII,

$$\Omega_1 = \sum_{n=1}^{12} \frac{12!}{n!\,(12-n)!}\Gamma_2^n \sum_{m=0}^{12-n} \frac{12!}{m!\,(12-n-m)!}\Gamma_3^m \Gamma_1^{12-m-n}$$

$$\times \frac{1-P_b}{(m-1)\,P_b+1} \cdot \frac{n}{(n-1)\,P_b+1},$$

$$\Omega_2 = \sum_{m=1}^{12} \frac{12!}{m!\,(12-m)!}\Gamma_3^m \sum_{n=0}^{12-m} \frac{(12-m)!}{n!\,(12-n-m)!}\Gamma_2^n \Gamma_1^{12-m-n}$$

$$\times \frac{1-P_b}{(n-1)\,P_b+1} \cdot \frac{m}{(m-1)\,P_b+1},$$

where $n$ is the number of CaM-pBS and $m$ is the number of T286-pBS, $P_b$ is the probability of $S$ site-mediated binding between CaMKII and NMDAR and

$$P_b = \frac{CaMKIIS{+}CaMKII^{*}S}{NMDAR_{Total}}.$$

Transfers of the $S$ site binding to the $T$ site

$$R_{11} = k_4 CaMKIIS, \tag{G.11}$$

$$R_{12} = k_4 CaMKII^{*}S. \tag{G.12}$$

Autophosphorylating CaMKIIS to CaMKII*S,

$$R_{13} = \frac{K_{cat} CaMKIIS.P.ATP}{K_{m1}{+}ATP}. \tag{G.13}$$

CaMKII turnover

$$R_{14} = k_5 CaMKIICaM, \tag{G.14}$$

$$R_{15} = k_5 CaMKII^{*}, \tag{G.15}$$

$$R_{16} = k_5 EiPSD, \tag{G.16}$$

$$R_{17} = k_5 CaMKIIS, \tag{G.17}$$

$$R_{18} = k_5 CaMKII^{*}S, \tag{G.18}$$

$$R_{19} = k_5 CaMKII^{*}T. \tag{G.19}$$

## G.2. ODEs

The time-dependent changes of the states of CaMKII are governed by 10 ODEs based on the above 19 reaction rates:

$Ca^{2+}$/CaM complex formation:

$$\frac{dCaCaM}{dt} = R_1 - R_2, \tag{G.20}$$

$$\frac{dCa_2CaM}{dt} = R_2 - R_3, \tag{G.21}$$

$$\frac{dCa_3CaM}{dt} = R_3 - R_4, \tag{G.22}$$

$$\frac{dCa_4CaM}{dt} = R_4 - R_5. \tag{G.23}$$

Changes in IS of CaMKII:

$$\frac{dCaMKIICaM}{dt} = R_5 - R_6 + R_7 - R_{14}, \tag{G.24}$$

$$\frac{dCaMKII^*}{dt} = R_6 - R_7 - R_{15}. \tag{G.25}$$

CaMKII-NMDAR formation and changes in HS of CaMKII:

$$\frac{dEiPSD}{dt} = R_8 - R_9 - R_{10} - R_{16}, \tag{G.26}$$

$$\frac{dCaMKIIS}{dt} = R_9 - R_{11} - R_{13} - R_{17}, \tag{G.27}$$

$$\frac{dCaMKII^*S}{dt} = R_{10} - R_{12} + R_{13} - R_{18}, \tag{G.28}$$

$$\frac{dCaMKII^*T}{dt} = R_{11} + R_{12} - R_{19}. \tag{G.29}$$

Constraints due to the conserved molecular numbers:

$$\begin{aligned} iCaMKII = {} & CaMKII_{Total} - CaMKIICaM - CaMKII^* \\ & - CaMKIIS - CaMKII^*S - CaMKII^*T, \end{aligned}$$

$$CaM = CaM_{Total} - CaCaM - Ca_2CaM - Ca_3CaM - Ca_4CaM - (CaMKII_{Total} - iCaMKII),$$

$$NMDAR = NMDAR_{Total} - CaMKIIS - CaMKII^*S - CaMKII^*T,$$

$$EoPSD = \frac{CaMKII_{Total}}{12} - EiPSD - (NMDAR_{Total} - NMDAR).$$

## G.3. Variables

Definition of the states/variables associated with inner state transition of CaMKII subunit of CaMKII.

| States (#) | Definition |
|---|---|
| $Ca^{2+}$ | Intracellular $Ca^{2+}$ ion |
| CaCaM | One $Ca^{2+}$ ion bound to CaM |
| Ca2CaM | Two $Ca^{2+}$ ions bound to CaM |
| Ca3CaM | Three $Ca^{2+}$ ions bound to CaM |
| Ca4CaM | $Ca^{2+}$/CaM complex |
| iCaMKII | Inhibited CaMKII subunit |
| CaMKIICaM | CaM bound CaMKII subunit |
| CaMKII* | Autophosphorylated CaMKII subunit |

## G.4. Parameters and Constants

Parameters and constants: their biological meaning, values and sources where the values are obtained.

| Symbol | Biological meaning | Value | Ref. |
|---|---|---|---|
| $k_{1f}$ | Binding rate of $Ca_4CaM$ and iCaMKII | $0.035\ \#^{-1}s^{-1}$ | [4] |
| $k_{1b}$ | Dissociation rate of CaMKIICaM into $Ca_4CaM$ and iCaMKII | $0.14\ s^{-1}$ | |
| $k_{2f}$ | Translocation rate of EoPSD into PSD | $0.0088\ s^{-1}$ | [7] |
| $k_{2b}$ | Dissociation rate of EiPSD out of PSD | $0.247\ s^{-1}$ | |
| $k_{3f}$ | Binding rate of elementary CaMKII to NMDAR at the S site | $0.008\ \#^{-1}s^{-1}$ | |
| $k_{3ba}$ | Dissociation rate of CaMKIIS into EiPSD and NMDAR | $0.38\ s^{-1}$ | |

(*Continued*)

(*Continued*)

| Symbol | Biological meaning | Value | Ref. |
|---|---|---|---|
| $k_{3bb}$ | Dissociation rate of CaMKII*S into EiPSD and NMDAR | $0.024\ s^{-1}$ | |
| $k_4$ | Transfer rate of NMDAR binding from S site to T site | $0.01\ s^{-1}$ | |
| $k_5$ | Turnover rate of CaMKII | $1/108000\ s^{-1}$ | [8] |
| $k_{6f}$ | Binding rate of $Ca^{2+}$ and CaM to form CaCaM | $0.0415\ \#^{-1}s^{-1}$ | [4] |
| $k_{6b}$ | Dissociation rate of CaCaM into $Ca^{2+}$ and CaM | $50\ s^{-1}$ | |
| $k_{7f}$ | Binding rate of $Ca^{2+}$ and CaCaM to form $Ca_2CaM$ | $1.45\ \#^{-1}s^{-1}$ | |
| $k_{7b}$ | Dissociation rate of $Ca_2CaM$ into $Ca^{2+}$ and CaCaM | $50\ s^{-1}$ | |
| $k_{8f}$ | Binding rate of $Ca^{2+}$ and $Ca_2CaM$ to form $Ca_3CaM$ | $0.2\ \#^{-1}s^{-1}$ | |
| $k_{8b}$ | Dissociation rate of $Ca_3CaM$ into $Ca^{2+}$ and $Ca_2CaM$ | $1250\ s^{-1}$ | |
| $k_{9f}$ | Binding rate of $Ca^{2+}$ and $Ca_3CaM$ to form $Ca_4CaM$ | $4.15\ \#^{-1}s^{-1}$ | |
| $k_{9b}$ | Dissociation rate of $Ca_4CaM$ into $Ca^{2+}$ and $Ca_3CaM$ | $1250\ s^{-1}$ | |
| $K_{cat1}$ | Autophosphorylation rate of CaMKIICaM into CaMKII* | $0.9\ s^{-1}$ | |
| $K_{cat2}$ | Dephosphorylation rate of CaMKII* by PP1 | $1.72\ s^{-1}$ | |
| $K_{m1}$ | Michaelis constant of the autophosphorylation | $1150\ \#^{-1}$ | |
| $K_{m2}$ | Michaelis constant of the dephosphorylation by PP1 | $660\ \#^{-1}$ | |
| $K_{m3}$ | Fraction of active CaMKII subunits at half maximal translocation rate | 0.39 | [7] |
| $K_{m4}$ | Fraction of autophosphorylated CaMKII subunits at half maximal dissociation rate | 0.019 | |
| $CaMKII_{Total}$ | Total number of CaMKII subunits | 1200 # | [11] |
| $CaM_{Total}$ | Total number of CaM molecule | Consistent with the $Ca^{2+}$ model | |
| $NMDAR_{Total}$ | Total number of NMDAR (or NR2B) | Consistent with the $Ca^{2+}$ model | |
| ATP | Total number of ATP | 240000 # | [13] |
| PP1 | Total number of PP1 | 145 | [7] |

## References

[1] Isaac J.T., Nicoll R.A. and Malenka R.C. (1995). Evidence for silent synapses: implications for the expression of LTP. *Neuron*, *15*(2), pp. 427–434.

[2] Durand G.M., Kovalchuk Y. and Konnerth A. (1996). Long-term potentiation and functional synapse induction in developing hippocampus. *Nature*, *381*(6577), pp. 71–75.

[3] Lüscher C. and Malenka R.C. (2012). NMDA receptor-dependent long-term potentiation and long-term depression (LTP/LTD). *Cold Spring Harbor Perspectives Biol.*, *4*(6), p. a005710.

[4] Chiba H., Schneider N.S., Matsuoka S. and Noma A. (2008). A simulation study on the activation of cardiac CaMKII $\delta$-isoform and its regulation by phosphatases. *Biophys. J.*, *95*(5), pp. 2139–2149.

[5] Barria A. and Malinow R. (2005). NMDA receptor subunit composition controls synaptic plasticity by regulating binding to CaMKII. *Neuron*, *48*(2), pp. 289–301.

[6] Paoletti P., Bellone C. and Zhou Q. (2013). NMDA receptor subunit diversity: impact on receptor properties, synaptic plasticity and disease. *Nat. Rev. Neurosci.*, *14*(6), pp. 383–400.

[7] He Y., Kulasiri D. and Samarasinghe S. (2015). Modelling the dynamics of CaMKII–NMDAR complex related to memory formation in synapses: The possible roles of threonine 286 autophosphorylation of CaMKII in long term potentiation. *J. Theoret. Biol.*, *365*, pp. 403–419.

[8] Ehlers M.D. (2003). Activity level controls postsynaptic composition and signaling via the ubiquitin-proteasome system. *Nature Neurosci.*, *6*(3), pp. 231–242.

[9] Liu L. *et al.* (2004). Role of NMDA receptor subtypes in governing the direction of hippocampal synaptic plasticity. *Science (N.Y.)*, *304*(5673), pp. 1021–1024.

[10] Massey P.V. *et al.* (2004). Differential roles of NR2A and NR2B-containing NMDA receptors in cortical long-term potentiation and long-term depression. *J. Neurosci.*, *24*(36), pp. 7821–7828.

[11] Ribrault C., Sekimoto K. and Triller A. (2011). From the stochasticity of molecular processes to the variability of synaptic transmission. *Nature Revi. Neurosci.*, *12*(7), pp. 375–387.

[12] Tang Y.-P. *et al.* (1999). Genetic enhancement of learning and memory in mice [10.1038/43432]. *Nature*, *401*(6748), pp. 63–69.

[13] Coultrap S.J. and Bayer K.U. (2012). CaMKII regulation in information processing and storage. *Trends Neurosci.*, *35*(10), pp. 607–618.

[14] Clayton D.A., Mesches M.H., Alvarez E., Bickford P.C. and Browning M.D. (2002). A hippocampal NR2B deficit can mimic age-related changes in long-term potentiation and spatial learning in the Fischer 344 rat. *J. Neurosci.*, *22*(9), pp. 3628–3637.

[15] Erreger K., Dravid S.M., Banke T.G., Wyllie D.J.A. and Traynelis S.F. (2005). Subunit-specific gating controls rat NR1/NR2A and NR1/NR2B NMDA channel kinetics and synaptic signalling profiles. *J. Physiol.*, *563*(Pt 2), pp. 345–358.

[16] Malenka R.C. and Nicoll A.R. (1999). Long-term potentiation — a decade of progress? *Science (N.Y.)*, *285*(5435), pp. 1870–1874.
[17] Lisman J., Yasuda R. and Raghavachari S. (2012). Mechanisms of CaMKII action in long-term potentiation. *Nature Revi. Neurosci.*, *13*(3), pp. 169–182.
[18] Lee S.J., Escobedo-Lozoya Y., Szatmari E.M. and Yasuda R. (2009). Activation of CaMKII in single dendritic spines during long-term potentiation. *Nature*, *458*(7236), pp. 299–304.
[19] Fukunaga K., Muller D. and Miyamoto E. (1995). Increased phosphorylation of Ca/Calmodulin-dependent protein kinase II and its endogenous substrates in the induction of long term potentiation. *J. Biol. Chem.*, *270*(11), pp. 6119–6124.
[20] Strack S. and Colbran R.J. (1998). Autophosphorylation-dependent targeting of calcium/calmodulin-dependent protein kinase II by the NR2B subunit of the N-Methyl-D-aspartate Receptor. *J. Biol. Chem.*, *273*(33), pp. 20689–20692.
[21] Merrill M.A., Chen Y., Strack S. and Hell J.W. (2005). Activity-driven postsynaptic translocation of CaMKII. *Trends Pharmacological Sci.*, *26*(12), pp. 645–653.
[22] Lisman J., Schulman H. and Cline H. (2002). The molecular basis of CaMKII function in synaptic and behavioural memory. *Nature Rev. Neurosci.*, *3*(3), pp. 175–190.
[23] Kennedy M.B. *et al.* (1990). Structure and regulation of type II calcium/calmodulin-dependent protein kinase in central nervous system neurons. *Cold Spring Harbor Symposia Quantitative Biology*, *55*, pp. 101–110.
[24] Hudmon A. and Schulman H. (2002). Structure–function of the multifunctional $Ca^{2+}$/calmodulin-dependent protein kinase II. *Biochem. J.*, *364*(3), pp. 593–611.
[25] Derkach V., Barria A. and Soderling T.R. (1999). Ca2+/calmodulin-kinase II enhances channel conductance of $\alpha$-amino-3-hydroxy-5-methyl-4-isoxazolepropionate type glutamate receptors. *Proc. Natl. Acad. Sci. USA*, *96*(6), pp. 3269–3274.
[26] Bayer K.U. *et al.* (2006). Transition from reversible to persistent binding of CaMKII to postsynaptic sites and NR2B. *J. Neurosci.*, *26*(4), pp. 1164–1174.
[27] Sanhueza M. and Lisman J. (2013). The CaMKII/NMDAR complex as a molecular memory. *Mol. Brain*, *6*(1), p. 1.
[28] Lisman J.E. and Zhabotinsky A.M. (2001). A model of synaptic memory: a CaMKII/PP1 switch that potentiates transmission by organizing an AMPA receptor anchoring assembly. *Neuron*, *31*(2), pp. 191–201.
[29] Foster K.A. *et al.* (2010). Distinct roles of NR2A and NR2B cytoplasmic tails in long-term potentiation. *J. Neurosci.*, *30*(7), pp. 2676–2685.
[30] Zhabotinsky A.M. (2000). Bistability in the $Ca^{2+}$/calmodulin–dependent protein kinase-phosphatase system. *Biophys. J.*, *79*(5), pp. 2211–2221.
[31] Pike F.G., Meredith R.M., Olding A.W.A. and Paulsen O. (1999). Postsynaptic bursting is essential for 'Hebbian' induction of associative long-term potentiation at excitatory synapses in rat hippocampus. *J. Physiol.*, *518*(Pt 2), pp. 571–576.

[32] Mansvelder H.D. and McGehee D.S. (2000). Long-term potentiation of excitatory inputs to brain reward areas by nicotine. *Neuron, 27*(2), pp. 349–357.
[33] Caporale N. and Dan Y. (2008). Spike timing-dependent plasticity: a Hebbian learning rule. *Ann. Rev. Neurosci., 31*, pp. 25–46.
[34] Raymond C.R. (2007). LTP forms 1, 2 and 3: different mechanisms for the 'long' in long-term potentiation. *Trends Neurosci., 30*(4), pp. 167–175.
[35] Lee H.-K., Barbarosie M., Kameyama K., Bear M.F. and Huganir R.L. (2000). Regulation of distinct AMPA receptor phosphorylation sites during bidirectional synaptic plasticity. *Nature, 405*(6789), pp. 955–959.
[36] Mullasseril P., Dosemeci A., Lisman J.E. and Griffith L.C. (2007). A structural mechanism for maintaining the 'on-state'of the CaMKII memory switch in the post-synaptic density. *J. Neurochem., 103*(1), pp. 357–364.
[37] Yashiro K. and Philpot B.D. (2008). Regulation of NMDA receptor subunit expression and its implications for LTD, LTP, and metaplasticity. *Neuropharmacology, 55*(7), pp. 1081–1094.
[38] Kim M.J., Dunah A.W., Wang Y.T. and Sheng M. (2005). Differential roles of NR2A- and NR2B-containing NMDA receptors in Ras-ERK signaling and AMPA receptor trafficking. *Neuron, 46*(5), pp. 745–760.
[39] Li S. *et al.* (2011). Soluble Abeta oligomers inhibit long-term potentiation through a mechanism involving excessive activation of extrasynaptic NR2B-containing NMDA receptors. *J. Neurosci., 31*(18), pp. 6627–6638.
[40] Snyder E.M. *et al.* (2005). Regulation of NMDA receptor trafficking by amyloid-beta. *Nature Neurosci., 8*(8), pp. 1051–1058.
[41] Gu Z., Liu W. and Yan Z. (2009). $\beta$-Amyloid impairs AMPA receptor trafficking and function by reducing $Ca(^{2+})$/calmodulin-dependent protein kinase II synaptic distribution. *J. Biol. Chem., 284*(16), pp. 10639–10649.
[42] Townsend M., Mehta T. and Selkoe D.J. (2007). Soluble A$\beta$ inhibits specific signal transduction cascades common to the insulin receptor pathway. *J. Biol. Chem., 282*(46), pp. 33305–33312.
[43] Koffie R.M., Hyman B.T. and Spires-Jones T.L. (2011). Alzheimer's disease: synapses gone cold. *Mol. Neurodegeneration, 6*(1), p. 1.

# Chapter 10

# Quantification of Synaptic Plasticity Towards Better Understanding of AD

## 10.1. Introduction

As we have seen before, neurons and synapses are interconnected in a complex nervous network for information processing and memory storage. The network is robust as the behavioral responses to environmental clues are rapid and precise, yet flexible to allow for the memory to be maintained for the desired duration [1]. The robustness and the flexibility of the nervous network are hypothesized to be contributed to by the "plastic" nature of the synaptic connections between two neurons [2]. The "plastic" nature denotes the modulable strength of the synapse dominated by the experience-dependent modifications in response to environmental clues [3, 4].

Synaptic plasticity can be hypothesized to be a process that dynamically regulates the strength of a synapse, which reflects its capacity to transmit the neuronal information, in response to environmental clues. As we have seen previously, the regulation is not limited to the properties of the synapse; it also controls the formation and removal of synapses in response to particular signals [5, 6]. Importantly, the formation of new synapses was observed in animal studies following learning activities to form new memories [7, 8]. Moreover, AD exhibits a strong disruption on the modulators of synaptic plasticity leading to synapse and memory losses [9, 10]. These observations revealed the essential role of synaptic plasticity in triggering memory formation. Thus, it is imperative to understand the roles of the main modulators as well as the molecular mechanisms of this remarkable emergent synaptic property [3].

With the abundant experimental data of the individual synaptic components, the modeling of synaptic plasticity enables the integration of the fragmented information of the synaptic functions into a system level view to provide insights into the roles and the interactions of the main synaptic modulators contributing to the behaviors of synaptic plasticity [11]. The existing models related to synaptic plasticity encapsulate a large amount of details of the synaptic interactions and help us to understand the structural and behavioral characteristics of synaptic plasticity [12]. The realistic modeling of synaptic plasticity is improved by an increasing number of synaptic components in each successive modeling exercise: for example, the Hayer and Bhalla model [13] contains 256 variables as the extreme case. However, little attention has been paid to the clarification of the essential modulators from the modeled synaptic components and the quantification of their collaborations in orchestrating synaptic plasticity.

In this chapter, we explain the main postsynaptic modulators of the N-methyl-D-aspartate receptor (NMDAR)-dependent synaptic plasticity and quantify the synaptic plasticity behaviors through the Buckingham $\pi$ theorem into a compact index, named synaptic plasticity index (SPI), in terms of the relationships among the clarified main postsynaptic modulators. The Buckingham $\pi$ theorem is a common method for dimensionless analysis of a given set of variables into a physically meaningful function without the need for the exact relationships among them. The theorem works by generating sets of dimensionless parameters based on the given variables in order to acquire balance in all the basic physical dimensions of which the given variables are composed (the details of the Buckingham $\pi$ theorem can be found in [14]). SPI, which links the synaptic strength with the dynamic interactions of the main postsynaptic modulators, is useful to investigate the general collaborations and the roles of the main postsynaptic modulators in the emergence of synaptic plasticity. Furthermore, due to its dimensionless nature, the index can be considered as a broad measure that can be used to predict the general design logics governing synaptic plasticity. Apart

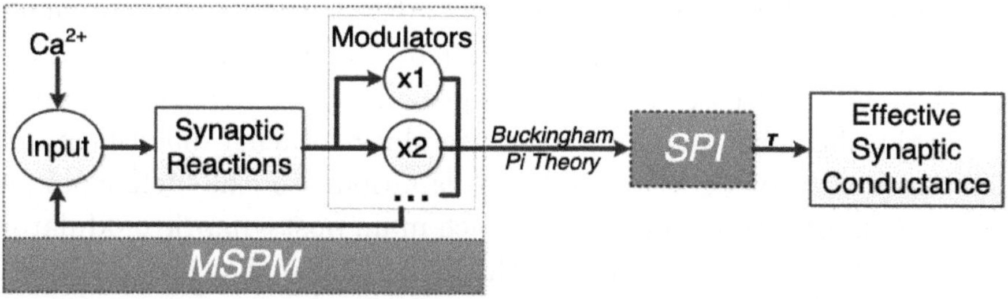


Fig. 10.1. Schematic Diagram for Calculation of SPI based on MSPM. MSPM captures the dynamics of the postsynaptic modulators in response to a $Ca^{2+}$ signal through the synaptic reactions driven simulation. The captured dynamics of the postsynaptic modulators are converted (based on the Buckingham $\pi$ theorem) into SPI. The resulting SPI can be used to predict effective synaptic conductance, an indicator of synaptic strength, through an expression in terms of $\tau$, which is derived from the AMPAR trafficking time courses.

from the contributions to investigate synaptic plasticity, SPI itself can be used as an alternative measure of synaptic plasticity.

We also develop a new $Ca^{2+}$/calmodulin-dependent mathematical model of synaptic plasticity, named MSPM (model of synaptic plasticity modulators), based on the main postsynaptic modulators as the variables to provide necessary data to calculate SPI (see the schematic diagram in Fig. 10.1) and we validate it against the behavioral predictions from the model. In developing this model, our guiding principle is the balance between the believability and the minimality, which are introduced as the key criteria of a good model by Casti [15]. In other words, MSPM should involve the least number of the synaptic components (the essential postsynaptic modulators), but mimics the behaviors of synaptic plasticity to the maximum extent. Interestingly, the new model shares a structure close to the conceptual models [16–18], which reveal the essential requirements for the bi-directional characteristics of synaptic plasticity, but exhibit very similar dynamics to the comprehensive synaptic signaling models [13, 19, 20] that are developed on and validated against the experimental results. Results of the validation show consistency between SPI and the model predictions showing that SPI can be used successfully to quantitatively describe the internal protein–protein interactions related to synaptic plasticity.

At the end of chapter, we analyze SPI on the roles of the main postsynaptic modulators and use SPI to understand the mutations in Alzheimer's disease. The analysis provides insights into the core of synaptic plasticity and shows the usefulness of such quantification. The results provide reasonable explanations to the behaviors of synaptic plasticity with respect to each main postsynaptic modulator and reveal the possible causes of the memory loss that are consistent with the $Ca^{2+}$ hypotheses of Alzheimer's disease [21, 22].

## 10.2. Main Postsynaptic Modulators Initiating NMDAR-Dependent Synaptic Plasticity

Synaptic plasticity integrates the dynamics as well as the interactions of the three regions of the synapse. The integration results in the alterations on the properties of the synapse, including the presynaptic neurotransmitter releasing probability and the postsynaptic response to the neurotransmitters [1, 4, 23]. All of these properties strongly influence the capacity to transmit neuronal information and serve as the key indicators of synaptic strength. Therefore, the core of synaptic plasticity relies on the mechanisms to modify these key indicators. These mechanisms take place across the whole synapse revealing the complexity of synaptic plasticity [1, 24]. However, the dominant expression site of synaptic plasticity seems to be dependent on the frequency or the pattern of the inducing stimulation [3, 4]. In this study, we keep our focus on the postsynaptic neuron because the synaptic plasticity activities were originally observed in the hippocampal neurons [2, 25, 26], and a number of studies show that the major form in this area, the NMDAR-dependent synaptic plasticity, is likely to express postsynaptically [3, 4].

A large amount of activities take place in the synapse. In the presynaptic neuron, the dominant activities are the synthesis of neurotransmitters and their release into the synaptic cleft in response to the neuronal signal (often to be an action potential). The released neurotransmitters diffuse across the synaptic cleft and bind to the postsynaptic receptors. The neurotransmitter-receptor binding activities in the synaptic cleft bridge the pre- and postsynaptic neurons, and hence link the postsynaptic reaction to the presynaptic response. In the postsynaptic neuron, the receptor property is tightly regulated in order to react precisely with respect to the presynaptic response. All these activities are dynamically up- or down-regulated to give flexibility to the synaptic strength that forms the base of synaptic plasticity [1, 23].

A-amino-3-hydroxy-5-methyl-4-isoxazole propionic acid receptor (AMPAR) is the major postsynaptic receptor in the hippocampus CA1 neurons to transmit neuronal information. Upon binding to the neurotransmitters, AMPAR allows cations (mostly $Na^+$ and $K^+$, a small amount of $Ca^{2+}$) to enter and depolarize the membrane of the postsynaptic neuron to generate an excitatory postsynaptic potential (EPSP). Strong EPSP beyond a threshold triggers an action potential that initiates the transmission process to the next neuron [1]. The magnitude of the EPSP is dependent on the postsynaptic depolarization, which is controlled through the regulation of the AMPAR property [23]. Hence, the regulatory mechanisms of the AMPAR property are essential to controlling synaptic plasticity in the hippocampal CA1 neurons.

Two major factors of the AMPAR property are important for regulating the postsynaptic response to the neurotransmitters: the activity and the single channel conductance. AMPAR is a mobile receptor; it normally recycles between the cytoplasm and the membrane of the postsynaptic neuron [23, 27–30]. AMPAR becomes active once stabilized in the postsynaptic density (PSD), which is a protein rich region in the postsynaptic neuron membrane. The stabilization of AMPAR in the PSD requires binding to PSD-docking proteins, such as *PSD-95* [31]. AMPARs in the PSD (PSD AMPAR) open the embedded ion channel to allow the cations to pass through the membrane [1, 27]. Therefore, the number of PSD AMPAR and their associated average single channel conductance determine the rate of the cations' flow as well as the strength of the resulting postsynaptic depolarization. These properties of AMPAR are regulated through phosphorylation at two sites of AMPAR, residues *ser831* and *ser845*, and the phosphorylation is regulated by the NMDAR-dependent postsynaptic process [23, 32]. The postsynaptic depolarization caused by cations' influx not only triggers the action potential, but also repels the $Mg^{2+}$ blocking of NMDAR [33]. Free from the $Mg^{2+}$ blocking, NMDAR becomes active and allows a transient $Ca^{2+}$ influx, when bound to a neurotransmitter, that leads to a rapid elevation in the intracellular $Ca^{2+}$ level [23, 33, 34]. Depending on the magnitude of the elevation, $Ca^{2+}$ selectively activates two different subsequent pathways mediated by calmodulin (*CaM*) leading to different modifications of the AMPAR residues [23, 35].

$Ca^{2+}$ selectively forms a switch that dynamically up- or down-regulates the synaptic strength with respect to the magnitude of the $Ca^{2+}$ elevations following the activation of NMDAR (Fig. 10.2(a)) [36–38].

A modest elevation activates the high $Ca^{2+}$ affinity calcineurin ($PP2B$), which then dephosphorylates inhibitor-1 ($I1$) [39, 40]. The phosphorylated $I1$ binds and inhibits protein phosphatase 1 ($PP1$), while the $PP2B$-dependent dephosphorylation of $I1$ frees the bound $PP1$ [41]. The released $PP1$ dephosphorylates both *ser831* and *ser845* of PSD AMPAR that results in the decrease of the single channel conductance and the enhanced internalization of PSD AMPAR via endocytosis, respectively [39, 42, 43]. In addition, recent evidences show that *PP2B* is required to enhance the endocytosis of AMPAR, but the exact mechanisms are unknown [43, 44]. All the $PP1$- and *PP2B*-dependent dephosphorylation of the AMPAR residues downregulates the synaptic strength, which is called depression. In contrast to depression, a high elevation of $Ca^{2+}$ levels triggers a kinase pathway that results in the upregulation of synaptic strength, which is called potentiation. In the kinase pathway, $Ca^{2+}/CAM$-dependent protein kinase II (*CaMKII*) and adenylyl cyclase ($AC$) are directly activated by $Ca^{2+}/CaM$ complex [40, 45]. *CaMKII* phosphorylates *ser831* of PSD AMPAR that increases the single channel conductance and $AC$ catalyzes adenosine triphosphate (*ATP*) into cyclic adenosine monophosphate ($cAMP$) [24, 45–51]. $cAMP$ binds to protein kinase A (*PKA*) regulatory domains and triggers the activation of *PKA* [24, 45]. The activated *PKA* phosphorylates AMPAR to prevent internalization and to enhance externalization into the membrane via exocytosis [23, 52–54].

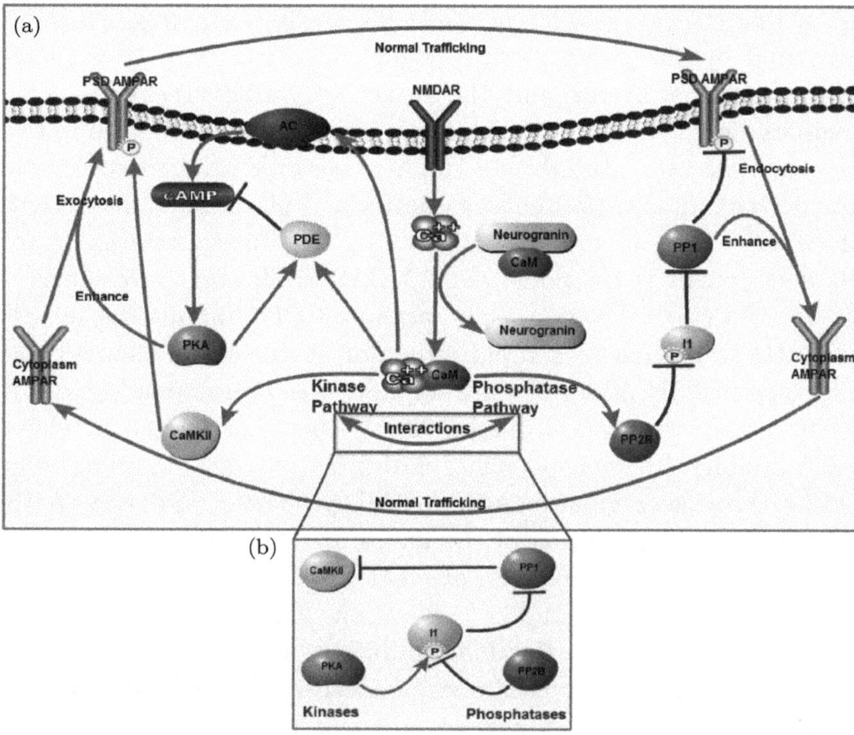


Fig. 10.2. Mechanisms initiating NMDAR-dependent synaptic plasticity. (a) AMPAR recycles between the cytoplasm and PSD of the postsynaptic neuron via exocytosis-based externalization and endocytosis-based internalization, and this trafficking can be controlled through a NMDAR-dependent process. (b) The two subpathways are also interacting with each other. PP1 directly represses the production of CaMKII, while PKA indirectly inhibits the production of PP1 through phosphorylating I1.

Through the activation of the kinase pathway, the single channel conductance and the number of PSD AMPAR are increased resulting in the upregulated synaptic strength.

Although activated by different intracellular $Ca^{2+}$ elevations, the direct interaction of $Ca^{2+}$ signals of both sub-pathways may interfere with each other, so that extra mechanisms of synaptic plasticity are required to differentiate the two pathways in order to prevent the interference between them (Fig. 10.2(b)). Firstly, the *CaM*-regulated proteins display different $Ca^{2+}$ sensitivities that are activated under different $Ca^{2+}$ levels [40]. Secondly, *cAMP* is regulated by phosphodiesterase (*PDE*), whose activity is promoted by both the activator and the target of *cAMP*, the $Ca^{2+}/CaM$ complex and *PKA* [24, 55]. This regulation forms a dual inhibitory process that strongly inhibits the activity of *cAMP* to ensure the low activity of *cAMP* under modest $Ca^{2+}$ elevation. Low activity of *cAMP* results in the low activity of *PKA*. Hence, internalization via endocytosis is the dominant AMPAR trafficking in response to the modest $Ca^{2+}$ elevation. Next, activated $PP1$ inhibits the activity of *CaMKII* leading to a rapid reduction in the single channel conductance of PSD AMPAR that amplifies the $PP1$-dependent downregulation of synaptic strength [40, 49, 50]. Lastly, the activated *PKA* inhibits $PP1$ via phosphorylating $I1$ [23, 40]. This inhibition blocks the downregulation of AMPAR properties and reverts the $PP1$-dependent inhibition of the kinase pathway. However, because of the dual regulation process of *cAMP*, *PKA*-dependent inhibition only happens at the high $Ca^{2+}$ elevation, where the $Ca^{2+}$ signal is strong enough to break the dual inhibition of *cAMP*.

The regulation of synaptic strength through AMPAR modifications does not last long; gene expressions are required to prolong and maintain the duration of AMPAR modifications [23]. Gene expressions are triggered by $Ca^{2+}$-dependent signals followed by the activation of postsynaptic transcription factors or pathways, such as *cAMP* response element-binding protein (*CREB*) and mitogen-activated protein kinase pathway (*MAPK*), and cause morphological changes in the synapse that accounts for the stabilized transmission capacity or memory storage [3, 4, 7]. However, gene expressions do not have a big impact on the initiation of synaptic plasticity. A large body of evidence shows the normal initiation of synaptic plasticity without the gene expression [3, 40]. Moreover, since the activation of gene expressions are the downstream events mediated by postsynaptic modulators, such as *PKA* and *CaMKII* [4, 40], gene expressions are represented through the dynamically changing levels of the corresponding postsynaptic modulators in our modeling approach. Hence, gene expressions are not considered as a main factor initiating synaptic plasticity.

Are all these synaptic components the main modulators of synaptic plasticity? The answer lies in the functionality of the synaptic components. As the main modulators, their functionalities must be essential in conducting the system behavior. The synaptic behavior of synaptic plasticity in the postsynaptic neuron appears as the altered AMPAR properties in order to regulate the synaptic strength. Hence, the proteins, which directly associate with AMPAR or decide the direction of modifying AMPAR properties, are the main modulators of synaptic plasticity. Clearly, not all the synaptic components fall into either of the two categories. According to the background information discussed above, the AMPAR association proteins are *PP*1, *PP*2*B*, *PKA* and *CaMKII*, and the decision proteins/ions are $Ca^{2+}$, *CaM* and *cAMP*. All of these main modulators severely impact the dynamics of synaptic plasticity, while the remaining synaptic components respond to the main modulators and mediate the related changes in the main modulators, resulting in the synaptic plasticity.

### 10.2.1. *Summary*

As previously discussed, the main modulators initiating synaptic plasticity can be determined from the detailed mechanisms of synaptic plasticity. The NMDAR-dependent $Ca^{2+}$ elevation and the associated *CaM* activation serve as the signal for synaptic plasticity. The signal selectively activates phosphatase and kinase pathways. The phosphatase pathway responds to modest $Ca^{2+}$ elevation and activates *PP*2*B* and *PP*1 in succession. *PP*2*B* and *PP*1 dephosphorylate the AMPAR residues leading to downregulation of synaptic strength. The kinase pathway responds to high $Ca^{2+}$ elevation and activates *cAMP*, *CaMKII* and *PKA*. *PKA* and *CaMKII* phosphorylate the AMPAR residues resulting in the upregulation of synaptic strength. The two pathways interact with each other and there are extra mechanisms that prevent the interference between them. Modest $Ca^{2+}$ elevation-activated *PP*1 inhibits the activity of *CaMKII*. Meanwhile, the dual inhibition of *cAMP* ensures low activity of the kinase pathway under the modest $Ca^{2+}$ elevation. Moreover, high $Ca^{2+}$ elevation-activated *PKA* inhibits the phosphatase pathway and reverts the downregulation of synaptic strength. As a conclusion,

$Ca^{2+}$, *CaM*, *cAMP*, *PKA*, *CaMKII*, *PP2B* and *PP1* are considered as the main modulators initiating synaptic plasticity.

## 10.3. Quantification of the NMDAR-Dependent Synaptic Plasticity

As discussed in the introduction, a mathematical model of synaptic plasticity would encapsulate the information related to synaptic interactions in a single quantitative index, which may be used to gain insights into the behavior of synaptic plasticity. One such representation expresses synaptic strength as a function of the intracellular $Ca^{2+}$ level that reveals the $Ca^{2+}$ dependence of synaptic plasticity [56]. However, this representation does not provide sufficient details of the postsynaptic mechanisms to reflect the inner protein–protein interactions of synaptic plasticity. A more detailed mechanistic-driven quantitative representation is required and we develop a detailed mechanistic index of synaptic plasticity, SPI, based on the dynamically changing levels of the postsynaptic modulators.

SPI focuses on describing the relationship between the main postsynaptic modulators and the AMPAR conductance ($G_{\text{AMPAR}}$), which is the cation influx capacity through all the PSD AMPARs within a unit time [13], of the postsynaptic neuron. $G_{\text{AMPAR}}$ determines many characteristics of the synaptic transmission, including the rate of the cation influx, the magnitude of EPSP and the generation of an action potential, that are all strong factors of the synaptic strength [1]. $G_{\text{AMPAR}}$ is a product of two main factors, the number and the average single channel conductance of AMPAR [57]. These factors are regulated through the competition between the *PKA*/*CaMKII*-dependent phosphorylation and the *PP2B*/*PP1*-dependent dephosphorylation of AMPAR that are the major mechanisms of synaptic plasticity as discussed previously [39–54]. Therefore, $G_{\text{AMPAR}}$ is a good measure of synaptic plasticity.

$G_{\text{AMPAR}}$ consists of two parts: (1) the basal conductance and (2) conductance enhancements. PSD AMPAR carries a base conductance with the magnitude dependent on the structural property of AMPAR. *PP1*-dependent undocking and endocytosis causes PSD

AMPAR to translocate into the cytoplasm and lose its functions as well as the conductance. Therefore, the number of PSD AMPAR is a determinant for the basal conductance. *CaMKII*-dependent phosphorylation of AMPAR changes the structure to allow more calcium influx that enhances the basal conductance. This *CaMKII*-dependent enhancement can be reverted through $PP1$-dependent dephosphorylation. Moreover, AMPAR, as a tetramer, contains two *ser831* residues that provide more flexibility for dynamic regulation of conductance. Hence, the *ser831* phosphorylation status of AMPAR is a determinant of the conductance enhancement. These determinants can be obtained directly from the monitored properties of AMPAR, or expressed as functions of the postsynaptic modulators in order to predict $G_{\mathrm{AMPAR}}$. The number of PSD AMPAR and the *ser831* phosphorylation status can be described by the following expressions in terms of the main postsynaptic modulators:

$$\begin{aligned} PSDAMPAR &= f_1([PKA], [PP1], [PP2B], \\ &\quad CytoplasmAMPAR), \qquad (10.1) \\ SinglePhosPSDAMPAR &= f_2([CaMKII], [PP1], \\ &\quad NonePhosPSDAMPAR), \qquad (10.2) \\ DoublePhosPSDAMPAR &= f_3([CaMKII], [PP1], \\ &\quad NonePhosPSDAMPAR), \qquad (10.3) \end{aligned}$$

where the quantities in square brackets are the concentrations of species in $\mu$M/L and other quantities are numbers of the corresponding AMPARs: $CytoplasmAMPAR$ is the number of the AMPAR in the cytoplasm, $NonePhosPSDAMPAR$ is the number of the PSD AMPAR without *ser831* phosphorylation, $SinglePhosPSDAMPAR$ is the number of the single *ser831* phosphorylated PSD AMPAR and $DoublePhosPSDAMPAR$ is the number of the double *ser831* phosphorylated PSD AMPAR. By applying the Buckingham $\pi$ theorem, the following dimensionless expressions are obtained (see Appendix H to this chapter for the details of the derivation):

$$\frac{PSDAMPAR}{CytoplasmAMPAR} = \emptyset \frac{[PKA]^{-b-c}}{[PP1]^{-b}[PP2B]^{-c}}, \quad (10.4)$$

$$\frac{SinglePhosPSDAMPAR}{NonePhosPSDAMPAR} = \varphi \left( \frac{[CaMKII]}{[PP1]} \right)^{d}, \quad (10.5)$$

$$\frac{DoublePhosPSDAMPAR}{NonePhosPSDAMPAR} = \theta \left( \frac{[CaMKII]}{[PP1]} \right)^{f}, \quad (10.6)$$

where $b$, $c$, $d$, $f$, $\emptyset$, $\varphi$ and $\theta$ are constants dependent on the cell-to-cell variability of synapses.

Since PSD AMPAR contains three types: without *ser831* phosphorylation, single *ser831* phosphorylated and double *ser831* phosphorylated, we can calculate the conductance for each type separately and sum their values into $G_{\text{AMPAR}}$. The conductance of the PSD AMPAR without *ser831* phosphorylation, $G_{NonePhosPSDAMPAR}$, is a product of their number and the average basal channel conductance ($\overline{G_{SingleChannel}}$):

$$G_{NonePhosPSDAMPAR} = NonePhosPSDAMPAR \cdot \overline{G_{SingleChannel}}. \quad (10.7)$$

Equation (10.7) can be normalized against the maximum possible level of $G_{\text{AMPAR}}$, at which AMPARs are all docked in PSD with double *ser831* phosphorylation, of the targeting synapse. We assume that PSD AMPAR with double *ser831* phosphorylation carries $G_{\max}$ single channel conductance, then the maximum possible level of $G_{\text{AMPAR}}$, max_$G_{\text{AMPAR}}$, is:

$$Max\_G_{AMPAR} = (PSDAMPAR + CytoplasmAMPAR) \cdot G_{\max}. \quad (10.8)$$

By combining Eqs. (10.7) and (10.8), the normalized $G_{NonePhosPSDAMPAR}$, $\hat{G}_{NonePhosPSDAMPAR}$ can be calculated:

$$\hat{G}_{NonePhosPSDAMPAR} = NonePhosPSDAMPAR \cdot \alpha, \quad (10.9)$$

where $\alpha = \overline{G_{SingleChannel}}/Max\_G_{AMPAR}$ is the normalized basal conductance coefficient. Similarly, we can calculate the normalized conductance of the single *ser831* phosphorylated PSD AMPAR, $\hat{G}_{NonePhosPSDAMPAR}$, as:

$$\hat{G}_{NonePhosPSDAMPAR} = SinglePhosPSDAMPAR \cdot (\alpha + \beta), \tag{10.10}$$

and the normalized conductance of the double *ser831* phosphorylated PSD AMPAR, $\hat{G}_{DoublePhosPSDAMPAR}$, as:

$$\hat{G}_{DoublePhosPSDAMPAR} = DoublePhosPSDAMPAR \cdot (\alpha + \gamma), \tag{10.11}$$

where $\beta = \overline{G_{SinglePhosEnhance}}/Max\_G_{\text{AMPAR}}$ is the normalized average single *ser831* phosphorylation conductance enhancement coefficient and $\gamma = \overline{G_{DoublePhosEnhance}}/Max\_G_{\text{AMPAR}}$ is the normalized average double *ser831* phosphorylation conductance enhancement coefficient. $\alpha$, $\beta$ and $\gamma$ are assumed as constant in the same neuron, but may be variable among different neurons.

The dimensionless form of $G_{\text{AMPAR}}$, normalized $G_{\text{AMPAR}}$, can be calculated as the sum of $\hat{G}_{NonePhosPSDAMPAR}$, $\hat{G}_{NonePhosPSDAMPAR}$ and $\hat{G}_{DoublePhosPSDAMPAR}$. Through simplification and rearrangement, the expression of the normalized $G_{\text{AMPAR}}$ is:

$$\begin{aligned} Normalized\ G_{AMPAR} = {} & \alpha \cdot PSDAMPAR \\ & + \beta \cdot SinglePhosPSDAMPAR \\ & + \gamma \cdot DoublePhosPSDAMPAR. \end{aligned} \tag{10.12}$$

By combining Eqs. (10.4), (10.5), (10.6) and (10.12), the inner decision, based on the determinants of $G_{\text{AMPAR}}$, with regard to the target of the $G_{\text{AMPAR}}$ modifications, SPI, is expressed as follows:

$$SPI = \emptyset \frac{[PKA]^{-b-c}}{\emptyset[PKA]^{-b-c} + [PP1]^{-b}[PP2B]^{-c}} \times \left( \alpha + \frac{\beta\varphi \left(\frac{[CaMKII]}{[PP1]}\right)^{d} + \gamma\theta \left(\frac{[CaMKII]}{[PP1]}\right)^{f}}{1 + \varphi\left(\frac{[CaMKII]}{[PP1]}\right)^{d} + \theta\left(\frac{[CaMKII]}{[PP1]}\right)^{f}} \right). \quad (10.13)$$

SPI is a future snapshot of the normalized $G_{\text{AMPAR}}$ if the current system status maintains long enough for the synapse to stabilize. Equation (10.13) expresses SPI as the product of the maximum SPI based on the current system status (Max_SPI: outside the bracket) and the current effective percentage (EP: inside the bracket). Max_SPI is a number between 0% (all AMPAR in the cytoplasm) and 100% (all AMPAR in PSD), representing the percentage of PSD AMPAR in the total AMPAR population. EP is also a number between $\alpha$ (no *ser831* phosphorylation on PSD AMPAR) and 100% (all PSD AMPARs are double *ser831* phosphorylated), determining the effective percentage of Max_SPI. The resulting SPI is an index between 0% and 100% representing weak to strong synaptic strength. SPI should increase rapidly towards 100% in response to potentiation and drop dramatically towards 0% in response to depression.

SPI is an instantaneous prediction of the normalized $G_{\text{AMPAR}}$ based on the inner protein–protein interactions at a given time, $t$, but the prediction takes time to convert to the authentic normalized $G_{\text{AMPAR}}$, because of the relatively slow diffusion rate of AMPAR between PSD and cytoplasm. Especially during synaptic plasticity, the continuous system status variations cause the continuous synaptic strength modifications leading to rapid changing values of SPI so that none of the predictions are able to stabilize in such situation. Moreover, the diffusion rate of AMPAR is variable with regard to different levels of the main postsynaptic modulators as discussed in the previous section. Therefore, SPI serves as the decision on the target to modify the normalized $G_{\text{AMPAR}}$ at a given time, $t$, and the decision varies with respect to the dynamic changing levels of the postsynaptic modulators. Hence, the normalized $G_{\text{AMPAR}}$

should reflect the integration of the temporal transients of SPI over a very small time interval, $\Delta t$, in response to the instantaneous inner protein–protein interactions. Since the AMPAR trafficking time courses follow the exponential increase or decay curves [58] that induce a delay in Max_SPI predictions, we formulate the normalized maximum $G_{\text{AMPAR}}$ in the PSD ($Max\_\hat{G}_{\text{AMPAR}}$) as exponential infinitesimal increments based on the Max_SPI as shown in Eq. (10.14):

$$\begin{aligned} Max\_\hat{G}_{AMPAR}(t+\Delta t) = Max\_\hat{G}_{AMPAR}(t) + ((Max\_SPI(t+\Delta t) \\ - Max\_\hat{G}_{AMPAR}(t)) \cdot (1 - e^{-\Delta t/\tau(t)}), \end{aligned} \tag{10.14}$$

where $Max\_SPI(t) = \frac{\emptyset[PKA]^{-b-c}}{\emptyset[PKA]^{-b-c}+[PP1]^{-b}[PP2B]^{-c}}$ and $\tau$ is derived from the AMPAR diffusion rate. Hence, the SPI prediction of the normalized $GAMPAR$, can be described by Eq. (10.15):

$$Normalized\ G_{AMPAR}(t+\Delta t) = Max\_\hat{G}_{AMPAR}(t+\Delta t) \cdot EP(t), \tag{10.15}$$

where EP is given by Eq. (10.16):

$$EP(t) = \alpha + \frac{\beta\varphi\left(\frac{[CaMKII]}{[PP1]}\right)^d + \gamma\theta\left(\frac{[CaMKII]}{[PP1]}\right)^f}{1+\varphi\left(\frac{[CaMKII]}{[PP1]}\right)^d + \theta\left(\frac{[CaMKII]}{[PP1]}\right)^f}. \tag{10.16}$$

SPI represents the internal decisions on the target of the synaptic strength modifications based on the instantaneous inner protein–protein interactions following the $Ca^{2+}$ signal. The normalized $G_{\text{AMPAR}}$ represents the effective AMPAR conductance in the PSD (an indicator of synaptic strength) that varies according to the internal decisions, but at a much slower pace due to the slower diffusion rate of AMPAR between PSD and cytoplasm. In the following sections, we use SPI (Eq. (10.13)) to refer to the internal decisions of synaptic strength modifications, and use normalized $G_{\text{AMPAR}}$ (Eq. (10.15)) to refer to the synaptic strength.

## 10.4. Development of MSPM and Validation of SPI

A large number of models have already been proposed regarding many aspects of synaptic plasticity [12]. Due to the focus on the postsynaptic neuron of this research, a $Ca^{2+}$-dependent mechanistic AMPAR trafficking model is the most suitable for the validation of SPI. Although many models [13, 16–20] satisfy the mechanistic requirement of the research, they are not suitable for our validation purposes. For example, Castellani *et al.* [16] modeled the $Ca^{2+}/CaM$-dependent bidirectional synaptic plasticity, but it is based on a theoretical mathematical framework rather than detailed synaptic reactions. The Hayer and Bhalla model [13] is a comprehensive mechanistic model based on the detailed synaptic reactions revealing the $Ca^{2+}/CaM$-dependent bidirectional aspects of synaptic plasticity precisely and providing useful insights into the bistability of long term synaptic plasticity, but it was proved to be too complex for our purposes. MSPM is a simpler one based on the postsynaptic modulators of synaptic plasticity as well as their associated synaptic reactions: MSPM is based on 11 differential equations whereas the Hayer and Bhalla model has 256 differential equations.

As mentioned previously, $Ca^{2+}$ signals selectively activate two distinct pathways based on the pattern of the incoming $Ca^{2+}$ elevation. Under the experimental conditions, a high $Ca^{2+}$ elevation is caused by high frequency (100–200 Hz) tetanic electrical stimulation resulting in potentiation, and a modest $Ca^{2+}$ elevation is caused by prolonged repetitive low-frequency (1 Hz) electrical stimulation resulting in depression [23]. These patterns of $Ca^{2+}$ elevation are well modeled as the exponential built-up and decay in the early models [59, 60]. In addition, the $Ca^{2+}$ binding activated CaM behaves as the mediator of the $Ca^{2+}$ signal that selectively activates subsequent pathways in response to different patterns of the $Ca^{2+}$ elevation [40]. Here, we use the Hayer and Bhalla model as the control for validating the behaviors of MSPM because of its excellent precision against experimental data. Therefore, we obtained the simulated levels of the $Ca^{2+}$ elevations and its binding partner, CaM, from the Hayer and Bhalla model and used them as the inputs for MSPM so that the inputs for both models are consistent.

MSPM is developed based on the $Ca^{2+}/CaM$-dependent interactions between the main postsynaptic modulators of synaptic plasticity discussed previously. There are a number of assumptions made in MSPM: (1) the modulators of synaptic plasticity only have two states, either inactive or active; (2) the total concentrations of the modulators, as a sum of the two states, are stable during the initiation of synaptic plasticity. Evidences show that the early phase synaptic plasticity does not require gene expression [3, 24], so that the composition of a neuron is stable at the initial stage. This assumption also allows MSPM to capture the active forms of the modulators only; (3) the synaptic processes are extremely fast, so that the degradations of the modulators do not contribute much to the system dynamics, especially in the initial phase of synaptic plasticity. Therefore, the degradations of the modulators are ignored; and (4) the different compartments between the cytoplasm and the surface of the postsynaptic neuron are ignored. All modeling variables are converted into the cytoplasm compartments and the volumes taken are 0.09 fL and 0.01 fL for the cytoplasm and the surface of the postsynaptic neurons, respectively. MSPM contains 11 species including different states of AMPAR *Glu1/2* subunit, AMPAR regulators (*PKA*, *CaMKII*, *PP*1 and *PP2B*) and system regulator, cAMP, which are all direct players of synaptic plasticity (see Fig. 10.3 for a brief description of the schematic diagram and its model equations). The involvement of only the postsynaptic modulators of synaptic plasticity ensures the precision of the validation of SPI.

MSPM is firstly validated against the experimental results of the AMPAR trafficking time courses in order to examine its biological validity. In order to mimic the experimental conditions, the same procedure of Hayer and Bhalla validation [13] is carried out: AMPAR direct interacted modulators (*PKA*, *CaMKII*, *PP2B* and *PP*1) are numerically fixed at their equilibrium levels and a pulse of AMPAR is introduced into either PSD or cytoplasm of the postsynaptic neuron. The AMPAR *Glu2/3* subunit trafficking is simulated based on the reaction from Hayer and Bhalla model [13]:

$$Cytoplasm\_Glu23 \underset{3.88889e-05}{\overset{0.0014}{\rightleftharpoons}} PSD\_Glu23.$$

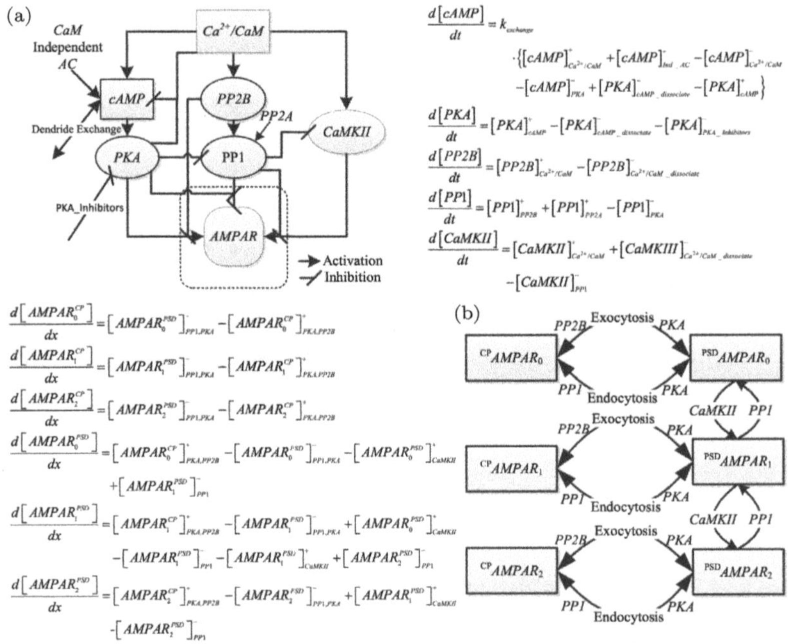


Fig. 10.3. Schematic diagrams and the equations of MSPM. (a) Synaptic protein–protein interactions in respose to $Ca^{2+}$ signals. Five proteins (*cAMP*, *PKA*, *PP2B*, *PP1* and *CaMKII*) are modeled and their equations are shown beside figure (a). In these equations, each term indicates a reaction and is described by a bracketed variable with a superscript sign and subscript words. The variable in the bracket indicates the domain of the reaction. +/− signs indicate activation or inhibition reactions, respectively. The subscript words indicate the proteins involved in the reactions (may be a substrate or an effector). (b) Extension of the dashed rectangle in figure (a): kinase and phosphatase-dependent modifications of AMPAR. Six different states of AMPAR are modeled and their states are described by superscript words and a subscript number. The superscript words indicate the location of the AMPAR (CP:cytoplasm) and the subscript number indicates the number of *ser831* phosphorylations. Their corresponding equations are shown beside figure (b). Equations are formatted similar to the equations of synaptic protein–protein interactions, except the +/− signs indicate phosphorylation and dephosphorylation reactions, respectively.

The resulting AMPAR *Glu2/3* subunit time course is combined with the AMPAR *Glu1/2* subunit time course from MSPM to represent the normal composition of AMPAR. The combined results are shown in Fig. 10.4, which shows a good agreement with the experimental observations.

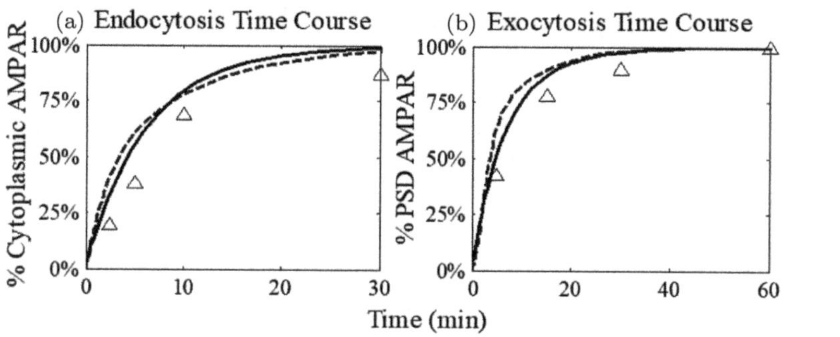


Fig. 10.4. Comparisons of the exocytosis and endocytosis time courses of AMPAR. The triangles represent the experimental observations of AMPAR trafficking [29]. The dashed lines represent the prediction of the AMPAR trafficking time courses from the Hayer and Bhalla model and the solid lines represent AMPAR trafficking time courses predicted from MSPM. (a) The endocytosis time course of AMPAR. This is obtained by examining the cytoplasmic changes following a pulse of AMPAR, 80 *Glu1/2* subunits and 26 *Glu2/3* subunits, into the PSD of the postsynaptic neuron. (b) The exocytosis time course of AMPAR. This is obtained by examining the PSD changes following a pulse of AMPAR, 80 *Glu1/2* and 26 *Glu2/3* subunits, into the cytoplasm of the postsynaptic neuron.

Next, the normalized $G_{\mathrm{AMPAR}}$ is obtained from MSPM by examining the number of PSD AMPAR (*Glu1/2* subunit) and their associated *ser831* phosphorylation status, and is compared against the same measure from the Hayer and Bhalla model to validate the system behaviors during synaptic plasticity. The behaviors of the normalized $G_{\mathrm{AMPAR}}$ calculated from both models are shown in Fig. 10.5 and a similar behavior is observed when the $Ca^{2+}$ signal is applied into the system at 10000s. However, MSPM exhibits a larger drop in the normalized $G_{\mathrm{AMPAR}}$ under depression condition that may be caused by the enhanced exocytosis, whereas the Hayer and Bhalla model does not show significant alteration in the PSD AMPAR numbers under depression. Overall, MSPM exhibits meaningful biological bidirectional behavior of synaptic plasticity that can be used to further validate SPI and to predict the general design logic governing synaptic plasticity.

The complicated synaptic interactions are likely to have unique design logic governing its functions, so that the appropriate derivation of $G_{\mathrm{AMPAR}}$ would maintain the same design logic of

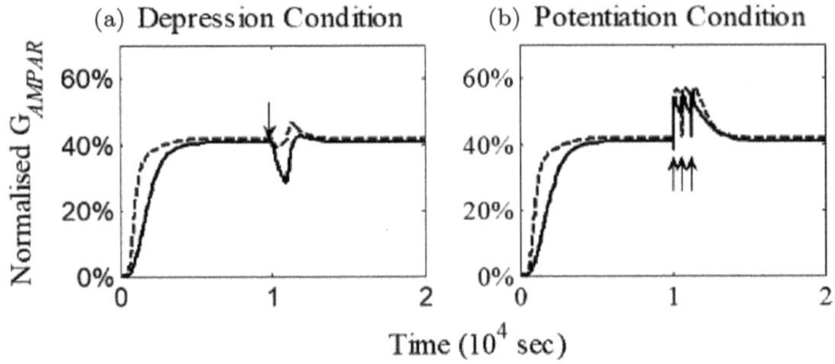


Fig. 10.5. Comparisons of model predictions of the normalized $G_{\text{AMPAR}}$. The solid lines and the dashed lines represent the normalized $G_{\text{AMPAR}}$ calculated from MSPM and the Hayer and Bhalla model, respectively. $G_{\text{AMPAR}}$ calculation is based on the following rules: (1) the basal conductance of each *PSD* AMPAR is taken as 1, (2) each *ser831* phosphorylation increases the conductance of the corresponding *PSD* AMPAR by 0.5. The calculated $G_{\text{AMPAR}}$ is normalized against its maximum, 160 in this case since the max number of AMPAR is fixed at 80 during the simulation. (a) The normalized $G_{\text{AMPAR}}$ is calculated under the depression induced by 900 stimulations at 1 Hz starting at 10000 s (black arrow). (b) The normalized $G_{\text{AMPAR}}$ is calculated under the potentiation induced by 3 bursts of 100 stimulations at 100 Hz separated by 600 s starting at 10000 s (black arrows).

the experimental observations of synaptic plasticity. The normalized $G_{\text{AMPAR}}$ maintains the design logic of the functionality of the synapse, and SPI prediction of the normalized $G_{\text{AMPAR}}$ is a theoretical framework derived from the biological information of synaptic properties. It is likely to have only one appropriate design logic for both measures to behave similarly and the target of this research is to identify that unique solution.

Figure 10.6 shows the comparison between the model and SPI predictions of the normalized $G_{\text{AMPAR}}$ based on MSPM.

Since we do not have sufficient data to determine the values of $\tau$ for the different levels of the postsynaptic modulators, we use an average value of 250s for $\tau$ in this comparison. Another reason to take $\tau$ as the average value in this study is because the synaptic plasticity signals, which change intracellular $Ca^{2+}$ levels, are highly transient in which the alterations in AMPAR trafficking rates are only effective in the short term. In order to obtain this average value of $\tau$, we first take an approximately value of $\tau$ from the AMPAR

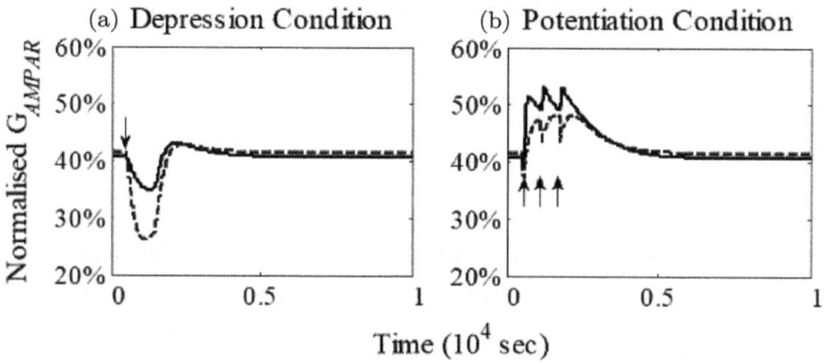


Fig. 10.6. Comparisons of model and SPI predictions of the normalized $G_{\text{AMPAR}}$. The solid lines and the dashed lines represent the MSPM prediction and the SPI prediction of the normalized $G_{\text{AMPAR}}$, respectively. Both predictions are based on the data provided by MSPM. (a) The predictions of the normalized $G_{\text{AMPAR}}$ under the depression induced by 900 stimulations at 1 Hz starting at 500 s (black arrow). (b) The predictions of the normalized $G_{\text{AMPAR}}$ under the depression induced by 3 bursts of 100 stimulations at 100 Hz separated by 600 s starting at 500 s (black arrows). Parameters used in calculating SPI predictions are $a = 1.025$, $b = 0.55$, $c = 0.985$, $d = 2.4325$, $\varphi = 8.75$, $\phi = 0.04$, $\theta = 9.15\text{e} - 4$, $\alpha = 0.5$, $\beta = 0.25$ and $\gamma = 0.5$.

tracking time courses shown in Fig. 10.4 based on the requirement for the exponential curve. We then take an interval of $\tau$ values from the lower 30% to the upper 30% of the approximately value and generate sets of the normalized $G_{\text{AMPAR}}$ based on these $\tau$ values. At the end, we analyze these sets of the normalized $G_{\text{AMPAR}}$ and identify the best one by comparing their correlation coefficient to the model prediction of the normalized $G_{\text{AMPAR}}$. As a result, the set with the average $\tau$ of 250 s gives the highest correlation coefficient to the model prediction. In Fig. 10.6, the behaviors of the two measures match well in pattern under both depression and potentiation conditions when $Ca^{2+}$ signals arrive at 500 s. However, the results show a significant difference in the amplitudes, due to the weaker elevation during the potentiation and stronger drop during the depression of SPI, that may be caused by the imprecise value of $\tau$. The imprecision of $\tau$ is due to two reasons: (1) There should be separate $\tau$ values regarding the exocytosis and endocytosis aspects of AMPAR, and (2) $\tau$ should be a variable regarding the dynamic levels of the main postsynaptic modulators rather than an average. A similar pattern revealed in the results is expected since MSPM is developed based on a similar design logic to SPI and this result improves the feasibility of identifying the design logics of synaptic plasticity from the experiments.

### 10.4.1. *Conclusions*

We present a biological meaningful index of synaptic plasticity, SPI, which encapsulates the inner protein–protein interactions in the emergence of synaptic plasticity. Along with SPI, we also present a simpler model of synaptic plasticity based on the clarified main postsynaptic modulators and it exhibits similar behaviors against much more complicated model. We then compare the SPI and the model predictions of the normalized $G_{\mathrm{AMPAR}}$ and the comparison reveals a well match in their patterns. Based on these observations, we conclude the following: (1) the central decision logic of the NMDAR-dependent synaptic plasticity is dependent on the levels of the essential postsynaptic modulators (*cAMP*, *PKA*, *CaMKII*, $PP1$ and $PP2B$) based on the current knowledge of synaptic plasticity; and (2) SPI can be used to confidently describe the behaviors of synaptic plasticity based on the collaborations among the AMPAR regulating proteins (*PKA*, *CaMKII*, $PP1$ and $PP2B$); and (3) both SPI and MSPM encapsulate the general design logic governing the NMDAR-dependent synaptic plasticity, but SPI may require accurate $\tau$ values to improve its precision. In spite of this weakness, SPI, as a single quantitative index, contains sufficient information of the inner protein–protein interactions of synaptic plasticity that is convenient and important for synaptic plasticity research. In the following sections, we investigate SPI with respect to the roles the postsynaptic modulators play in synaptic plasticity and use SPI to understand the mutations following Alzheimer's disease.

## 10.5. The Roles of the Main Postsynaptic Modulators in Conducting NMDAR-Dependent Synaptic Plasticity

The functions of the main postsynaptic modulators are discussed previously, but the knowledge regarding the impacts of these functions on synaptic plasticity is rather limited. The understandings of the impacts in terms of the strength, the pattern and the difference between the modulators remain unknown. With SPI, the direct and detailed images of the collaborations between the main postsynaptic modulators in the emergence of synaptic plasticity can be obtained which may provide insights into the distinct ways by which the main postsynaptic modulators function towards synaptic plasticity.

We set up a local sensitivity analysis of SPI in order to investigate the patterns of SPI in response to changing levels of the main postsynaptic modulators. One modulator is perturbed at a time from negligible to its maximum level (sum of the inactive and active forms) in MSPM (see material and methods for the details), and other modulators are fixed at their equilibrium levels, which give the SPI equilibrium level of 40% according to MSPM. The corresponding SPI in response to the perturbations is shown in Fig. 10.7.

The results propose a clear relationship between the main postsynaptic modulators and the synaptic plasticity. In relation to

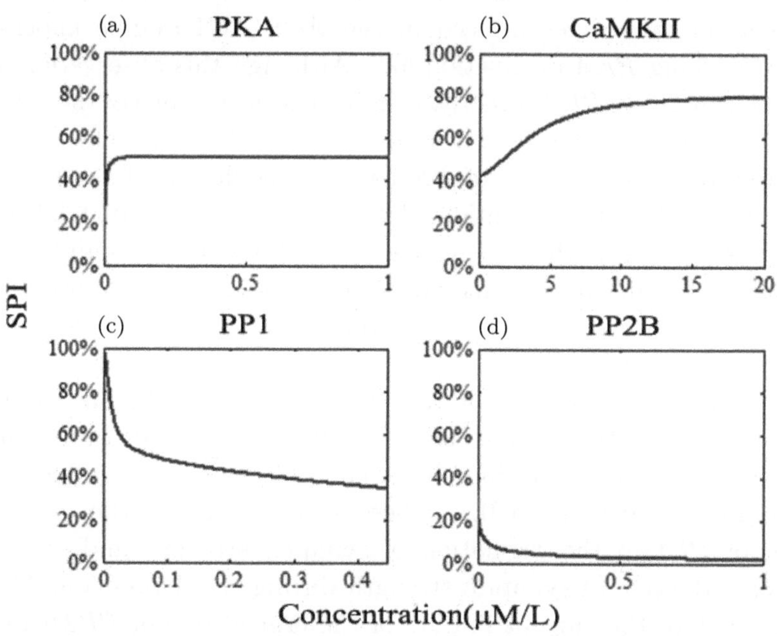


Fig. 10.7. Local sensitivity of SPI. (a) SPI of the perturbed PKA levels, (b) SPI of the perturbed CaMKII levels, (c) SPI of the perturbed PP1 levels, and (d) SPI of the perturbed PP2B levels.

the kinases, the increase of SPI in response to *PKA* (Fig. 10.7(a)) is more rapid compared to its response to *CaMKII* (Fig. 10.7(b)). A slight increase of *PKA* leads to rapid elevation of SPI, and the same amount of elevation of SPI requires a much larger increase of *CaMKII*. This result proposes an explanation to the observations that the induction of potentiation based on multiple trains of high-frequency stimulations does not require *PKA* [61], but a single train of high frequency stimulations must require *PKA* to induce potentiation [62, 63]. This is because *CaMKII* functions inside the PSD of postsynaptic neurons and a single train of high frequency stimulation is too weak, both in terms of its strength and duration, to drive and maintain a large amount of *CaMKII* into the PSD to induce potentiation. Hence, a single train of high frequency stimulation requires the assistance of *PKA* to inhibit the phosphatase pathway and to enhance the exocytosis of AMPAR in order to induce potentiation. Moreover, low levels of *PKA* cause SPI to reach as low as 0% blocking the induction of depression. It was observed that depression cannot be induced in the $AC8/AC1$ double knock-out mice blocking *PKA* production [63]. Although this observation may not be caused by *PKA* directly, it indicates the essence of the *cAMP-PKA* pathway in the regulation of synaptic strength. These results suggest the essential role of *PKA* in the regulation of the synaptic strength and in the initiation of synaptic plasticity. In relation to the phosphatases, their influence on SPI are more complex. SPI decreases rapidly in response to the increasing levels of both $PP2B$ and $PP1$. However, there is a difference between the decreases: SPI in response to perturbed $PP1$ decreases from 100% to approximately 40% (Fig. 10.7(c)) and SPI in response to perturbed $PP2B$ decreases from approximately 40% to 0% (Fig. 10.7(d)). Hence, $PP1$ reduces SPI towards its equilibrium level and $PP2B$ further reduces SPI from its equilibrium to 0%. These results confirm the dominant roles of $PP1$ in the regulation of synaptic strength and of $PP2B$ in the reduction of synaptic strength during the depression. This is consistent to the current finding of the crucial role of $PP2B$ in the induction of long-term depression, but unclear role in the regulation of synaptic strength [44].

## 10.6. SPI in Alzheimer's Disease

As we see in previous chapters, AD is strongly associated with the increased formation of amyloid beta ($\beta$-amyloid) oligomers [64], which is formed and accumulated outside of the neuron. The observations from the mice with Alzheimer's disease revealed the significant calcium overloads associated with the deposition of $\beta$-amyloid. The calcium overload is progressed by two mechanisms: (1) enhancing $Ca^{2+}$ influx: the enhancement could be an activation of $Ca^{2+}$ voltage channels or $\beta$-amyloid itself forming ion-conducting pores [21, 22]; and (2) altering release of $Ca^{2+}$ from internal stores, such as the endoplasmic reticulum (ER) [65, 66].

The intracellular $Ca^{2+}$ is very important because it serves as the signal for many synaptic functions including synaptic plasticity. Based on the extreme precision of synaptic plasticity, Berridge [22] proposed a $Ca^{2+}$ hypothesis which states that Alzheimer's disease prevents new memory formation through calcium overload to induce continuous depression to remove existing memory storage. Even a slight calcium overload causes a large impact on the precision of synaptic plasticity and prevents the induction of the potentiation, which initiates memory consolidation.

In this chapter, we use SPI to obtain insights into the synaptic functional disruptions caused by calcium overloads. A high $Ca^{2+}$ resting level was observed in mice with Alzheimer's disease [67], recorded 0.247 $\mu$M/L, which is more than twice the normal level, approximately 0.11 $\mu$M/L. In this investigation, four different $Ca^{2+}$ resting levels: 0.08 $\mu$M/L, 0.16 $\mu$M/L, 0.24 $\mu$M/L and 0.5 $\mu$M/L are used, where 0.08 $\mu$M/L represents the normal $Ca^{2+}$ resting level and the other three are remodeled $Ca^{2+}$ resting levels. Within the remodeled $Ca^{2+}$ resting levels, 0.16 $\mu$M/L and 0.24 $\mu$M/L are slightly overloaded and 0.5 $\mu$M/L is heavily overloaded closing to the potentiation region.

SPI of the 4 different $Ca^{2+}$ resting levels under potentiation conditions are shown in Fig. 10.8. Their behaviors exhibit different amplitudes in response to the altered levels of the main postsynaptic modulators, but no significant difference is noticed in their patterns of expressions. The baselines of SPI in response to the calcium

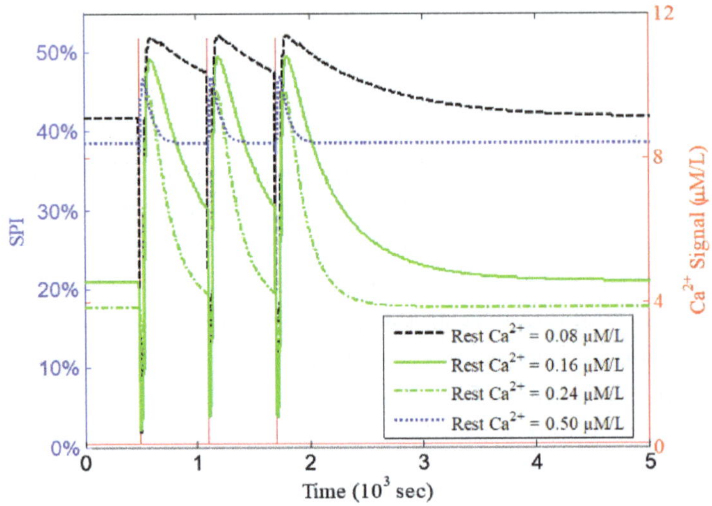


Fig. 10.8. SPI of four different $Ca^{2+}$ resting levels under potentiation condition. Potentiation is induced by 3 bursts of 100 stimulations at 100 Hz separated by 600s starting at 500s. The solid lines represent the $Ca^{2+}$ signal in response to the stimulations. SPI of the normal neuron (0.08 $\mu$M/L) is represented by the dash line. SPI values of the remodeled systems are represented as solid and dash-dot and dot lines. Green lines (0.16 $\mu$M/L and 0.24 $\mu$M/L) show SPI of slight calcium overloads and blue line (0.50 $\mu$M/L) shows the SPI of heavy calcium overloads.

overloads firstly decrease with respect to the slight calcium overloads, and then increase with respect to the heavy calcium overloads close to the potentiation region. These results reflect the $Ca^{2+}$ selectively as discussed previously. However, the amplitudes of SPI diminish as the $Ca^{2+}$ resting level increase, which may be caused by the increasing level of *PP2B* during the moderate $Ca^{2+}$ elevation as shown in the experiments [68]. The potentiation percentage of 0.50 $\mu$M/L is much less (118% of its baseline) compared to that of 0.08 $\mu$M/L (125% of its baseline). These results suggest that the general decision logic of synaptic plasticity is maintained with respect to slight calcium overloads, but may be disrupted with respect to heavy calcium overloads. This may distinguish the stages of Alzheimer's disease from memory loss (early) to neuronal apoptosis (end).

SPI predictions of normalized $G_{\mathrm{AMPAR}}$ are shown in Fig. 10.9 and reveal significant differences among the patterns of the potentiation

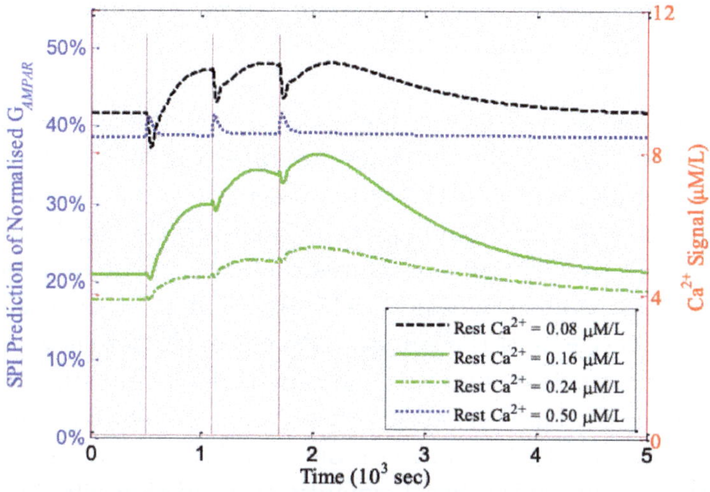


Fig. 10.9. SPI predictions of the normalized $G_{\text{AMPAR}}$ of four different $Ca^{2+}$ resting levels under potentiation condition. Potentiation is induced by 3 bursts of 100 stimulations at 100 Hz separated by 600 s starting at 500 s. The red lines represent the $Ca^{2+}$ signal in response to the stimulations. The normalized $G_{\text{AMPAR}}$ predictions of the normal neuron (0.08 $\mu$M/L) is represented by the black dash line. The normalized $G_{\text{AMPAR}}$ predictions of the remodeled systems is represented as green (solid and dash-dot) and blue (dot) lines. Green lines (0.16 $\mu$M/L and 0.24 $\mu$M/L) show the normalized $G_{\text{AMPAR}}$ of slight calcium overloads and blue line (0.50 $\mu$M/L) shows the normalized $G_{\text{AMPAR}}$ predictions of heavy calcium overloads. $\tau$ used are 250 s, 650 s, 1650 s and 2150 s for 0.08 $\mu$M/L, 0.16 $\mu$M/L, 0.24 $\mu$M/L and 0.50 $\mu$M/L of $Ca^{2+}$ resting levels, respectively.

of the $Ca^{2+}$ resting levels. The potentiation can be easily induced in the normal neuron as shown by the three peaks rapidly following each burst of the stimulations. At slight calcium overloads, the peaks are smaller and take more time to be reached. Especially in the neurons with 0.24 $\mu$M/L $Ca^{2+}$ resting level, the peaks are less than half of the normal neuron's and take more than 3 bursts of stimulation to reach. In the neuron with 0.5 $\mu$M/L $Ca^{2+}$ resting level, the potentiation percentage is very small (less than 110% of the baseline) and attenuates rapidly indicating a failure in the potentiation. These results suggest that slight calcium overloads set synaptic strength into the depression region of the normal neuron, which induces continuous removal of memory storage. Heavy calcium overloads completely blocks the potentiation to prevent memory formation. This is consistent with the $Ca^{2+}$ hypothesis of Alzheimer's disease.

The mutations of Alzheimer's disease in response to the changing levels of $Ca^{2+}$ resting concentration are probably contributed by the dynamic regulation of AMPAR as discussed previously. The changing basal levels of synaptic strength in response to the $Ca^{2+}$ overloads as shown in Fig. 10.9 indicate the restructuring of the ratio between PSD AMPAR and cytoplasmic AMPAR that influences the total capacity of $G_{\mathrm{AMPAR}}$ of the synapse. Recent evidences show that the injection of $\beta$-amyloid oligomers a few hours before investigation does not influence the basal synaptic transmission [64, 68], but long-term exposure of $\beta$-amyloid oligomers induces AMPAR removal and spine losses that influence the basal synaptic transmission [69, 70]. These observations may indicate the different stages of $\beta$-amyloid oligomers toxicity, which are represented from $Ca^{2+}$-independent depression of synaptic strength to $Ca^{2+}$ dynamics remodeling and $Ca^{2+}$-dependent restructuring of the dendrite spine. One reasonable explanation is that the $Ca^{2+}$-independent depression of synaptic strength slowly and progressively remodels the property of the synapse resulting in the $Ca^{2+}$ dynamics remodeling, leading to synapse shrinkage and memory loss. These $Ca^{2+}$-independent and -dependent processes repeat until reaching the threshold of $Ca^{2+}$ to induce neuronal apoptosis. The dendrite spine restructuring also causes the abnormal potentiation percentages during slight $Ca^{2+}$ overloads, so that the cell-to-cell variability parameters of SPI need to be adjusted to reflect such alteration. In addition, the approximation of the $\tau$ value may be another factor to cause the abnormality in the potentiation percentage.

## 10.7. Methods

MSPM is developed based on the detailed biological reactions and their associated reaction rates retrieved from the DOQCS databases [71]. Through careful analyses and simplifications, a set of 11 ODEs are developed and presented in the following (see the details of the model derivations in Appendix H):

$$\frac{d[cAMP]}{dt} = k_{\text{exchange}} \cdot \left\{ \frac{K_{cat1}[ATP][Ca_4CaM][AC1]_{\text{Total}}}{(k_{m1} + [Ca_4CaM])(k_{m2} + [ATP])} \right.$$

$$+ \frac{K_{\text{cat}2}[ATP]([AC2]^e + \exp(-0.11 \cdot t)}{([AC2]^0 - [AC2]^e))}{(k_{m3} + [ATP])}$$

$$- \frac{K_{\text{cat}3}[cAMP][cAMP \cdot PDE]_{\text{Total}}}{k_{m4} + [cAMP]}$$

$$\times \left(1 + \frac{[PKA]}{k_{m5} + [PKA]}\right)$$

$$- \frac{K_{\text{cat}4}[cAMP][PDE1]_{\text{Total}}}{k_{m6} + [cAMP]}$$

$$\left(1 + \frac{5[Ca_4CaM]}{k_{m7} + [Ca_4CaM]}\right)$$

$$-4k_1[R_2C_2][cAMP]^4 + 4k_2[PKA]^2$$

$$\left. ([R_2]_{\text{Total}} - [R_2C_2]) \right\},$$

$$\frac{d[PKA]}{dt} = (2k_1[R_2C_2][cAMP]^4 - 2k_2[PKA]^2([R_2C_2]_{\text{Total}} - R_2C_2))$$

$$\cdot \frac{(k_{m8} + [PKA])^2}{(k_{m8} + [PKA])^2 + k_{M8}[PKA \cdot Inhibitor]_{\text{Total}}},$$

$$\frac{d[PP2B]}{dt} = k_3[CaCaM]\frac{([PP2B]_{\text{Total}} - [PP2B])[Ca]^{n1}}{(k_{m9} + [Ca]^{n1})} - k_4[PP2B],$$

$$\frac{d[PP1]}{dt} = \frac{K_{\text{cat}5}[PP2B]([PP1]_{\text{Total}} - [PP1])}{k_{m10} + ([PP1]_{\text{Total}} - [PP1])}$$

$$+ \frac{K_{\text{cat}5}[PP2A]([PP1]_{\text{Total}} - [PP1])}{k_{m11} + ([PP1]_{\text{Total}} - [PP1])}$$

$$- \frac{K_{\text{cat}7}[PKA]([I1]_{\text{Total}} - [PP1]_{\text{Total}} + [PP1])}{k_{m12} + ([I1]_{\text{Total}} - [PP1]_{\text{Total}} + [PP1]},$$

$$\frac{d[CaMKII]}{dt} = k_5[CaCaM]([CaMKII]_{\text{Total}} - [CaMKII]) - \frac{V_{m1}[CaMKII]^{n2}}{k_{m13} + [CaMKII]^{n2}} - \frac{K_{\text{cat8}}[PP1][CaMKII]^{n3}}{k_{m14} + [CaMKII]^{n3}},$$

$$\frac{d[AMPAR]^{\text{None}}_{\text{Cytoplasm}}}{dt} = -K_{\text{exo}}[AMPAR]^{\text{None}}_{\text{Cytoplasm}} + K_{\text{end}}[AMPAR]^{\text{None}}_{PSD},$$

$$\frac{d[AMPAR]^{\text{Single}}_{\text{Cytoplasm}}}{dt} = -K_{\text{exo}}[AMPAR]^{\text{Single}}_{\text{Cytoplasm}} + K_{\text{end}}[AMPAR]^{\text{Single}}_{PSD},$$

$$\frac{d[AMPAR]^{\text{Double}}_{\text{Cytoplasm}}}{dt} = -K_{\text{exo}}[AMPAR]^{\text{Double}}_{\text{Cytoplasm}} + K_{\text{end}}[AMPAR]^{\text{Double}}_{PSD},$$

$$\frac{d[AMPAR]^{\text{None}}_{PSD}}{dt} = K_{\text{exo}}[AMPAR]^{\text{None}}_{\text{Cytoplasm}} - K_{\text{end}}[AMPAR]^{None}_{PSD} - \frac{K_{\text{cat9}}[ActiveCaMKII][AMPAR]^{\text{None}}_{PSD}}{k_{m19} + [AMPAR]^{\text{None}}_{PSD}} + \frac{K_{\text{cat10}}[PP1][AMPAR]^{\text{Single}}_{PSD}}{k_{m20} + [AMPAR]^{\text{Single}}_{PSD}},$$

$$\frac{d[AMPAR]^{\text{Single}}_{PSD}}{dt} = K_{\text{exo}}[AMPAR]^{\text{Single}}_{Cytoplasm} - K_{\text{end}}[AMPAR]^{\text{Single}}_{PSD} + \frac{K_{\text{cat9}}[ActiveCaMKII][AMPAR]^{\text{None}}_{PSD}}{k_{m19} + [AMPAR]^{\text{None}}_{PSD}}$$

$$- \frac{K_{\text{cat10}}[PP1][AMPAR]_{PSD}^{\text{Single}}}{k_{m20} + [AMPAR]_{PSD}^{\text{Single}}}$$

$$- \frac{K_{\text{cat9}}[ActiveCaMKII][AMPAR]_{PSD}^{\text{Single}}}{k_{m19} + [AMPAR]_{PSD}^{\text{Single}}}$$

$$+ \frac{K_{\text{cat10}}[PP1][AMPAR]_{PSD}^{\text{Double}}}{k_{m20} + [AMPAR]_{PSD}^{\text{Double}}},$$

$$\frac{d[AMPAR]_{PSD}^{\text{Double}}}{dt} = K_{\text{exo}}[AMPAR]_{\text{Cytoplasm}}^{\text{Double}}$$

$$+ K_{\text{end}}[AMPAR]_{PSD}^{\text{Double}}$$

$$+ \frac{K_{\text{cat9}}[ActiveCaMKII][AMPAR]_{PSD}^{\text{Single}}}{k_{m19} + [AMPAR]_{PSD}^{\text{Single}}}$$

$$- \frac{K_{\text{cat10}}[PP1][AMPAR]_{PSD}^{\text{Double}}}{k_{m20} + [AMPAR]_{PSD}^{\text{Double}}},$$

where

$$K_{\text{exo}} = \frac{V_{m2}k_{m16}[PKA]^{n4}}{(k_{m15} + [PKA]^{n4})(k_{m16} + [PP2B])},$$

$$K_{\text{end}} = \frac{V_{m3}k_{m18}[AMPAR]_{PSD}^{\text{None}}[PP1]^{n5}}{(k_{m17} + [PP1]^{n5})(k_{m18} + [PKA])}.$$

The model shown in this chapter is solved by using Matlab with the parameter values and initial conditions shown in Tables 10.1 and 10.2. The parameters governing the synaptic inner protein–protein interactions are retrieved from the DOQCS databases and the parameters of the AMPAR trafficking are estimated to fit the experimental results of the AMPAR trafficking time courses shown in Fig. 10.4.

Table 10.1. Parameter values of MSPM.

| Parameters | Value | Parameters | Value | Parameters | Value |
|---|---|---|---|---|---|
| km1 | 0.02 $\mu$M | n1 | 2 | Kcat1 | 4.5 $\mu$M/s |
| km2 | 300 $\mu$M | n2 | 0.98 | Kcat2 | 2 $\mu$M/s |
| km3 | 300 $\mu$M | n3 | 2 | Kcat3 | 10 $\mu$M/s |
| km4 | 19.841 $\mu$M | n4 | 2 | Kcat4 | 1.667 $\mu$M/s |
| km5 | 0.0085 $\mu$M | n5 | 2 | Kcat5 | 0.34 $\mu$M/s |
| km6 | 39.7 $\mu$M | $k_{\text{diffuse}}$ | 0.018 $\mu$M/s | Kcat6 | 2 $\mu$M/s |
| km7 | 0.0069 $\mu$M | k1 | 65 $\mu$M/s | Kcat7 | 9 $\mu$M/s |
| km8 | 0.0167 $\mu$M | k2 | 2 $\mu$M/s | Kcat8 | 0.0115 $\mu$M/s |
| km9 | 0.3 $\mu$M | k3 | 599.99 $\mu$M/s | Kcat9 | 0.5 $\mu$M/s |
| km10 | 4.9708 $\mu$M | k4 | 1 $\mu$M/s | Kcat10 | 0.35 $\mu$M/s |
| km11 | 16 $\mu$M | k5 | 0.01 $\mu$M/s | | |
| km12 | 7.5001 $\mu$M | Vm1 | 0.202 $\mu$M/s | | |
| km13 | 6.2612 $\mu$M | Vm2 | 0.75 $\mu$M/s | | |
| km14 | 0.0494 $\mu$M | Vm3 | 0.003 $\mu$M/s | | |
| km15 | 0.04 $\mu$M | | | | |
| km16 | 0.0015 $\mu$M | | | | |
| km17 | 0.198 $\mu$M | | | | |
| km18 | 0.2 $\mu$M | | | | |
| km19 | 10.00001 $\mu$M | | | | |
| km20 | 0.22222 $\mu$M | | | | |

Table 10.2. Buffer levels, initial conditions and total levels of MSPM.

| Buffer | | Initial conditions | | Total levels | |
|---|---|---|---|---|---|
| [ATP] | 2000 $\mu$M | [cAMP] | 0 $\mu$M | [AC1] | 0.074074 $\mu$M |
| | | | | [AC2] | 0.074074 $\mu$M |
| | | [PKA] | 0 $\mu$M | [cAMP.PDE] | 0.55556 $\mu$M |
| | | | | [PDE1] | 2.5926 $\mu$M |
| | | [PP2B] | 0 $\mu$M | [$R_2C_2$] | 0.5 $\mu$M |
| [PP2A] | 0.1111 $\mu$M | | | [PKA.Inhibitor] | 0.25926 $\mu$M |
| | | [PP1] | 0.4444 $\mu$M | [PP2B] | 1 $\mu$M |
| | | | | [PP1] | 0.4444 $\mu$M |
| | | [CaMKII] | 0 $\mu$M | [I1] | 0.4444 $\mu$M |
| | | | | [CaMKII] | 20 $\mu$M |

## Appendix H

### H.1. SPI Derivation

Based on the background information, we have the following equations describing the relationships governing the synaptic properties:

$$\frac{PSDAMPAR}{CytoplasmAMPAR} = f_1(PKA, PP1, PP2B),$$

$$\frac{SinglePhosPSDAMPAR}{NonePhosPSDAMPAR} = f_2(CamKII, PP1),$$

$$\frac{DoublePhosPSDAMPAR}{NonePhosPSDAMPAR} = f_3(CamKII, PP1).$$

Here, kinases (*PKA*, *CaMKII*) and phosphatases (*PP*1, *PP*2*B*) are positively and negatively correlated with the corresponding synaptic properties, respectively.

Applying the Buckingham $\pi$ theorem gives us the following expressions:

$$\frac{PSDAMPAR}{CytoplasmAMPAR} = \phi[PKA]^a[PP1]^b[PP2B]^c,$$

$$\frac{SinglePhosPSDAMPAR}{NonePhosPSDAMPAR} = \varphi[CaMKII]^d[PP1]^e,$$

$$\frac{DoublePhosPSDAMPAR}{NonePhosPSDAMPAR} = \theta[CaMKII]^f[PP1]^g.$$

where $a$, $b$, $c$, $d$, $e$, $f$, $g$, $\varphi$, $\phi$ and $\theta$ are constants dependent on the cell-to-cell variability of synapses. Based on the two physical dimensions, mole (M) and liter (L), we have the following equalities of the exponents:

$$a + b + c = 0,$$

$$-a - b - c = 0,$$

$$d + e = 0,$$

$$-d - e = 0,$$

$$f + g = 0,$$

$$-f - g = 0.$$

Solving the system gives us the following values for the exponents:

$$a = -b - c,$$
$$e = -d,$$
$$g = -f,$$

where $b, c < 0$ and $e, f > 0$ indicating the positive relationship between $PKA/CaMKII$ and the synaptic strength and the negative relationship between $PP2B/PP1$ and the synaptic strength. Now we have:

$$\frac{PSDAMPAR}{CytoplasmAMPAR} = \phi \frac{[PKA]^{-b-c}}{[PP1]^{-b}[PP2B]^{-c}}, \tag{H.1}$$

$$\frac{SinglePhosPSDAMPAR}{NonePhosPSDAMPAR} = \varphi \left(\frac{[CaMKII]}{[PP1]}\right)^d, \tag{H.2}$$

$$\frac{DoublePhosPSDAMPAR}{NonePhosPSDAMPAR} = \theta \left(\frac{[CaMKII]}{[PP1]}\right)^f. \tag{H.3}$$

Based on Eq. (H.1), the PSD AMPAR/Total AMPAR ratio can be derived as:

$$\frac{PSDAMPAR}{TotalAMPAR} = \frac{[PKA]^{-b-c}}{\phi[PKA]^{-b-c} + [PP1]^{-b}[PP2B]^{-c}}. \tag{H.4}$$

This expression determines the maximum level of SPI.

PSD AMPAR consists of three states: none *ser831* phosphorylation, a single *ser831* phosphorylation and double *ser831* phosphorylation. If the number of the none phosphorylated PSD AMPAR is 1, then the total number of PSD AMPAR can be derived based on Eqs. (H.2) and (H.3):

$$1 + \varphi \left(\frac{[CaMKII]}{[PP1]}\right)^d + \theta \left(\frac{[CaMKII]}{[PP1]}\right)^f. \tag{H.5}$$

Based on Eq. (H.5), the ratio of none phosphorylated PSD AMPAR to the PSD AMPAR can be expressed as

$$\frac{NonePhosPSDAMPAR}{PSDAMPAR} = \frac{1}{1+\varphi\left(\frac{[CaMKII]}{[PP1]}\right)^d + \theta\left(\frac{[CaMKII]}{[PP1]}\right)^f}. \tag{H.6}$$

Similarly, the ratios of single and double phosphorylated PSD AMPAR to the PSD AMPAR are expressed as

$$\frac{SinglePhosPSDAMPAR}{PSDAMPAR} = \frac{\varphi\left(\frac{[CaMKII]}{[PP1]}\right)^d}{1+\varphi\left(\frac{[CaMKII]}{[PP1]}\right)^d + \theta\left(\frac{[CaMKII]}{[PP1]}\right)^f}, \tag{H.7}$$

and

$$\frac{DoublePhosPSDAMPAR}{PSDAMPAR} = \frac{\theta\left(\frac{[CaMKII]}{[PP1]}\right)^f}{1+\varphi\left(\frac{[CaMKII]}{[PP1]}\right)^d + \theta\left(\frac{[CaMKII]}{[PP1]}\right)^f}. \tag{H.8}$$

By combining Eqs. (H.4), (H.7) and (H.8), SPI is represented by Eq. (H.9) as a product of its maximum level and effective percentage:

$$SPI = \frac{\phi[PKA]^{-b-c}}{\phi[PKA]^{-b-c} + [PP1]^{-b}[PP2B]^{-c}} \times \left(\alpha + \frac{\beta\varphi\left(\frac{[CaMKII]}{[PP1]}\right)^d + \gamma\theta\left(\frac{[CaMKII]}{[PP1]}\right)^f}{1+\varphi\left(\frac{[CaMKII]}{[PP1]}\right)^d + \theta\left(\frac{[CaMKII]}{[PP1]}\right)^f}\right), \tag{H.9}$$

where $\alpha$ is the normalized base conductance coefficient, $\beta$ is the normalized single *ser831* phosphorylation conductance enhancement coefficient and $\gamma$ is the normalized double *ser831* phosphorylation conductance enhancement coefficient.

## H.2. Model Formulation

Biological systems are modeled as systems of biochemical reactions that describe the major forms of protein activities: (1) activation/inhibition and (2) synthesis/degradation. Due to the rapid synaptic reactions during the initiation of synaptic plasticity, we assume that the short timeframe does not fit the timescale of type (2) reaction rates, so all the biochemical reactions in the MSPM are type (1) activities. In addition, most of the reactions associated with synaptic plasticity are the enzymatic-dependent phosphorylation/dephosphorylation activities. Therefore, we model most of the biochemical reactions based on Michaelis-Menten kinetics.

Consider the following situation where AMPAR is phosphorylated by *CaMKII*:

$$AMPAR + CAMKII \underset{k_{\text{off}}}{\overset{k_{\text{on}}}{\rightleftharpoons}} AMPAR,$$

$$CAMKII \xrightarrow{K_{\text{cat}}} AMPAR^{*} + CaMKII.$$

The dynamics of the components involved in the phosphorylation process can be described by the following ODEs based on the mass action law:

$$\begin{aligned}
\frac{d[AMPAR]}{dt} &= -k_f[AMPAR][CaMKII] \\
&\quad +k_b[AMPAR \cdot CaMKII], \qquad \text{(H.10)} \\
\frac{d[CaMKII]}{dt} &= -k_f[AMPAR][CaMKII] \\
&\quad + k_b[AMPAR \cdot CaMKII] \\
&\quad + K_{\text{cat}}[AMPAR \cdot CaMKII], \qquad \text{(H.11)} \\
\frac{d[AMPAR \cdot CaMKII]}{dt} &= k_f[AMPAR][CaMKII] \\
&\quad - k_b[AMPAR \cdot CaMKII] \\
&\quad - K_{\text{cat}}[AMPAR \cdot CaMKII], \qquad \text{(H.12)}
\end{aligned}$$

and

$$\frac{d[AMPAR^*]}{dt} = K_{\text{cat}}[AMPAR \cdot CaMKII]. \tag{H.13}$$

If $k_f$ and $k_b$ are large, indicating very rapid collision between AMPAR and *CaMKII* in forming the $AMPAR \cdot CaMKII$ complex as well as the fast rate of reverting $AMPAR \cdot CaMKII$ back into AMPAR and *CaMKII*, then it can be assumed that AMPAR, *CaMKII* and $AMPAR \cdot CaMKII$ complex mix instantaneously so that their levels shift to the equilibrium immediately. As the zero flux at the equilibrium, the level of $AMPAR \cdot CaMKII$ complex can be approximated by the following expression:

$$[AMPAR \cdot CaMKII] = k_f/k_b[AMPAR \cdot CaMKII]. \tag{H.14}$$

Equation (H.14) can be modified to:

$$[AMPAR \cdot CaMKII] = k_f/k_b[AMPAR]([CaMKII]^0 - [AMPAR \cdot CaMKII]). \tag{H.15}$$

Rearrange Eq. (H.15):

$$[AMPAR \cdot CaMKII] = \frac{[AMPAR][CaMKII]^0}{k_d[AMPAR]}, \tag{H.16}$$

where $k_d = k_f/k_b$. Substituting Eq. (H.16) into Eq. (H.13), the dynamic of the phosphorylated AMPAR is:

$$\frac{d[AMPAR^*]}{dt} = \frac{K_{\text{cat}}[CaMKII]^0[AMPAR]}{k_d + [AMPAR]}.$$

The key signals of synaptic plasticity, $Ca^{2+}$ and the $Ca^{2+}/CaM$ complex, are modeled in a number of synaptic plasticity models [13, 59, 71], so that it is not necessary to remodel the interactions between $Ca^{2+}$ and *CaM* again in MSPM. The simulated $Ca^{2+}$ and $Ca^{2+}/CaM$ dynamics are retrieved from the existing models and are used as the inputs of MSPM. By doing this, MSPM could focus more on the interactions between the modulators initiating synaptic plasticity and it is more convenient to compare with other synaptic plasticity models with simulated $Ca^{2+}$ and $Ca^{2+}/CaM$ dynamics.

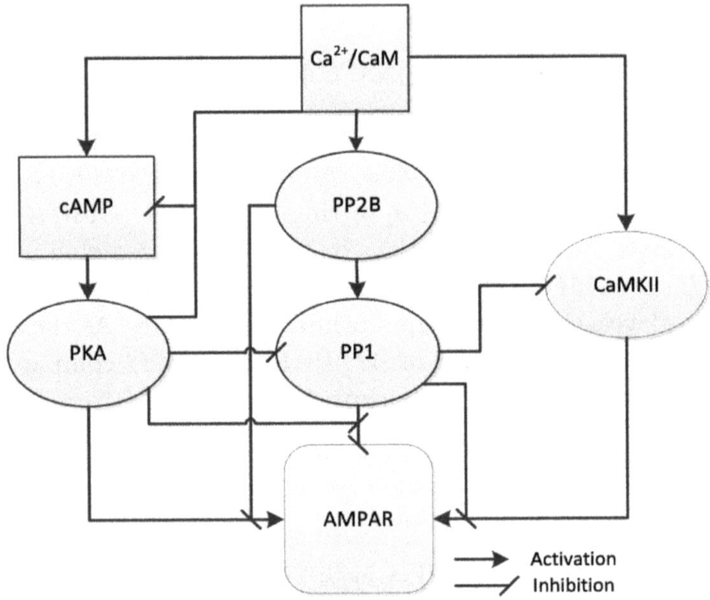


Fig. H.1. Schematic diagram of synaptic interactions in the emergence of synaptic plasticity.

DOQCS databases [60] provided the main pathways as well as the reaction rates of the biochemical reactions involved in synaptic plasticity. The MSPM model is developed based on the proposed design logic discussed in the main text (Fig. H.1) and the pathway information was provided by the DOQCS databases. MSPM is further simplified into an operational model with the main modulators of synaptic plasticity only, in order to track the basal dynamics of synaptic plasticity. The detailed model formation steps are described in the following sections of this document.

## H.3. cAMP and PKA

The cyclic adenosine monophosphate ($cAMP$) and $PKA$ pathway involve the activations of $cAMP$-associated adenylyl cyclase ($AC$) and phosphodiesterase ($PDE$), which lead to the regulated activity of $cAMP$ as well as its target protein, $PKA$. These interactions are presented by the following biological reactions:

$$AC2 \leftrightarrow AC2^*,$$

$$Ca_4CaM + AC1 \leftrightarrow Ca_4CaM \cdot AC1,$$

$$ATP \xrightarrow{Ca_4CaM \cdot AC1, AC2^*} cAMP,$$

$$cAMP \cdot PDE^* \underset{PKA}{\Leftrightarrow} cAMP \cdot PDE,$$

$$PDE1 + Ca_4CaM \leftrightarrow CaM \cdot PDE1,$$

$$cAMP \xrightarrow{cAMP \cdot PDE, CaM \cdot PDE1} AMP,$$

$$R_2C_2 + 4cAMP \leftrightarrow 2PKA + R_2 \cdot cAMP_4,$$

$$PKA + PKA \cdot Inhibitor \leftrightarrow Inhibited \cdot PKA.$$

Here, $cAMP$ and $PKA$ indicate the active forms of the corresponding proteins, while all other notations indicate the inactive forms.

The activation of adenylyl cyclase type 2 ($AC2$), the $\mathrm{Ca}^{2+}/CaM$ independent $AC$, does not require assistance of other components and its level is very low, which only causes significant changes to the level of $cAMP$ at low system activity. Therefore, the dynamics of $AC2$ can be fitted into a simple function in terms of the simulation time, $t$:

$$[AC2] = ([AC2]^e + \exp(-0.11 \cdot t)([AC2]^0 - [AC2]^e)),$$

where $[AC2]^e$ and $[AC2]^0$ are the equilibrium and initial levels of $AC2$, respectively, and $t$ is the simulation time.

The reaction rate of $\mathrm{Ca}^{2+}/CaM$-dependent activation of adenylyl cyclase type 1 ($AC1$) is very big (long arrow) indicating a rapid formation of $AC1$ in the presence of $\mathrm{Ca}^{2+}/CaM$ complex. Therefore, we assume that the level of $AC1$ changes rapidly in relation to the level of $\mathrm{Ca}^{2+}/CaM$ complex ($Ca_4CaM$) and a Hill function could describe the activation of $AC1$:

$$[Ca_4CaM \cdot AC1] = \frac{[AC1]_{\mathrm{Total}}[Ca_4CaM]}{(k_{m(AC1 \cdot Ca_4CaM)} + [Ca_4CaM]},$$

where $[AC1]_{\text{Total}}$ is the total level of $AC1$ and $k_{m(AC1 \cdot Ca_4CaM)}$ is the level of $Ca_4CaM$ at which half of the $AC1$ is converted to the active form, $Ca_4CaM \cdot AC1$.

Same as the $AC1$ formation, $CaM \cdot PDE1$ and $cAMP \cdot PDE^*$ formations are rapid, and are largely dependent on the levels of $\text{Ca}^{2+}/CaM$ complex and $PKA$, respectively. Therefore, we describe their dynamics by Hill functions in terms of the $\text{Ca}^{2+}/CaM$ complex and $PKA$:

$$[CaM \cdot PDE1] = \frac{[PDE1]_{\text{Total}}[Ca_4CaM]}{(k_{m(PDE1 \cdot Ca_4CaM)} + [Ca_4CaM])},$$

$$[cAMP \cdot PDE^*] = \frac{[cAMP \cdot PDE]_{\text{Total}}[PKA]}{(k_{m(cAMP \cdot PDE \cdot PKA)} + [PKA])},$$

where $[PDE1]_{\text{Total}}$ and $[cAMP \cdot PDE]_{\text{Total}}$ are the total levels of $PDE1$ and $cAMP \cdot PDE$, respectively. $k_{m(PDE1 \cdot Ca_4CaM)}$ is the level of $Ca_4CaM$ at which half of the $PDE1$ is converted to $CaM \cdot PDE1$, and $k_{m(cAMP \cdot PDE \cdot PKA)}$ is the level of $PKA$ at which half of the $cAMP \cdot PDE$ is converted to $cAMP \cdot PDE^*$.

Initially, $PDE1$ and $cAMP \cdot PDE$ repress $cAMP$ activity at very low levels. Upon being catalyzed by the $\text{Ca}^{2+}/CaM$ complex or $PKA$ into more active forms ($CaM \cdot PDE1$ or $cAMP \cdot PDE^*$), the inhibition of cAMP increases approximately five-fold and one-fold, respectively (according to the reaction rates provided by the database). Therefore, we assume that the basal inhibition rates of $cAMP$ associated with $PDE1$ and $cAMP \cdot PDE$ are 1, whereas the $CaM \cdot PDE1$- and/or $cAMP \cdot PDE^*$-dependent inhibition rates of $cAMP$ could reach the maxima of 6 and 2, respectively. Hence, the $cAMP$ dynamics can be described by the following ODE:

$$\frac{d[cAMP]}{dt} = k_{\text{diffuse}} \left( \frac{K_{\text{cat}(AC1 \cdot ATP)}[ATP][Ca_4CaM]}{(k_{m(AC1 \cdot Ca_4CaM)} + [Ca_4CaM]) (k_{m(AC1 \cdot ATP)} + [ATP])} \right.$$

$$+ \frac{K_{\text{cat}(AC2 \cdot ATP)}[ATP]([AC2]^e + \exp(-0.11 \cdot t)([AC2]^0 - [AC2]^e))}{(k_{m(AC2 \cdot ATP)} + [ATP])}$$

$$
\begin{aligned}
&- \frac{K_{\text{cat}(cAMP\cdot PDE\cdot cAMP)}[cAMP][cAMP\cdot PDE]_{\text{Total}}}{(k_{m(cAMP\cdot PDE\cdot cAMP)} + [cAMP])} \\
&\times \left(1 + \frac{[PKA]}{k_{m(cAMP\cdot PDE\cdot PKA)} + [PKA]}\right) \\
&- \frac{K_{\text{cat}(PDE1\cdot cAMP)}[cAMP][PDE1]_{\text{Total}}}{k_{m(PDE1\cdot cAMP)} + [cAMP]} \\
&\times \left(1 + \frac{5[Ca_4CaM]}{k_{m(PDE1\cdot Ca_4CaM)} + [Ca_4CaM]}\right) \\
&- 4k_{f(PKA\cdot\text{activation})}[R_2C_2][cAMP]^4 \\
&\left. + 4k_{b(PKA\cdot\text{activation})}[PKA]^2([R_2]_{\text{Total}} - [R_2C_2]) \right),
\end{aligned}
$$

where $[R_2C_2]_{\text{Total}}$ is the total level of $R_2$, which is the regulatory subunit of *PKA*. $k_{\text{diffuse}}$ is the diffusion rate of *cAMP* from the cytoplasm to the dendrite.

Similarly, the dynamics of *PKA* are described by the following ODE:

$$
\begin{aligned}
\frac{d[PKA]}{dt} &= 2k_{f(PKA\cdot\text{activation})}[R_2C_2][cAMP]^4 \\
&\quad - 2k_{b(PKA\cdot\text{activation})}[PKA]^2 \\
&\quad \times([R_2C_2]_{\text{Total}} - [R_2C_2]_{\text{Total}} - [R_2C_2]) \\
&\quad - k_{f(PKA\cdot\text{activation})}[PKA][PKA\cdot Inhibitor] \\
&\quad + k_{b(PKA\cdot inhibition)}[Inhibited\cdot PKA].
\end{aligned}
$$

The inhibition of *PKA* is a rapid reaction, so it is assumed that the inhibition of *PKA* occurs immediately after the production of *PKA* and the balance between *PKA* and its inhibitor immediately reaches the equilibrium. Based on this assumption, the level of inhibited *PKA* can be expressed by a function of the level of *PKA*:

$$
[Inhibited\cdot PKA] = \frac{[PKA\cdot Inhibitor]_{\text{Total}}[PKA]}{k_{m(PKA\cdot Inhibitor)} + [PKA]},
$$

where $[PKA \cdot Inhibitor]_{\text{Total}}$ is the total level of $PKA$ inhibitor and $k_{m(PKA \cdot Inhibitor)}$ is the level of $PKA$ at which half of the $PKA$ inhibitor is converted to Inhibited $PKA$. The total levels of $PKA$ are also conserved:

$$[R_2C_2] = [R_2C_2]_{\text{Total}} - \frac{[PKA]}{2} - \frac{[PKA \cdot Inhibitor]_{\text{Total}}[PKA]}{k_{m(PKA \cdot Inhibitor)} + [PKA]}.$$

Next, the following time derivatives can be obtained:

$$\begin{aligned}\frac{d}{dt}[Inhibited \cdot PKA] &= \frac{d}{dt}\left(\frac{[PKA \cdot Inhibitor]_{\text{Total}}[PKA]}{k_{m(PKA \cdot Inhibitor)} + [PKA]}\right)\\ &= \frac{k_{m(PKA \cdot Inhibitor)}[PKA \cdot Inhibitor]_{\text{Total}}\frac{d[PKA]}{dt}}{(k_{m(PKA \cdot Inhibitor)} + [PKA])^2}.\end{aligned}$$

Based on the time derivatives, the ODE of $PKA$ can be modified to

$$\begin{aligned}\frac{d[PKA]}{dt} &= (2k_{f(PKA\cdot\text{activation})}[R_2C_2][cAMP]^4\\ &\quad - 2k_{b(PKA\cdot\text{activation})}[PKA]^2([R_2C_2]_{\text{Total}} - R_2C_2))\\ &\quad - \frac{k_{m(PKA \cdot Inhibitor)}[PKA \cdot Inhibitor]_{\text{Total}}\frac{d[PKA]}{dt}}{(k_{m(PKA \cdot Inhibitor)} + [PKA])^2}.\end{aligned}$$

Arranging the above equation, the dynamics of $PKA$ is now:

$$\begin{aligned}\frac{d[PKA]}{dt} &= (2k_{f(PKA\cdot\text{activation})}[R_2C_2][cAMP]^4\\ &\quad - 2k_{b(PKA\cdot\text{activation})}[PKA]^2([R_2C_2]_{\text{Total}} - R_2C_2))\\ &\quad \times g\frac{(k_{m(PKA \cdot Inhibitor)} + [PKA]^2}{(k_{m(PKA \cdot Inhibitor)} + [PKA]^2 + k_{m(PKA \cdot Inhibitor)}[PKA \cdot Inhibitor]_{\text{Total}}},\end{aligned}$$

and

$$[R_2C_2] = [R_2C_2]_{\text{Total}} - \frac{[PKA]}{2} - \frac{[PKA \cdot Inhibitor]_{\text{Total}}[PKA]}{k_{m(PKA \cdot Inhibitor)} + [PKA]}.$$

## H.4. PP2B

The total level of all the forms of $PP2B$ is conserved:

$PP2B$ is activated through a signal cascade (shown as the below biological reactions) involving both $Ca^{2+}$ and $Ca^{2+}$/CaM complex. $Ca_4 \cdot CaM \cdot Ca_4 \cdot PP2B$ is the active form of $PP2B$ in the biological pathway.

$$Basal \cdot PP2B + 2Ca \leftrightarrow Ca_2PP2B + 2Ca,$$

$$Ca_2PP2B + 2Ca \leftrightarrow Ca_4 \cdot PP2B + Ca_4 \cdot CaM,$$

$$Ca_4 \cdot PP2B + Ca_4 \cdot CaM \longrightarrow Ca_4CaM \cdot Ca_4 \cdot PP2B.$$

Here, we use $PP2B$ to indicate the active form of $PP2B$ for simplicity. The total level of all the forms of $PP2B$ is conserved:

$$[Basal{\cdot}PP2B]+[Ca_2{\cdot}PP2B]+[Ca_4{\cdot}PP2B]+[PP2B] = [PP2B]_{\text{Total}}.$$

Inactive $PP2B$ has three forms, $Basal \cdot PP2B$, $Ca_2 \cdot PP2B$ and $Ca_4 \cdot PP2B$. We assume that the ratio between the three inactive forms is dependent on the $Ca^{2+}$ signal, so that $Ca_4 \cdot PP2B$ can be expressed as a function of the level of $Ca^{2+}$ by applying the Hill function:

$$[Ca_4 \cdot PP2B] = \frac{([PP2B]_{\text{Total}} - [PP2B])[Ca^{2+}]^n}{(k_{m(Ca_4 \cdot PP2B)} + [Ca^{2+}]^n)},$$

where $k_{m(Ca_4 \cdot PP2B)}$ is the $n$th power of the level of $Ca^{2+}$ at which half of the inactive form of $PP2B$ is $Ca_4 \cdot PP2B$.

Finally, the kinetics of the active form $PP2B$ is described as the following ODE based on the mass action law:

$$\frac{d[PP2B]}{dt} = k_{f(PP2B)}[CaCaM]\frac{([PP2B]_{\text{Total}} - [PP2B])[Ca]^n}{(k_{m(Ca_4 \cdot PP2B)} + [Ca]^n)} - k_{b(PP2B)}[PP2B].$$

## H.5. PP1

The $PP1$ pathway describes the state changes of $PP1$ between the inactive form, $I1^*PP1$, and the active form, $PP1$. These changes are shown as the following biological reactions:

$$I1 \cdot PP1 \rightarrow PP1 + I1,$$

$$I1 \xrightarrow{PKA} I1^*,$$

$$I1^* \xrightarrow{PP2A, PP2B} I1,$$

$$PP1 + I1^* \leftrightarrow I1^*PP1,$$

$$I1^*PP1 \xrightarrow{PP2A, PP2B} I1 \cdot PP1.$$

$PP1$ binds to the phosphorylated inhibitor 1 ($I1$) and becomes inactive, while the phosphorylation and dephosphorylation of $I1$ requires $PKA$- or $PP2A/PP2B$-dependent catalysis, respectively. The fast-binding rate between $PP1$ and the phosphorylated $I1$ consumes most of the phosphorylated $I1$ and results in the low level of the stand-alone phosphorylated $I1$ throughout the whole process. Hence, we assume that $I1$ firstly phosphorylates by $PKA$ and then the phosphorylated $I1$ rapidly binds to $PP1$ in which the negligible level of phosphorylated $I1$ is left over. Therefore, the phosphorylated $I1$ only exists as $I1^*PP1$ complex. In addition, we assume that $I1$ and $PP1$ bind or unbind immediately after the phosphorylation or the dephosphorylation of $I1$, respectively. Now, the interactions between $PP1$ and $I1$ can be simplified to the following biological reactions:

$$I1^*PP1 \xrightarrow{PP2A, PP2B} I1 + PP1 \xrightarrow{PKA} I1^*PP1.$$

Since we assumed that all the phosphorylated $I1$ rapidly binds and inhibits $PP1$, the inhibition rate of $PP1$ is approximately equal to the production rate of the phosphorylated $I1$. However, this assumption is only valid when the total level of $I1$ is equal to or less than that of $PP1$ (in this case, it is true according to the DOQCS databases).

The competitive interactions between $PKA$ and $PP2B/PP2A$ to alter the activity of $I1$ are the core of the $PP1$ pathway. Due to the catalytic characteristic of these interactions to $I1$, Michaelis–Menten kinetics are used to approximate the dynamics of $PP1$:

$$\frac{d[PP1]}{dt} = \frac{K_{cat(PP2B\cdot PP1)}[PP2B]([PP1]_{\text{Total}} - [PP1])}{k_{m(PP2B\cdot PP1)} + ([PP1]_{\text{Total}} - [PP1])} + \frac{K_{cat(PP2A\cdot PP1)}[PP2A]([PP1]_{\text{Total}} - [PP1])}{k_{m(PP2A\cdot PP1)} + ([PP1]_{\text{Total}} - [PP1])} - \frac{K_{cat(PKA\cdot PP1)}[PKA]([I1]_{\text{Total}} - [PP1]_{\text{Total}} + [PP1])}{k_{m(PKA\cdot PP1)} + ([I1]_{\text{Total}} - [PP1]_{\text{Total}} + [PP1])},$$

where $[PP1]_{\text{Total}}$ is the total level of $PP1$, and $[I1]_{\text{Total}}$ is the total level of $I1$. Constants $k_{m(PP2B\cdot PP1)}$, $k_{m(PP2A\cdot PP1)}$ and $k_{m(PKA\cdot PP1)}$ are the levels of substrates, $I1^*PP1$ ($PP2B$- and $PP2A$-dependent catalysis, respectively) and $I1$, at which the reaction speeds reach half of their maxima. Since $PP2A$ is not sensitive to the $Ca^{2+}$ signal, the level of $PP2A$ is assumed to be a constant.

## H.6. CaMKII

Activation of *CaMKII* requires a sequence of activities: (1) binding to a $Ca^{2+}/CaM$ molecule; and (2) diffusing to and docking in a protein rich zone, called the postsynaptic density (PSD). *CaMKII* must undergo both activities to become active, but either activity can happen first. This activation process is described by the following set of biological reactions:

$$CaMKII \cdot CaM + NMDAR \rightarrow CaMKII \cdot CaM \cdot PSD,$$

$$CaMKII \cdot thr286^* \cdot CaM + NMDAR \rightarrow CaMKII \cdot thr286 \cdot CaM \cdot PSD,$$

$$Ca_4CaM \cdot PSD + CaMKII \cdot PSD \rightarrow CaMKII \cdot CaM \cdot PSD,$$

where docking in the PSD is represented as binding to a NMDAR receptor.

*CaMKII* is deactivated when *CaMKII* dissociates from the $Ca^{2+}/CaM$ binding and it occurs under either of the two situations: (1) decreased $Ca^{2+}$ level; and (2) presence of the $PP1$-dependent regulation. These processes are described by the following set of reactions:

$$CaMKII \cdot CaM \cdot PSD \rightarrow Ca_4CaM \cdot PSD + CaMKII \cdot PSD,$$

$$CaMKII \cdot thr286 \cdot PSD \xrightarrow{PP1 \cdot PSD} CaMKII \cdot PSD,$$

$$CaMKII^{***} \cdot PSD \xrightarrow{PP1 \cdot PSD} CaMKII \cdot thr305 \cdot PSD,$$

where $CaMKII^{***}$ indicates *CaMKII* which is phosphorylated at both *thr*286 and *thr*305.

Also, different inactive states of *CaMKII* are interchangable without altering the activity of *CaMKII*. These interactions are described by the following set of reactions:

$$Ca_4CaM + CaMKII \leftrightarrow CaMKII \cdot CaM,$$

$$Ca_4CaM + CaMKII \cdot thr286 \leftrightarrow CaMKII \cdot thr286^* \cdot CaM,$$

$$Ca_4CaM \cdot thr305 + NMDAR \leftrightarrow CaMKII \cdot thr305 \cdot PSD,$$

$$Ca_4CaM + NMDAR \leftrightarrow CaMKII \cdot PSD,$$

$$CaMKII \cdot thr286 \cdot PSD + Ca_4CaM \cdot PSD$$

$$\leftrightarrow CaMKII \cdot thr286 \cdot CaM \cdot PSD,$$

$$CaMKII \cdot thr286 \xrightarrow{tot \cdot CaM \cdot CaMKII} CaMKII^{***},$$

$$CaMKII \cdot CaM \xrightarrow{tot \cdot CaM \cdot CaMKII} CaMKII \cdot thr286^* \cdot CaM,$$

$$CaMKII \cdot thr286 \xrightarrow{tot \cdot autonomous \cdot CaMKII} CaMKII^{***},$$

$$CaMKII \cdot CaM \xrightarrow{tot \cdot autonomous \cdot CaMKII} CaMKII \cdot thr286^* \cdot CaM,$$

$$CaMKII \cdot thr286 \cdot PSD \xrightarrow{tot \cdot CaM \cdot CaMKII \cdot PSD} CaMKII^{***} PSD,$$

$$CaMKII \cdot CaM \cdot PSD \xrightarrow{tot \cdot CaM \cdot CaMKII \cdot PSD} CaMKII$$

$$\cdot thr286^* \cdot CaM \cdot PSD,$$

$$CaMKII \cdot thr286 \cdot PSD \xrightarrow{tot \cdot autonomous \cdot CaMKII \cdot PSD}$$

$$CaMKII^{***} PSD,$$

$$CaMKII \cdot CaM \cdot PSD \xrightarrow{tot \cdot autonomous \cdot CaMKII \cdot PSD} CaMKII$$

$$\cdot thr286^* \cdot CaM \cdot PSD,$$

$$CaMKII \cdot thr286^{*} \cdot CaM \xrightarrow{PP1} CaMKII \cdot CaM,$$

$$CaMKII \cdot thr286 \xrightarrow{PP1} CaMKII,$$

$$CaMKII \cdot thr305 \xrightarrow{PP1} CaMKII,$$

$$CaMKII^{***} \xrightarrow{PP1} CaMKII \cdot thr286,$$

$$CaMKII^{***} \xrightarrow{PP1} CaMKII \cdot thr305,$$

$$CaMKII \cdot thr286^{*} \cdot CaM \cdot PSD \xrightarrow{PP1 \cdot PSD} CaMKII \cdot CaM \cdot PSD,$$

$$CaMKII^{***} \cdot PSD \xrightarrow{PP1 \cdot PSD} CaMKII \cdot thr286 \cdot PSD.$$

MSPM only considers two states of *CaMKII*, active and inactive. *CaMKII* becomes active when binds to a $Ca^{2+}/CaM$ molecule and diffuses into PSD. The *CaMKII* docking in the PSD (PSD *CaMKII*) is shown as a binding to NMDAR in the above reactions. We assume that the level of NMDAR is sufficient for *CaMKII* binding so that the activation rate of *CaMKII* is purely dependent on the level of the $Ca^{2+}/CaM$ complex. The binding to $Ca^{2+}/CaM$ complex may break down when either $Ca^{2+}$ level goes down or $PP1$-dependent inhibition is applied. Hence, the dynamics of *CaMKII* consists of three parts and can be described by the following ODE:

$$\frac{d[CaMKII]}{dt} = k_{(CaMKII\cdot\text{activation})}[Ca_4CaM]([CaMKII]_{\text{Total}} - [CaMKII]) - \frac{V_{m(dissociate)}[CaMKII]^{n1}}{k_{m(dissociate)} + [CaMKII]^{n1}} - \frac{K_{\text{cat}(CaMKII\cdot PP1)}[PP1][CaMKII]^{n2}}{k_{m(CaMKII\cdot PP1} + [CaMKII]^{n2}},$$

where $[CaMKII]_{\text{Total}}$ is the total level of *CaMKII*. $k_{m(\text{dissociate})}$ is $n1$th power of the level of *CaMKII* at which the dissociate rate of the *CaMKII* and $Ca^{2+}/CaM$ binding reaches the half of the maximum, $V_{m(\text{dissociate})}$, and $k_{m(CaMKII\cdot PP1)}$ is $n2$th power of the level of *CaMKII* at which the rate of $PP1$-dependent regulation of *CaMKII* reaches half of its maximum.

## H.7. AMPAR Trafficking

AMPAR contains two *PKA*-dependent phosphorylation sites (*ser845*) and two *CaMKII*-dependent phosphorylation sites (*ser831*) that are the targets of the complicated kinase and phosphatase competitive network. Here, we are only interested in two aspects of AMPAR: its activity and conductance. Hence, it is not necessary for MSPM to capture the detailed composition of AMPAR and the required information of AMPAR is the localization of AMPAR and the *CaMKII*-dependent phosphorylation status of PSD AMPAR. Now, we can formulate a conceptual AMPAR trafficking diagram with six distinct states of AMPAR (Fig. H.2): each describes a unique combination of the two required pieces of information: localization

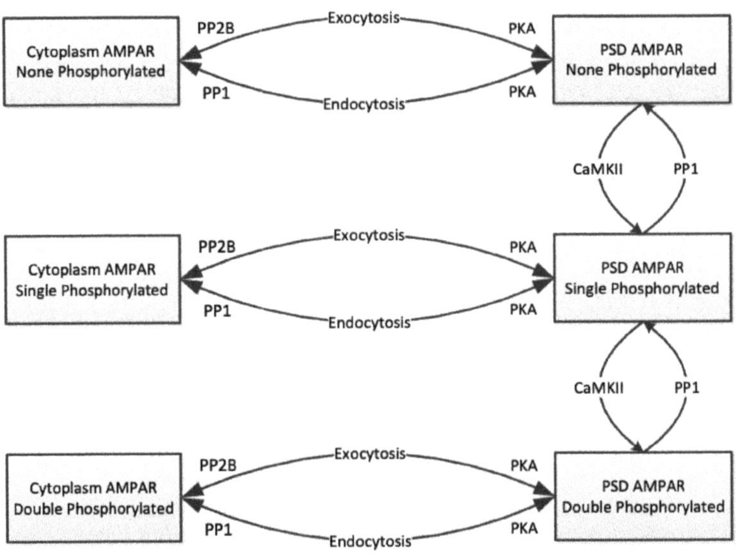


Fig. H.2. Conceptual diagram of AMPAR trafficking. Six distinct states of AMPAR are indicated by six rectangles. The arrow indicates the interchange between different states. The localization movements are indicated by the horizontal arrows that are regulated by PKA, PP1 and PP2B through enhanced exocytosis or endocytosis. The phosphorylation processes of PSD AMPAR are indicated by the vertical arrows, where CaMKII governs the phosphorylation and PP1 is responsible for the dephosphorylation.

(cytoplasm or PSD) and *CaMKII*-dependent *ser831* phosphorylation status (none, single or double).

The localization of AMPAR is the result of the competitive regulation between kinases (PKA) and phosphatases (PP1 and PP2B). We assume that AMPAR recycles at a maximum externalization rate, $V_{m(Externalization)}$, and a maximum internalization rate, $V_{m(Internalization)}$. The PKA-driven process enhances the externalization rate and can be explained by the Hill function, $[PKA]^{n1}/(k_{m(enhance-ex)} + [PKA]^{n1})$, where $k_{m(enhance-ex)}$ is the $n1$th power of the level of PKA at which the externalization rate of AMPAR increases to half of its maximum. The PP2B-driven process reduces the externalization rate and can be explained by the Hill function, $k_{m(reduce-ex)}/(k_{m(reduce-ex)} + [PP2B]^{n2})$, where $k_{m(reduce-ex)}$ is the $n2$th power of the level of $PP2B$ at which the externalization rate of AMPAR reduces to half of its maximum. Similarly, the enhancement and reduction of internalization of AMPAR can be explained by $PP1$-driven enhancement, $[PP1]^{n3}/(k_{m(enhance-in)} + [PP1])^{n3}$ $k_{m(enhance-in)}/(k_{m(enhance-in)} + [PP2B]^{n2})$, and $PKA$-driven reduction, $k_{m(reduce-in)}/(k_{m(reduce-in)} + [PKA]^{n4})$.

The conductance-related *CaMKII*-dependent phosphorylation and $PP1$-dependent dephosphorylation of AMPAR are modeled with Michaelis-Menten kinetics. Finally, the dynamics of the six states of AMPAR are described by the following ODEs:

$$\frac{d[AMPAR]^{None}_{Cytoplasm}}{dt} = -K_{exo}[AMPAR]^{None}_{Cytoplasm} + K_{end}[AMPAR]^{None}_{PSD},$$

$$\frac{d[AMPAR]^{Single}_{Cytoplasm}}{dt} = -K_{exo}[AMPAR]^{Single}_{Cytoplasm} + K_{end}[AMPAR]^{Single}_{PSD},$$

$$\frac{d[AMPAR]^{Double}_{Cytoplasm}}{dt} = -K_{exo}[AMPAR]^{Double}_{Cytoplasm} + K_{end}[AMPAR]^{Double}_{PSD},$$

$$\frac{d[AMPAR]_{PSD}^{None}}{dt} = K_{\text{exo}}[AMPAR]_{\text{Cytoplasm}}^{\text{None}} - K_{\text{end}}[AMPAR]_{PSD}^{\text{None}} - \frac{K_{\text{cat}(phosphorylation)}[ActiveCaMKII][AMPAR]_{PSD}^{None}}{k_{m(phosphorylation)} + [AMPAR]_{PSD}^{\text{None}}} + \frac{K_{\text{cat}(dephosphorylation)}[PP1][AMPAR]_{PSD}^{\text{Single}}}{k_{m(dephosphorylation)} + [AMPAR]_{PSD}^{\text{Single}}},$$

$$\frac{d[AMPAR]_{PSD}^{\text{Single}}}{dt} = K_{\text{exo}}[AMPAR]_{\text{Cytoplasm}}^{\text{Single}} - K_{\text{end}}[AMPAR]_{PSD}^{\text{Single}} + \frac{K_{cat(phosphorylation)}[ActiveCaMKII][AMPAR]_{PSD}^{\text{None}}}{k_{m(phosphorylation)} + [AMPAR]_{PSD}^{\text{None}}} - \frac{K_{cat(dephosphorylation)}[PP1][AMPAR]_{PSD}^{\text{Single}}}{k_{m(dephosphorylation)} + [AMPAR]_{PSD}^{\text{Single}}} - \frac{K_{cat(phosphorylation)}[PP1][AMPAR]_{PSD}^{\text{Single}}}{k_{m(phosphorylation)} + [AMPAR]_{PSD}^{\text{Single}}} + \frac{K_{cat(dephosphorylation)}[PP1][AMPAR]_{PSD}^{\text{Double}}}{k_{m(dephosphorylation)} + [AMPAR]_{PSD}^{\text{Double}}},$$

$$\frac{d[AMPAR]_{PSD}^{\text{Double}}}{dt} = K_{\text{exo}}[AMPAR]_{\text{Cytoplasm}}^{\text{Double}} + K_{\text{end}}[AMPAR]_{PSD}^{\text{Double}} + \frac{K_{\text{cat}(phosphorylation)}[ActiveCaMKII][AMPAR]_{PSD}^{\text{Single}}}{k_{m(phosphorylation)} + [AMPAR]_{PSD}^{\text{Single}}} - \frac{K_{\text{cat}(dephosphorylation)}[PP1][AMPAR]_{PSD}^{\text{Double}}}{k_{m(dephosphorylation)} + [AMPAR]_{PSD}^{\text{Double}}},$$

where $[ActiveCaMKII] = [CaMKII] + 0.2222$,

$$K_{exo} = \frac{V_{m(Externalization)}k_{m(reduce-ex)}[PKA]^{n1}}{(k_{m(enhance-ex)} + [PKA]^{n1})(k_{m(reduce-ex)} + [PP2B]^{n2})},$$

$$K_{end} = \frac{V_{m(Internalization)}k_{m(reduce-in)}[AMPAR]^{\text{None}}_{PSD}[PP1]^{n3}}{(k_{m(enhance-in)} + [PP1]^{n3})(k_{m(reduce-in)} + [PKA]^{n4})}.$$

## References


[1] Bear M.F., Connors B.W. and Paradiso M.A. (2007). *Neuroscience: Exploring the Brain.* Lippincott Williams & Wilkins, Philadelphia, PA.

[2] Bliss T.V. and Collingridge G.L. (1993). A synaptic model of memory: long-term potentiation in the hippocampus. *Nature*, *361*, pp. 31–39.

[3] Kandel E.R. (2009). The biology of memory: a forty-year perspective. *J. Neurosci.* *29*, pp. 12748–12756.

[4] Mayford M., Siegelbaum S.A. and Kandel E.R. (2012). Synapses and memory storage. *Cold Spring Harb. Perspect Biol.* *4*.

[5] Engert F. and Bonhoeffer T. (1999). Dendritic spine changes associated with hippocampal long-term synaptic plasticity. *Nature*, *399*, pp. 66–70.

[6] Morgado-Bernal I. (2011). Learning and memory consolidation: linking molecular and behavioral data. *Neurosci.*, *176*, pp. 12–19.

[7] Lamprecht R. and LeDoux J. (2004). Structural plasticity and memory. *Nat. Rev. Neurosci.*, *5*, pp. 45–54.

[8] Deng W., Aimone J.B. and Gage F.H. (2010). New neurons and new memories: how does adult hippocampal neurogenesis affect learning and memory? *Nat. Rev. Neurosci.*, *11*, pp. 339–350.

[9] Trommer B.L. *et al.* (2005). ApoE isoform-specific effects on LTP: blockade by oligomeric amyloid-beta1-42. *Neurobiol. Dis.* *18*, pp. 75–82.

[10] Viola K.L., Velasco P.T. and Klein W.L. (2008). Why Alzheimer's is a disease of memory: the attack on synapses by A beta oligomers (ADDLs). *J. Nutr. Health Aging*, *12*, pp. 51S–57S.

[11] Kotaleski J.H. and Blackwell K.T. (2010). Modeling the molecular mechanisms of synaptic plasticity using systems biology approaches. *Nat. Rev. Neurosci.*, *11*, pp. 239–251.

[12] Manninen T., Hituri K., Hellgren Kotaleski J., Blackwell K.T. and Linne M.-L. (2010). Postsynaptic signal transduction models for long-term potentiation and depression. *Frontier Comput. Neurosci.*, *4*.

[13] Hayer A. and Bhalla U.S. (2005). Molecular Switches at the Synapse Emerge from Receptor and Kinase Traffic. *PLoS Comput. Biol.*, *1*, p. e20.

[14] Yarin L.P. (2011). *The Pi-Theorem: Applications to Fluid Mechanics and Heat and Mass Transfer*, Springer Science & Business Media.

[15] Casti J. (1992). *Reality Rules*, 2 Volume Set, John Wiley & Sons.

[16] Castellani G.C., Quinlan E.M., Bersani F., Cooper L.N. and Shouval H.Z. (2005). A model of bidirectional synaptic plasticity: From signaling network to channel conductance. *Learning Memory*, *12*, pp. 423–432.

[17] Castellani G.C., Bazzani A. and Cooper L.N. (2009). Toward a microscopic model of bidirectional synaptic plasticity. *Proc. Nat. Acad. Sci. USA*, *106*, pp. 14091–14095.

[18] D'Alcantara P., Schiffmann S.N. and Swillens S. (2003). Bidirectional synaptic plasticity as a consequence of interdependent $Ca^{2+}$-controlled phosphorylation and dephosphorylation pathways. *European J. Neurosci.*, *17*, pp. 2521–2528.

[19] Bhalla U.S. and Iyengar R. (1999). Emergent properties of networks of biological signaling pathways. *Science*, *283*, pp. 381–387.

[20] Bhalla U.S. (2002). Biochemical signaling networks decode temporal patterns of synaptic input. *J. Comput. Neurosci.*, *13*, pp. 49–62.

[21] Berridge M.J. (2010). Calcium hypothesis of Alzheimer's disease. *Eur. J. Physiol.*, *459*, pp. 441–449.

[22] Berridge M.J. (2011). Calcium signalling and Alzheimer's disease. *Neurochem. Res.*, *36*, pp. 1149–1156.

[23] Citri A. and Malenka R.C. (2008). Synaptic plasticity: multiple forms, functions, and mechanisms. *Neuropsychopharmacology*, *33*, pp. 18–41.

[24] Stanton P.K., Bramham C. and Scharfman H.E. (2010). *Synaptic Plasticity and Transsynaptic Signaling*, Springer.

[25] Bliss T.V.P. and Lømo T. (1973). Long-lasting potentiation of synaptic transmission in the dentate area of the anaesthetized rabbit following stimulation of the perforant path. *J. Physiol.*, *232*, pp. 331–356.

[26] Dudek S.M. and Bear M.F. (1992). Homosynaptic long-term depression in area CA1 of hippocampus and effects of N-methyl-D-aspartate receptor blockade. *Proc. Natl. Acad. Sci.*, *89*, pp. 4363–4367.

[27] Collingridge G.L., Isaac J.T. and Wang Y.T. (2004). Receptor trafficking and synaptic plasticity. *Nat. Rev. Neurosci.*, *5*, pp. 952–962.

[28] Malinow R. and Malenka R.C. (2002). AMPA receptor trafficking and synaptic plasticity. *Ann. Rev. Neurosci.*, *25*, pp. 103–126.

[29] Derkach V.A., Oh M.C., Guire E.S. and Soderling T.R. (2007). Regulatory mechanisms of AMPA receptors in synaptic plasticity. *Nat. Rev. Neurosci.*, *8*, pp. 101–113.

[30] Shepherd J.D. and Huganir R.L. (2007). The cell biology of synaptic plasticity: AMPA receptor trafficking. *Annu. Rev. Cell. Dev. Biol.*, *23*, pp. 613–643.

[31] Ehrlich I., Klein M., Rumpel S. and Malinow R. (2007). PSD-95 is required for activity-driven synapse stabilization. *Proc. Natl. Acad. Sci.*, *104*, pp. 4176–4181.

[32] Lee H.-K. *et al.* (2003). Phosphorylation of the AMPA receptor GluR1 subunit is required for synaptic plasticity and retention of spatial memory. *Cell*, *112*, pp. 631–643.

[33] Kullmann D.M. and Lamsa K.P. (2007). Long-term synaptic plasticity in hippocampal interneurons. *Nat. Rev. Neurosci.*, *8*, pp. 687–699.

[34] Chatterjea D., Hamid E., Leonard J.P. and Alford S. (2010). Phosphorylation-state-dependent regulation of NMDA receptor short-term plasticity modifies hippocampal dendritic $Ca^{2+}$ *Transient. J. Neurophysiol.* *104*, pp. 2203–2213.

[35] Malenka R.C. and Bear M.F. (2004). LTP and LTD: an embarrassment of riches. *Neuron*, *44*, pp. 5–21.
[36] Lisman J. (1989). A mechanism for the Hebb and the anti-Hebb processes underlying learning and memory. *Proc. Natl. Acad. Sci. USA*, *86*, pp. 9574–9578.
[37] Bear M.F., Cooper L.N. and Ebner F.F. (1987). A physiological basis for a theory of synapse modification. *Science*, *237*, pp. 42–48.
[38] Yang S.-N., Tang Y.-G. and Zucker R.S. (1999). Selective Induction of LTP and LTD by Postsynaptic [$Ca^{2+}$] Elevation. *J. Neurophysiol.*, *81*, pp. 781–787.
[39] Beattie E.C. *et al.* (2000). Regulation of AMPA receptor endocytosis by a signaling mechanism shared with LTD. *Nat. Neurosci.*, *3*, pp. 1291–1300.
[40] Xia Z. and Storm D.R. (2005). The role of calmodulin as a signal integrator for synaptic plasticity. *Nat. Rev. Neurosci.*, *6*, pp. 267–276.
[41] Mulkey R.M., Endo S., Shenolikar S. and Malenka R.C. (1994). Involvement of a calcineurin/inhibitor-1 phosphatase cascade in hippocampal long-term depression. *Nature*, *369*, pp. 486–488.
[42] Carroll R.C., Beattie E.C., von Zastrow M. and Malenka R.C. (2001). Role of AMPA receptor endocytosis in synaptic plasticity. *Nat. Rev. Neurosci.*, *2*, pp. 315–324.
[43] Snyder G.L. *et al.* (2003). Regulation of AMPA receptor dephosphorylation by glutamate receptor agonists. *Neuropharmacology*, *45*, pp. 703–713.
[44] Jurado S., Biou V. and Malenka R.C. (2010). A calcineurin/AKAP complex is required for NMDA receptor-dependent long-term depression. *Nat. Neurosci.*, *13*, pp. 1053–1055.
[45] Banke T.G. *et al.* (2000). Control of GluR1 AMPA receptor function by cAMP-dependent protein kinase. *J. Neurosci.*, *20*, pp. 89–102.
[46] Barria A., Muller D., Derkach V., Griffith L.C. and Soderling T.R. (1997). Regulatory phosphorylation of AMPA-type glutamate receptors by CaM-KII during long-term potentiation. *Science*, *276*, pp. 2042–2045.
[47] Lisman J., Schulman H. and Cline H. (2002). The molecular basis of CaMKII function in synaptic and behavioral memory. *Nat. Rev. Neurosci.*, *3*, pp. 175–190.
[48] Blitzer R.D. *et al.* (1998). Gating of CaMKII by cAMP-regulated protein phosphatase activity during LTP. *Science*, *280*, pp. 1940–1942.
[49] Colbran R.J. and Brown A.M. (2004). Calcium/calmodulin-dependent protein kinase II and synaptic plasticity. *Curr. Opin. Neurobiol.*, *14*, pp. 318–327.
[50] Lee S.-J.R., Escobedo-Lozoya Y., Szatmari E.M. and Yasuda R. (2009). Activation of CaMKII in single dendritic spines during long-term potentiation. *Nature*, *458*, pp. 299–304.
[51] Wu Y. *et al.* (2002). Calmodulin kinase II and arrhythmias in a mouse model of cardiac hypertrophy. *Circulation*, *106*, pp. 1288–1293.
[52] Esteban J.A. *et al.* (2003). PKA phosphorylation of AMPA receptor subunits controls synaptic trafficking underlying plasticity. *Nat. Neurosci. 6*, pp. 136–143.

[53] Zheng Z. and Keifer J. (2009). PKA has a critical role in synaptic delivery of GluR1- and GluR4-containing AMPARs during initial stages of acquisition of *in vitro* classical conditioning. *J. Neurophysiol.*, *101*, pp. 2539–2549.
[54] Lee H.-K., Takamiya K., He K., Song L. and Huganir R.L. (2010). Specific roles of AMPA receptor subunit GluR1 (GluA1) phosphorylation sites in regulating synaptic plasticity in the CA1 region of hippocampus. *J. Neurophysiol.*, *103*, pp. 479–489.
[55] Lugnier C. (2006). Cyclic nucleotide phosphodiesterase (PDE) superfamily: A new target for the development of specific therapeutic agents. *Pharmacology Therapeutics*, *109*, pp. 366–398.
[56] Shouval H.Z., Castellani G.C., Blais B.S., Yeung L.C. and Cooper L.N. (2002). Converging evidence for a simplified biophysical model of synaptic plasticity. *Biol. Cybern.*, *87*, pp. 383–391.
[57] Benke T.A., Luthi A, Isaac J.T.R. and Collingridge G.L. (1998). Modulation of AMPA receptor unitary conductance by synaptic activity. *Nature*, *393*, pp. 793–797.
[58] Ehlers M.D. (2000). Reinsertion or Degradation of AMPA receptors determined by activity-dependent endocytic sorting. *Neuron.*, *28*, pp. 511–525.
[59] Helmchen F., Imoto K. and Sakmann B. (1996). $Ca^{2+}$ buffering and action potential-evoked $Ca^{2+}$ signaling in dendrites of pyramidal neurons. *Biophys. J.*, *70*, pp. 1069–1081.
[60] Zhabotinsky A.M. (2000). Bistability in the $Ca^{2+}$/calmodulin-dependent protein kinase-phosphatase system. *Biophys. J.*, *79*, pp. 2211–2221.
[61] Wong S.T. *et al.* (1999). Calcium-stimulated adenylyl cyclase activity is critical for hippocampus-dependent long-term memory and late phase LTP. *Neuron.*, *23*, pp. 787–798.
[62] Otmakhova N.A., Otmakhov N., Mortenson L.H. and Lisman J.E. (2000). Inhibition of the cAMP pathway decreases early long-term potentiation at CA1 hippocampal synapses. *J. Neurosci.*, *20*, pp. 4446–4451.
[63] Zhang M., Storm D.R. and Wang H. (2011). Bidirectional synaptic plasticity and spatial memory flexibility require $Ca^{2+}$-stimulated adenylyl cyclases. *J. Neurosci.*, *31*, pp. 10174–10183.
[64] Walsh D.M. *et al.* (2002). Naturally secreted oligomers of amyloid $\beta$ protein potently inhibit hippocampal long-term potentiation *in vivo*. *Nature*, *416*, pp. 535–539.
[65] Stutzmann G.E. *et al.* (2006). Enhanced ryanodine receptor recruitment contributes to $Ca^{2+}$ disruptions in young, adult, and aged Alzheimer's disease mice. *J. Neurosci.*, *26*, pp. 5180–5189.
[66] Cheung K.-H. *et al.* (2008). Mechanism of $Ca^{2+}$ disruption in Alzheimer's disease by presenilin regulation of InsP3 receptor channel gating. *Neuron.*, *58*, pp. 871–883.
[67] Lopez J.R. *et al.* (2008). Increased intraneuronal resting [$Ca^{2+}$] in adult Alzheimer's disease mice. *J. Neurochem.*, *105*, pp. 262–271.
[68] Chen Q.-S., Wei W.-Z., Shimahara T. and Xie C.-W. (2002). Alzheimer amyloid $\beta$-peptide inhibits the late phase of long-term potentiation through

calcineurin-dependent mechanisms in the hippocampal dentate gyrus. *Neurobiol. Learn. Mem.*, *77*, pp. 354–371.
[69] Hsieh H., Boehm J., Sato C., Iwatsubo T., Tomita T., Sisodia S. and Malinow R. (2006). AMPAR removal underlies A$\beta$-induced synaptic depression and dendritic spine loss. *Neuron*, *52*(5), pp. 831–843.
[70] Kuchibhotla K.V. *et al.* (2008). A$\beta$ plaques lead to aberrant regulation of calcium homeostasis in vivo resulting in structural and functional disruption of neuronal networks. *Neuron*, *59*(2), pp. 214–225.
[71] Sivakumaran S., Hariharaputran S., Mishra J. and Bhalla U.S. (2003). The Database of Quantitative Cellular Signaling: management and analysis of chemical kinetic models of signaling networks. *Bioinformatics*, *19*, pp. 408–415.

# Index

Lightning Source UK Ltd.
Milton Keynes UK
UKHW020931190122
397385UK00001B/46

9 781800 610118